FUNDAMENTALS OF THE PETROLEUM INDUSTRY

by Robert O. Anderson

FUNDAMENTALS OF THE PETROLEUM INDUSTRY

by Robert O. Anderson

University of Oklahoma Press: Norman

Library of Congress Cataloging in Publication Data

Anderson, Robert O., 1917-
   Fundamentals of the petroleum industry.

   Includes index.
   1. Petroleum industry and trade — History. I. Title.
HD9560.5.A55   1984   338.2′7282′09   84-40271
ISBN 0-8061-1909-8
ISBN 0-8061-1916-0 (pbk.)

CONTENTS

PREFACE

I write of an industry in which I have spent more than 40 years. I have had the good fortune to see it from every viewpoint, having been involved in all aspects, first as an independent and more recently as head of one of the largest companies.

Most vividly I remember the thousands of hard-working and dedicated people of this industry, together with their families, in all the far corners of the world. In no other industry can one match their enthusiasm and belief in their work. But the industry and its people have seen many changes in the past four decades. The flamboyant characters of the early oil patches have given way to trained university graduates. They may be less colorful, but their stewardship of the industry augurs well for the future.

Unfortunately, very few people within or outside the oil industry fully understand its complexities. Nor do they appreciate it as the vast and efficient worldwide producer and distributor of energy that it is. In a large oil company many people spend their entire careers in a single department or branch of the industry without ever grasping the nature of the whole.

Perhaps this book will provide, in some small way, an insight not only for those on the inside but also for the increasing numbers of individuals in both the private and the public sectors who have a growing interest in the affairs of this great and vital industry. I take much of my inspiration from an earlier book entitled *Fundamentals of the Petroleum Industry,* by Dorsey Hager, which was a bible for many of those who entered the industry years ago. Hager's work was of great value to me as a neophyte in

the business. I hope that this book will serve similarly as a guide to those who follow.

Chapter notes have been collected at the end of the text, and the interested reader may refer to them for additional information and documentation. The notes refer to chapter and page number. There is no attempt to document by original source every fact and assertion. The notes tell the interested reader where he or she can find more information and occasionally expand upon the text itself.

Many persons have assisted in the making of this book. I thank those senior managers of Atlantic Richfield Company who contributed their own special knowledge and expertise. Special thanks go to Leslye Borden, who helped in the editing and research; to Tom Yerxa, who designed the book; and to Marshall Lumsden and A. Donald Anderson, who coordinated its preparation.

In particular, I extend my gratitude and appreciation for the help and cooperation of Peter J. Brennan, whose contributions as researcher, technical consultant, and editor helped greatly to shape this work in its present form.

Robert O. Anderson
Roswell, New Mexico
1983

INTRODUCTION

The industry has come of age during my lifetime. Roughly two billion dollars a day now change hands in worldwide petroleum transactions. It is the world's first trillion-dollar industry in terms of annual dollar sales. It employs most of the world's shipping as the carrier of oil between continents and nations.

Whether nations are producers or consumers, oil and oil-related taxes are the major sources of income for many of the world's people. Petroleum is the largest single item in the balance of payments and exchanges between nations. It is thus easy to understand why oil is so intimately involved in governmental processes at all levels throughout the world.

During my time in the petroleum industry, world consumption in each decade has exceeded the entire world consumption of energy up to that time. In short, in 40 years there has been a sixteenfold increase in daily world consumption of petroleum. An even faster growth in capital requirements has accompanied this geometric increase in consumption as our dwindling reserves of low-cost energy are replaced by newer and much more expensive substitutes.

The world petroleum situation today is complex and confusing. It is difficult to sort out and understand conflicting stories and opinions without reviewing the circumstances that led up to the present global predicament. Indeed, it is impossible to understand where we are without recalling how we got here.

Optimism has always characterized this industry. Back when Henry Ford began mass-producing the popular motorcar, there was no doubt that the industry could provide enough fuel to put the world on wheels. The United States confidently fueled World War I with almost no effort. We had ample refining capacity to go with our rich petroleum reserves. By 1920, people were talking of oil resources running out, and the price of gasoline reached 25 cents a gallon. On a constant-dollar basis that would compare with today's European prices. Then two huge oil fields were discovered within the city limits of Los Angeles, Kettleman Hills and Midway. The shortage went away.

Soon afterward a single field discovered in East Texas in 1932 produced so much oil that, in the conventional wisdom of the time, the world seemed to be afloat in crude. The only problem was how to get rid of the surplus. Production from the East Texas field carried us through World War II.

In the postwar period the recovery of Europe was built not entirely on the Marshall Plan but on the Marshall Plan augmented by a sea of cheap energy, much of it from the new fields of the Middle East. Major refinery complexes were developed, first in Italy and later in Rotterdam. Also in the mid-1950s came the development of the supertanker, which enabled Middle East oil to become more competitive in world trade. This new refining capacity and these cheap energy sources displaced coal and subsidized economic growth.

In late 1956 and early 1957, the first cargoes of Middle East crude arrived in the United States. The impact was particularly sharp on the Pacific Coast, where supplies had been extremely tight. My independent company at Norwalk, California, received the first cargo of Middle East crude to be delivered on the coast, and most of the major West Coast companies followed suit.

By 1957, the U.S. domestic industry was fighting for survival against less expensive imports, and a crude-oil import bill was passed by Congress to protect the U.S. industry. A typical new oil discovery in Texas was restricted to producing 25 barrels a day (bpd) compared to an average Middle East well of 30,000 to 40,000 bpd, (The Texas Railroad Commission, which regulated the oil industry in that state, presumably limited production to conserve and prevent waste but also to prevent domestic prices from dropping to a level that would encourage premature abandonments and inhibit further exploration.)

At all times U.S. government policy was directed toward the single goal of holding prices on energy to the lowest possible level, regardless of replacement cost. There was an absolute conviction that the world was swimming in oil and that energy shortages within less than a century or so were unthinkable. Texas alone was assumed to have massive shut-in petroleum supplies, that is, wells with valves closed off to stop production.

In Washington, D.C., official estimates of shut-in reserves in the United States ran to five or six million bpd. Not only would these reserves have been more than adequate to protect our national security, but also they would have been large enough to discourage higher production in the Middle East or Venezuela, lest the United States unleash them in retaliation.

In many ways Americans enjoyed the best of both worlds. We had large internal supplies and cheap, unlimited energy from abroad. The United States not only controlled most of the world's production but was virtually the sole source of technology and reserve estimates. It was easy to convince ourselves that we were living in a world without energy limits.

But during the 1960s, U.S. oil production and reserves had quietly been deteriorating, though even by 1969 few had yet perceived the drastic drop. This was not the only misperception, even among so-called experts. On August 10, 1969, for example, the Tulsa *Daily World* quoted the staff of the White House Task Force on Oil Import Controls, which, working without data from the companies involved, estimated that North Slope crude oil in Alaska could be sold at the well head for *36 cents a barrel* and still yield a 16 percent rate of return.

The article accused the principal holders of the big North Slope discovery at Prudhoe Bay of withholding data pending a state lease sale in the fall. It also mentioned an analysis by M. A. Adelman, professor of economics in the Massachusetts Institute of Technology, who estimated the capital cost at Prudhoe Bay at 10.6 cents a barrel (these costs will ultimately reach more than $2.50 per barrel). "At current prices," Adelman's paper estimated, "delivered cost of North Slope crude will be $1.11 a barrel in Los Angeles, $1.36 in Chicago, 96 cents on the East Coast via the Northwest Passage and $1.81 via the Panama Canal; 96 cents in Yokohama, and 76 cents in Northern Europe."

In fact, as of this writing, lifting costs alone of North Slope crude (before capital expense, transportation, taxes, and the like) amount to around $1.13 a barrel. It is difficult to believe that a supposedly competent and well-informed group could have so seriously underestimated costs and thereby misled the public on an issue of this magnitude.

In retrospect, we can regard this period as the point at which dominance over world oil production passed rather quickly from the United States to the Middle East nations. Sometime in the early 1970s the state of Texas, which had carried this country through two wars, reached its maximum level of production and dropped all producing restrictions. The net increase was less than 100,000 bpd. Our supposedly vast hidden reserves had evaporated overnight.

Throughout 1972 and into mid-1973, it became increasingly apparent that world demand was beginning to exceed the world's ability to produce. Informed observers were concerned that the world could be moving toward a long period of chronic deficiencies of crude oil and products. During the summer of 1973, a few spot shortages of gasoline occurred (particularly in the Rocky Mountain states), underlining the growing seriousness of the situation.

Earlier in the summer, Libya and Algeria had stunned the world oil industry by arbitrarily increasing their posted prices to approximately $7 a barrel. There was some temporary buyer resistance, but the price move inspired OPEC to call a meeting for early October to discuss a significant increase in OPEC pricing. Suddenly the world community became acquainted not only with the acronym OPEC but also with the entire concept of energy supplies. At end of 1973, OPEC, led by the shah of Iran, redoubled its previous arbitrary pricing to roughly $11 a barrel, where it remained for nearly five years.

In the wake of these increases, world consumption fell some three million bpd, a result of market pressures that drove down demand in most of the free world. The United States, however, elected to subsidize and regulate rather than face normal market forces. As a result, U.S. consumption and imports continued to rise during the period when OPEC policies should have created the greatest pressure to reduce consumption.

With the collapse of Iran as a producing nation in early 1979, almost ten percent of the free world's supply was lost, creating an instant and chaotic supply situation. Virtually every other producing country stepped up its operations to maximum levels. The net result was still a worldwide shortage of some two or three million bpd, which was made up for a time from existing inventories. Then OPEC raised oil prices again. And again.

Saudi Arabia was a flywheel for OPEC, finely balancing supply and demand for five years, but it can no longer play that role, even if it wants to. In fact, no government, or group of governments, can provide effective control or balance in a situation like today's, in which we live with the threat of impacts even greater than those of 1973-74, impacts that could destroy the economies of most of the Third World nations. Even the OPEC cartel can no longer control the numbers.

Nevertheless, it is becoming evident that almost three-fourths of the oil produced in the world today is produced by governments, not oil companies. Governments and private concerns do business in a very different way. Governments do not have the pressures, the energies, the tempo, the dynamics of the private sector. Nowhere is this fact more vividly demonstrated than in the contrast between petroleum practices of the Soviet Union and those of the United States. Typical drilling speeds in the Soviet Union, for example, are five to ten times slower than those in the United States.

As the world industry moves into the hands of nationalized or government companies, these differences will become critical. We are already witnessing major slowdowns in development and exploration because host governments do not see their national interests built on the same time frame as the ones on which private industry operates. More and more, world supply becomes vulnerable to world politics.

Through most of the 1970s and into the early 1980s, the U.S. oil industry operated under rigid government controls. These controls had little or no relationship to future supplies or prices. They were largely dedicated to keeping prices low. This policy, or lack of policy, was a setback for the nation and its future generations. Politics rather than statesmanship was and, unfortunately, still is the guiding principle in governmental actions both here and abroad.

Today, thanks to the repeal of government price controls, the shortages have vanished. The price of oil, through the forces of supply and demand, is closer to its real day-to-day value in the marketplace. At the same time, we have learned some valuable lessons from the artificially created shortages of the 1970s and early 1980s. For one, conservation has proved to be an even more

effective force in managing our supply of energy than we had thought. By 1983 the consumption of crude oil in the United States had declined by approximately 3.76 million bpd from a high of 18.85 million bpd in 1978.

The petroleum industry can never return to the past. The unfettered competition of the early years is probably gone forever, but it was in that spirit that oilmen explored for, found, and produced the oil that fueled the enormous growth of our economy since the industry began in 1859.

FUNDAMENTALS OF THE PETROLEUM INDUSTRY

CHAPTER 1 –
THE BEGINNINGS

"Drake's Folly," they called it. People snickered at the man who was drilling a hole in solid rock to find a substance that was oozing from the ground all about, a substance, moreover, that was a nuisance, a contaminant of brine wells.

Edwin Laurentine Drake was a retired railroad conductor, failed speculator, and self-styled "colonel." On August 27, 1859, near Titusville, Pennsylvania. Drake's primitive cable-drilling rig struck oil at the negligible depth of 69½ feet and flowed 20 barrels a day. Colonel Drake was lucky. He not only tapped the largest field ever found at such a shallow depth but found a sweet and paraffinic oil that was ideal for the needs of the day, needs that were just beginning to grow. He got $20 a barrel for it, a price not seen again until modern times.

Had Drake's well been 6,950 feet down or its oil sulfurous and asphaltic, the petroleum industry would have been longer in coming and the colonel forgotten. As it was, he went from failed speculator to successful oil producer and back to failed speculator. He died poor, the first and far from the last oilman to follow that cycle.

Although many date the modern oil industry from that 1859 well, the industry is really a creature of the twentieth century. Its growth paralleled the rise of the internal-combustion engine. It really came of age with the discovery of the great East Texas oil fields in the early years of the century. Perhaps the twenty-first and succeeding centuries will run on some other energy source. The Age of Oil, unlike the Stone Age, the Bronze Age, and the Iron Age, may not last long, but the Oil Age has already changed the world as did no epoch before it.

Oil's ascendancy may be brief. Its history is not. Oil predates man. It was there in the earth before the first man ever trod the surface above it or stepped into a sticky pool of pitch. If ancient animals behaved as modern ones do, perhaps ancestors of our modern bears made use of petroleum deposits. In western Washington today, for instance, wallowing bears have worn smooth the sides of exposed oil seepages.

Evidence of man's first use of petroleum is

3

Edwin Laurentine Drake (right) stands in front of his well with one of his investors, Peter Wilson, a Titusville druggist. Drake's discovery of oil showed the world that petroleum could be drilled for in a purposeful manner and produced in substantial quantities— 20 barrels a day.

archaeological rather than written. Some of the earliest archaeological discoveries, from ancient Sumer, are of mosaic inlaid in asphalt nearly 6,000 years old. Ancient peoples used asphalt to cement stones together in their buildings and in the construction of the pyramids. They used it to waterproof cisterns and silos and to caulk their boats.

Even the dead benefited. Egyptian morticians used bitumen to make headdresses for mummies and wrapped the mummies in linen soaked in asphalt. While this treatment undoubtedly kept the mummies in an excellent state of preservation, it also made them flammable. Invaders regularly used mummies for fuel.

Although there are petroleum seepages and natural gas vents all over the world, the first written records of petroleum and man's use of it come from where it is most abundant, the Middle East.

The earliest region of petroleum manifestation was on the west coast of the Caspian Sea, in what is now the Baku region of the USSR. The area was part of Persia until 1806 and remains today a major production center. Several thousand years before Christ, oil and gas seepages spewed eternal flames that were venerated. The inhabitants built temples in homage to the fires. The temples became fixtures of Zoroastrianism, the long-dominant religion in Persia, in which fire, earth, and water represent the principal god, Ahura Mazda, later a brand name for electric light bulbs and an automobile.

Travelers reported that such a temple still existed in the region as late as 1870. Indeed, the religion that reveres fire is today found in Iran, modern Persia, and particularly in the Parsee community of Bombay, India.

The burning gas springs and naturally occurring pools of petroleum have been sources of curiosity and commerce to travelers and visitors for millennia. Herodotus (ca. 450 B.C.), the source of most of our modern knowledge of the world before Alexander the Great, made several references to petroleum products. He wrote that the builders of the walls of Babylon made fired bricks and then "used hot bitumen for cement." Where did they get the bitumen? "Eight day's journey from Babylon there is another city, called Is, on a smaller river that runs into the Euphrates; and this river, also named

Is, brings down lumps of bitumen; and hence the bitumen was fetched for the walls of Babylon." That was a respectable quantity of bitumen to transport a considerable distance from an area now inside Iran.

In another reference, this well-traveled raconteur mentions a well 40 furlongs from Arderrica, also in Iran, from which "men get asphalt and brine and oil in this way: they draw liquid by means of a crane; but instead of a bucket they tie half of a skin on the end of the arm. A man lowers it into the well and scoops up liquid which he pours into a receptacle, and from there to another container, where it is separated into three channels. The asphalt and brine solidify at once; the oil, which the Persians call rhadinake, is black and has a sickly smell." Crude petroleum and asphalt have changed not at all in 2,500 years; in some respects, neither have the recovery and separation methods.

Herodotus mentions an island off North Africa where girls using feathers coated with pitch picked up gold from the mud of a lake. On another island off Greece were pools of pitch "that smells like asphalt but better in all respects than the pitch of Pieria. They pour it into a pit that has been dug close by the pool, and when they have enough, they scoop it from the pit into jars."

Ecbatana, summer capital of the great King Darius in Media, now Iran, had eternal fires and appropriate temples. Alexander the Great visited the city in pursuit of Darius. There is no record that Alexander put to practical use the petroleum he found there, though it is hard to believe that so inventive and practical a man would not have devised a military application for oil. Perhaps his army's flaming arrows were tipped with pitch.

For Alexander's amusement, a train of oil was laid upon the street and ignited. Flame quickly swept the street and just as quickly went out, without damage to the roadway. In a further entertainment, a young boy singer was doused with oil by one of Alexander's counselors, who thought the flame would have no more effect on the boy than it did on the street. The records say that the boy jumped into a pool and saved his life.

Eight hundred years later, in A.D. 533-34, the Byzantine general Belisarius made novel and effective use of greased pigs. In a campaign against the Vandals in North

4

Game board decorated with shells set in bitumen.

Fire temple painted on a Samara-type bowl.

Greek fire, a mixture of pitch, oil, and other chemicals, was the secret weapon of the Byzantines in their naval battles. Hurled off the bow of war galleys, it was ignited and spread by contact with water.

Africa, he unleashed a herd of the petroleum-anointed and flaming animals into the enemy ranks, throwing the Vandals into great confusion. Belisarius's small army routed the foe, captured their king, and recovered the golden vessels of King Solomon that the Vandals had taken in the sack of Rome 80 years before.

Later, in the seventh century A.D., the Byzantines of Constantinople, now Istanbul, developed the most significant military use of petroleum up to that time and for some time to come, the famous Greek Fire. This sticky, combustible mixture of pitch, crude petroleum, crude naphtha, charcoal, sulfur, phosphorus, potassium nitrate, and other materials was the Byzantine secret weapon that successfully defended Constantinople until its fall to the superior artillery of the Turks in 1456.

In the West, the earliest written references to petroleum are in the Bible. In Genesis, God said to Noah: "Make thee an ark of gopher wood. . . . Pitch it within and without with pitch." The Vale of Siddim was said to be full of "slime pits," interpreted to mean bitumen, pitch, or tar pits, into which fell the fleeing kings of Sodom and Gomorrah. In 1918, commenting on the morality of oil-boom towns, the *Oil and Gas Journal* quipped that Sodom and Gomorrah were, after all, oil-field towns.

In Deuteronomy, the Lord made Jacob "to suck honey out of the rock, and oil from the flinty rock." In the Book of Job, when poor Job is relating how good life used to be, he says, "The rock poured me out rivers of oil." In Maccabees, Nehemiah sent priests in Persia to the eternal fire, but there they found no fire but "thick water." So he commanded them to pour the "water" on wood and stones, and "there was a great fire kindled," and "Nehemiah called this thing Nephthar, . . . but many men call it Nephthai."

Some say this reference is the origin of the word *naphtha,* but more recent authorities trace the derivation through Greek back to the Persian *neft,* signifying moist. After all, Persia was the source of most ancient knowledge about petroleum. In the early days, naphtha generally designated petroleum products of Persian or Iranian origin, and only later was this word applied to a particular range of petroleum spirits, whatever their origin.

These early references are not mere curiosities. In areas like Baku and Ecbatana, near what is now Hamadan, in Iran, the presence of oil and gas was and is self-evident. In other areas, time and use have eroded or obscured the surface evidence of hydrocarbons. Modern petroleum prospectors have found these ancient references valuable in again locating the underground sources of forgotten pools and seeps.

There is ample evidence that both the primitive and the civilized inhabitants of the Americas used petroleum in much the same way as ancient peoples did. The Indians of Central America and Mexico used asphalt as cement, and the Toltecs set mosaics in asphalt in much the same way as did the ancient Sumerians. North American Indians, such as the Senecas and the Iroquois, used oil as medicine as well as to paint their bodies and make ceremonial fires. Excavations in California showed that ancient Indians used asphalt for sticking decorative beads to articles and for waterproofing baskets. They even used tar to caulk their boats. Near Lake Tulare, aborigines placed asphalt on the eyes of their dead.

Columbus noted Indians using oil as medicine. Travelers found and recorded evidence of oil on the coast of Peru as early as 1669, and the Spaniards even granted oil concessions in that country in 1692. Producers dug oil from pits — the usual production method then — at Santa Elena, Ecuador, in 1700. A Spanish friar writing in 1604 noted at Saint Helena a fountain of pitch that the mariners used to pitch their ropes and tackle.

In the North, a Franciscan friar is said to have found an oil spring near what is now Cuba, New York, in 1627. Early travelers such as the Russian naturalist Peter Kalin in 1748 reported finding oil in Pennsylvania. George Washington, traveling to wilderness Virginia in 1776 with his comrade-in-arms, General Andrew Lewis, found an oil spring on which Washington filed a claim. Said Washington in an addendum to his will listing his property: "This tract was taken by General Lewis and myself on account of the bituminous spring which it contains of so inflammable a nature as to burn as freely as spirits and is nearly as difficult to extinguish." The site is now at Burning Springs, West Virginia.

On the other side of the world, firm references to petroleum are nearly as ancient as those of the Middle

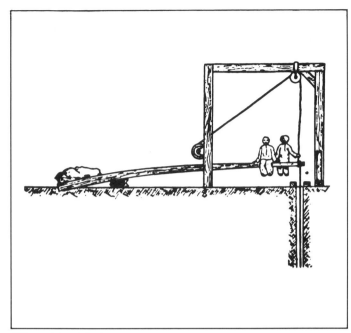

Chinese spring-pole rig.

Medicinal benefits of petroleum were known in medieval Bavaria.

East. Early Chinese references are 3,000 years old and indicate that the Chinese, like our Pennsylvania farmers, were really looking for brine from which to make salt. When they found brine, they often found petroleum, too, which they separated, saved, and used in lamps.

The Chinese were also pioneer well drillers. Two thousand years ago, they had developed drilling techniques to sink wells to 3,000 feet. While the rest of the world was digging pits and bailing out seepage, the Chinese were drilling deep holes with strings of tools similar to modern cable tool rigs.

Sometimes the Chinese found gas. They built pipeline distribution systems of bamboo tubing to bring the gas into their palaces for heating and light.

Japanese references come later, the earliest being a report of A.D. 674 that the emperor Teuchi was offered gifts of burning earth and burning water — asphalt and naphtha.

Closer to modern times, references to petroleum become more numerous and, presumably, more accu-

rate. In the thirteenth century, Marco Polo mentioned the oil trade at Baku, apparently by then an established commerce; "On the confines toward Georgine, there is a fountain from which oil springs in great abundance, inasmuch as a hundred shiploads might be taken from it at one time. This oil is not good to use as food, but is good to burn, and is also used to anoint camels that have the mange. People come from vast distances to fetch it, for in all countries around there is no other oil."

Oil was also known and recorded in Prussia and Bavaria in the Middle Ages; a medicinal oil called Saint Quirinus oil was sold in Bavaria in 1436. In Galicia, another modern center for Russian petroleum production, oil was described as "earth balm" in 1506 and in 1750 reportedly was used as a medicinal agent in the treatment of cattle, as well as for lighting and lubrication.

The northern Italian town of Salsomaggiore, near which natural gas issued in 1226, adopted a burning salamander as the town crest. Leonardo da Vinci's sketchbooks reveal his interest in drilling machines. One sketch shows a hand-operated shallow-hole drilling machine that remarkably resembles a modern cable rig. Pierre Belon, writing in 1553, refers to the Monte Zibio oil, another Italian seepage. In Genoa, Parma, and other cities, oil from pit wells dug 40 to 60 feet deep was used for lighting as early as 1640.

Perhaps the greatest oil commerce in the Far East occurred in Burma, where as early as 1735 many hundreds of hand-dug pit wells yielded 650,000 barrels a year. The Burmese did a brisk business in oil trade among their neighbors, most of their product going for medicinal purposes.

Despite oil's obvious flammability and value as a fuel, illuminant, and lubricant, early users failed to exploit it fully, paying inordinate attention to its supposed medicinal values. Baku's 28,000-barrel production in 1819, for example, was refined for use as cordials and medicine. And when European Americans began to take note of the impurity produced along with their brine, the more enterprising promoted it as a cure-all.

One of these was Samuel M. Kier, a Pittsburgh druggist who owned some brine wells that incidentally produced about 1,000 barrels of petroleum a year. Kier bottled the oil, labeled it "Rock Oil," advertised it widely as a panacea for just about everything, and sold it for $1.50 a gallon in bulk.

He was not particularly successful, but the oil remedy became popularly known as Seneca Oil, after the tribe whose medicine men had endorsed the oil's magical medicinal properties. In time, Seneca Oil became corrupted to the more familiar "snake oil," the generic term for nostrums of great promise but little value.

Undaunted, Kier devised a crude still for making lamp oil. The oil gave light, but with such smoke and smell that that venture, too, failed.

About the same time, James Young, of Glasgow, Scotland, introduced a process to make lamp oil through distillation of coal or oil shale. The process traveled to the United States, and by 1855 several factories were making "coal oil" for lamps.

Abraham Gesner, a Canadian geologist, developed a similar process for producing oil from coal and licensed U.S. plants to manufacture it. Gesner called his product "kerosene," from the Greek words for oil and wax.

In a few years, many plants were making up to 23,000 gallons of coal oil and kerosene a year from natural asphalts, soft coal, and shale. It was smelly, smoky, and expensive but cheaper than the vegetable, whale, and animal oils that still dominated the lamp-lighting markets. Whale oil, the leading illuminating oil, was $2.50 a gallon. Economics and entrepreneurs would spur the development of cheaper and often better substitutes.

In 1857, Colonel A. C. Ferris, a lamp-oil maker, ran some of Kier's Seneca Oil through his processing plant. It yielded such a superior product that he immediately sent men everywhere to buy all the crude oil they could find at $20 a barrel. The world was ready for an ample, dependable supply of petroleum of consistent quality to feed the fast-growing illuminating-oil market, as well as industry's growing need for lubricants and, soon, fuels. At $20 a barrel, oil was worth looking for.

Drake invested in the Pennsylvania Rock Oil Company, which got oil by skimming it from surface deposits and by taking it as a by-product of brine wells. Drake, sent to Pennsylvania by the company to look over its properties, became interested in the drilling techniques used to produce brine and, incidentally, oil. He wondered whether the same techniques could be used to produce oil exclusively.

When Pennsylvania Rock Oil Company went bankrupt, Drake obtained a lease on its lands and formed the Seneca Oil Company. He assembled a salt-drilling rig and, after drilling through 30 feet of rock, hit the pay zone at 69½ feet on August 27, 1859. At last, all the essential elements had come together: markets, supply, and the technology to produce both the raw material and the marketable product. The oil industry was born.

People stood in line to buy Drake's oil at $20 a barrel. The lure of profits in the fledgling industry soon had the Pennsylvania hills dotted with rigs. In other nations, the industry developed parallel to that in the United States. Drilled wells began replacing mined pits even in the prolific areas of central Europe, Russia, and far-off Burma.

A vibrant new industry that promised huge profits and saw demand growing as fast as supply produced feverish speculation and, naturally, had both winners and losers. One of the big losers, unfortunately, was Edwin L. Drake, who lost heavily in oil speculations in New York and died in broken health and destitute in 1880. Even then, an oil play was not a sure thing. Many a wealthy and highly successful oilman has ended up in a pauper's grave, a victim of the inherently high-risk aspect of the industry and one large gamble too many.

Growth of the industry after 1859 was explosive at home and abroad. West Virginia, New York, and Ohio became oil states in a very few years. In the West, California and Colorado, barely settled, were oil producers in the early 1860s; Ventura County, California, in 1861 and Florence, Colorado, in 1862 — long before the Indians and the buffalo had vanished from the plains. A well was drilled at Paola, Kansas, in 1860. There were wells in Kentucky as early as 1861, in Tennessee in 1866, and in Wyoming in 1867. Until 1900, however, the only important oil states were still east of the Mississippi River.

From 1859 to 1900 the main petroleum product throughout the world was kerosene for illuminating oil. The main object of refining was to extract as much kerosene as possible from the crude. Lubricants and some fuel oil became increasingly important toward the close of the century, but bitumen or asphalt was considered a useless nuisance, and gasoline was burned off as unwanted.

The internal-combustion engine, which became the largest consumer of petroleum products, was invented not long after the birth of the oil industry. It was many years, however, before the engine was a significant consumer of petroleum products, particularly in the personal automobile. Until then, the industry existed largely to light the lamps of the world.

There were many lamps. From the Drake well's 20 barrels a day, production in the United States alone increased in the next 41 years to 173,830 barrels a day in 1900. Even then the United States was not the largest producer. Russia led with 206,310 barrels a day. Total world production was about 400,000 barrels a day, the other significant producers being Indonesia (then the Dutch East Indies), Poland, Rumania, Burma, Japan, and Canada. The United States soon leaped ahead, however.

A man with a dream and a man with a theory joined

Samuel Kier was one of the most successful marketers of petroleum as medicine. He sold nearly 240,000 half-pints of Rock Oil.

PETROLEUM, OR ROCK OIL.

A NATURAL REMEDY!

PROCURED FROM A WELL IN ALLEGHENY COUNTY, PA

Four hundred feet below the Earth's Surface!

PUT UP AND SOLD BY

SAMUEL M. KIER,

CANAL BASIN, SEVENTH STREET, PITTSBURGH, PA.

The healthful balm from Nature's secret spring,
The bloom of health, and life to man will bring,
As from her depths the magic liquid flows,
To calm our suff'rings, and assuage our woes.

CAUTION.—As many persons are now going about and vending an article of a spurious character, calling it Petroleum, or Rock Oil, we would caution the public against all preparations bearing that name not having the name of S. M. Kier written on the label of the bottle.

PETROLEUM.—It is necessary, upon the introduction of a new medicine to the notice of the public, that something should be said in relation to its powers in healing disease, and the manner in which it acts. Man's organization is a complicated one; and to understand the functions of each organ, requires the study of years. But to understand that certain remedies produce certain impressions upon these organs, may be learned by experience in a short time. It is by observation in watching the effects of various medicines, that we are enabled to increase the number of curative agents; and when we have discovered a new medicine and attested its merits, it is our duty to bring it before the public, so that the benefits to be derived from it may be more generally diffused, but have no right to hold back a remedy whose powers are calculated to remove pain and to alleviate human suffering and disease. THE PETROLEUM HAS BEEN FULLY TESTED! About one year ago it was placed before the public as A REMEDY OF WONDERFUL EFFICACY. Every

Kier used advertisements like this, with testimonials from satisfied customers, to publicize and sell his Rock Oil cure-all.

with a new technology to change the oil industry completely. The man with the dream, a one-armed brickmaker, Patillo Higgins, wanted to build a model city atop what he was convinced was a pool of oil. He teamed up with the man with a theory, Anthony Lucas, an Austrian mining engineer who believed that oil was associated with salt domes. They tried to find oil in a geologic formation near Beaumont, Texas.

Higgins had gone north from Texas to learn more about brickmaking. He found the kilns of Pennsylvania, fired by oil and gas, more efficient than the kilns he used, which were fired by coal and wood. He also studied the oil business, learned the surface signs that oilmen look for, and became convinced that there was oil under a slight hillock on the flat plain near Beaumont.

In 1892, Higgins formed a company to look for the oil. Six years and three efforts later, he was deeply in debt and had found nothing. In a final effort, he advertised for help and found Lucas. Lucas provided the technology Higgins needed. Lucas chose a steam-driven rotary drilling rig rather than the cable tools that dominated the drilling industry in that period.

Rotaries were not all that new. The early rigs were primitive affairs, driven by animal power. In the 1880s, they had been used in South Dakota to explore for water. In 1894, rotary drillers looking for water in Corsicana, Texas, found oil instead. By the time Lucas adopted this equipment for drilling in Beaumont, more than a hundred oil wells had already been drilled with rotary rigs. Lucas's contribution was to put together an essentially modern rig complete with boilers, engine, draw works, rotary table, pumps, drill pipe, and bits. In addition, he pioneered the use of drilling mud to flush out the bore and support its walls in the soft formations of the Gulf Coast.

Equally important, Higgins and Lucas hired as drillers the Hamill brothers, Jim, Al, and Curt, of Corsicana. Their practical ingenuity and determination had as much to do with the developers' success as did their own vision, theory, and technology.

The fifth Higgins well, the second for Lucas on the same formation, was spudded on October 27, 1900. Nearly three difficult months later, on January 10, 1901,

the hole was down 1,020 feet, and the crew was changing the drill string. Suddenly there was a tremendous roar. Mud and tools erupted from the hole as the crew fled for their lives. The uproar subsided, and the men cautiously returned to survey the damage, prudently shutting down the boiler. Then the well roared again. A jet of gases ejected the last of the mud, and a column of green crude soared 200 feet above the remains of the drilling rig. The great Spindletop, the first gusher ever seen in the West, had come in. The oil business would never be the same again.

Spindletop flowed 84,000 barrels a day. That single well increased total world production by more than 20 percent and increased production in the United States by nearly 50 percent. The well flowed for nine days and deposited 800,000 barrels of crude in a lake that extended nearly a mile from the well. The Hamills finally capped Spindletop with a makeshift arrangement of fittings and valves, the first "christmas tree."

Within days, Higgins's dream of a model city had

Soon after "Captain" Anthony Lucas came to Beaumont, Texas, he struck oil and advanced to "Colonel." He made "General" when people realized that Spindletop was sitting on an ocean of oil.

Special trains came to Spindletop Heights from as far away as New York, Philadelphia, and Saint Louis. Capped by the first christmas-tree valves, the wells were opened up once in a while so that crowds could see spectacular gushers.

disappeared under a forest of wooden drilling derricks as wildcatters and speculators rushed in from all over, bringing in their wake shanties and shacks, as well as the con men, pimps and prostitutes, roustabouts, grocers, bankers, and lawyers that follow every boom.

Spindletop temporarily had a disastrous effect on the oil market. The $20 a barrel that Drake got for his oil was long a thing of the past, but Texas oil was bringing two dollars a barrel. As the boom developed, that quickly fell to two or three cents. Three years later, the field began to play out, and the price recovered to as much as 83 cents a barrel.

Spindletop is now a depleted field, the derricks and shanties gone with the pimps and the promoters, the dream of a model city drowned under a river of oil, 18 million barrels the first year. But in its own way, like Drake's well, Spindletop represented the conjunction of need, theory, technology, and the men to make them work at a critical time in the development of the industry and society.

Shortly after Spindletop, the gasoline previously burned and the asphalt formerly dumped found uses. With the development of the internal-combustion engine, the rivers of oil found new markets. Henry Ford, working in a small shop in Detroit, had a dream that changed the world. Large numbers of cars created a huge demand for inexpensive energy, which stimulated the search for still more oil. The first powered airplane flew two years later. Electrification became widespread. Locomotives and ships began burning oil rather than coal. Thanks to the proving of Lucas's theories, more new oil soon was found. The discovery and exploitation of this oil fostered the continued proliferation of energy-consuming technology and created the consumer society of today.

11

CHAPTER 2 –
U.S. TO 1931

One aspect of the story of the American oil industry begins a few years before Drake's 1859 well with, appropriately enough, a scientific study.

In 1855, Benjamin H. Silliman, Jr., a professor of chemistry in Yale University, New Haven, undertook a study of Pennsylvania oils for two New York investors, George H. Bissell and Jonathan G. Eveleth. Bissell and Eveleth, who were interested in Seneca Oil, had bought the Titusville Lumber Company, including 105 acres of land on Oil Creek and two 99-year leases on 2,200 acres on Oil Creek and in Venango County. They then formed the Pennsylvania Rock Oil Company and hired Silliman to look over samples of the oil from seeps on their property.

Silliman discovered that simple distillation by heating permitted the oil to be separated into a variety of individual products. "In short," he reported, "your company has in its possession a raw material from which, by a simple and not expensive process, they may manufacture very valuable products." Silliman inspired Bissel and Eveleth to think in terms far broader than mere medical potions and to think of drilling specifically for oil rather than skimming it from ponds and creeks.

Some financial matters needed attention first. For reasons of taxation and regulation, New York State was not the best locale for a corporation. Therefore, Bissell and Eveleth closed the Pennsylvania Rock Oil Company of New York and sold out to Asahel Pierpont and William A. Ives, of New Haven, Connecticut. These gentlemen then formed a new company under Connecticut law, also named the Pennsylvania Rock Oil Company (Connecticut). The principals, who owned the assets of the old New York company, including the Titusville Lumber Company properties, then leased the 1,200 acres to the Connecticut company, which was capitalized at $300,000 in 12,000 shares of $25 each.

Thus all the new Pennsylvania Rock Oil Company had, besides an already tangled financial and legal history, was a lease on 1,200 acres in northwestern Pennsylvania and a scientific report handed down from the predecessor company. Things were to get more tangled before anyone saw oil.

The company entered into an agreement with Drake, who was a stockholder and resident of New Hampshire, to explore the properties and drill for oil there. Drake would pay the company 12 cents for each barrel of oil produced. In the spring of 1858, Drake made over his lease to Pierpont, Ives, and the rest as equal partners, and the group set themselves up as the Seneca Oil Company with Drake as president and managing director.

When Drake struck oil in 1859, he founded an industry that was, in truth, waiting to be founded. The industrial revolution was gathering steam, and the common man needed to see and to read after sunset. The whaling industry, which at its peak employed 70,000 men in 7,000 ships, was now in decline, but demand for light and lubricants was growing.

Why had no one exploited rock oil before? A contemporary writer explained: "Divine Providence orders that what is not needed is not known, . . and Providence never gave us 'oil out of the flinty rock' before, because public necessity had not arisen, nor had facilities for preparing and transporting it been provided." This same author, writing nine months after the Drake discovery, further observed, "What the briny sea now refuses, the flinty rock affords."

Drake is credited with bringing in the first well specifically drilled to find oil, though there is controversy over that, too. Canada claims the honor for J. H. Williams, a Hamilton, Ontario, resident of Scottish birth, who is reputed to have mined, refined, and marketed his oil, in perhaps the world's first integrated petroleum operation, before Drake spudded his Titusville well.

Factors that benefit and plague the oil industry to the present day prevailed in microcosm in Titusville in 1859. For a while, people bought Drake's oil at $20 a barrel. After all, Seneca Oil was a valuable commodity. As far back as 1797, the estate of a local merchant listed three 10-gallon kegs of Seneca Oil at $50 a keg. It was said to be effective for rheumatism and the mange. But day after day, as Drake's well continued to pump out 20 barrels or so, other drillers arrived to share in the bonanza. When

Professor Benjamin Silliman, hired by Bissell and Eveleth, evaluated uses for petroleum. His "Report on Rock Oil" was a turning point in the history of oil.

13

only two other wells were competing with his, Drake complained about overproduction and suggested that the producers tailor their output to market demand. His sensitivity to supply-and-demand imbalances brought into focus a problem endemic in the industry.

That winter, however, supply far exceeded demand. Prices had collapsed to $10 a barrel by January 1860 and to ten cents a barrel by December 1861. While Drake longed for production controls, he was hustling his product to market in places like Erie, Chicago, Cincinnati, and Pittsburgh. He even gave oil away, as much as 100 barrels, following the advice of a local writer: "It will be money in the pockets of producers and sellers, for a while, to give away much oil, that its merits may be tested."

In Pittsburgh, Drake met George M. Mowbray, a chemist associated with the firm of Schieffelin Brothers of New York. On March 12, 1860, Drake and Mowbray signed what was probably the first oil-marketing agreement, in which Schieffelin would market Seneca Oil in return for a 7.5 percent commission on the proceeds.

Meanwhile, drillers were bringing in new wells in the Titusville area in a pattern that became familiar in the American oil industry. In February 1860, one well came in just 100 yards from the Drake well, on land leased on the James Parker farm. This well cost $3,000. An Ohio man bought a one-twenty-fourth interest in the project for $10,000. Another partner had an interest in the world's first refinery, on Oil Creek, which he sold for $25,000. Fortunes were already being made.

In March 1860, another well came in on the other side of Oil Creek, half a mile away, flowing 75 barrels per day from 144 feet. The well soon ceased to flow and was abandoned, but was later deepened to 550 feet, then deepened a third time and resumed producing in 1873.

Not everyone formed a company, went in partnership, or sold out. James Evans, of Franklin, Pennsylvania, a blacksmith, made his own set of light drilling tools and, with the help of two sturdy sons and a sturdier spring pole, brought in a 25-barrel-per-day oil well 75 feet down from the bottom of the family water well. This produced a heavy oil known afterwards as Franklin crude, particularly useful for lubricants.

Evans refused all offers for his well and did not even sell interests in it. He lived comfortably on the income, produced only when it was convenient and expedient, and never at the maximum rate, which contemporaries figured would have yielded him $20,000 or $30,000 every three months. Instead, he pumped a daily average of only 75 barrels during that period, which netted him $10,000. Evans's greedier colleagues regarded him as a fool, but his was perhaps the first example of a practice that became known as prorationing.

A contemporary account, published in June 1860, captures some of the excitement of that era: "Titusville was named after Jonathan Titus, who, 60 years ago, cut his way into the wilderness from eastern Pennsylvania. For many years, few followed. But now, never was an army more eager to get possession of a citadel than a host now are to gain the occupancy of some, at least, of this oil land. It is now the rendezvous of strangers eager for speculation. The capitalists, as well as that large class of men not so rich as ready to venture, are streaming in from all quarters. . . . They barter prices in claims and shares, buy and sell sites, and report the depth, show or yield of wells. . . .

"Those who leave today tell others of the well they saw yielding 50 barrels pure oil a day. . . . The story sends back more tomorrow, who must see before they can credit. . . . Never was a hive of bees in time of swarming more astir, or making a greater buzz."

All the basic elements of today's oil industry were being worked out and largely settled during these early years: land leasing, exploration, drilling technology, well logging, refining, transportation and marketing, and secondary recovery from depleted fields. Even disasters were not long in coming. Drake's well caught fire, but with damage only to the structure and his dignity. In 1861, however, a flowing well on Oil Creek caught fire and turned into a three-day conflagration that killed 19 people.

By the end of 1859, there were already 15 plants refining petroleum. Most coal-oil refiners, of whom there were some 53 in the country, representing an investment of $6 million, took one look at the new petroleum industry and prudently decided to switch rather than fight. The economics were irresistible. For about $200, a man could build a petroleum refinery to process five barrels a day; for $4,000, he could refine more than the richest well could produce. By contrast, an economically feasible coal-oil plant cost between $75,000 and $100,000, and, although the end products were similar, making oil from coal involved several additional and expensive steps. "In refining oil," observed one Titusville commentator, "you spend but little, yet profit much."

By today's standards, drilling along Oil Creek was not expensive. Three men could make three feet a day with primitive spring-pole drills that literally hammered their way through the rock. Including wages and supplies, they could drill a well for between $1,000 and $1,500. Companies were formed by the dozens. Farmers mortgaged their farms for $2 a month to get in on the bonanza. A producer could raise, barrel, and freight oil from Titusville to New York for 12.5 cents a gallon. At that rate, a 20-barrel well produced a daily profit of $254.

A typical story was that of J. W. Sherman, of Cleve-

14

At the J. M. Henderson Oil Brokerage in Titusville, buyers and sellers transacted business without the convenience of modern computer systems.

land, who leased the Foster farm in Pennsylvania, ran out of money, sold a one-sixteenth interest for a horse which proved useless, sold another one-sixteenth to buy a boiler and engine which proved better, sold another one-sixteenth for $180 cash to buy fuel and a shotgun, and finally brought in a 2,000-barrel-a-day well that paid out over $2 million in its lifetime. Almost identical stories were repeated over the next several decades of oil development.

The public sector also cashed in. This marked the start of government's endless fascination with oil profits. In 1860, in Franklin, Pennsylvania, the Farmers' and Mechanics' well on school property gave over a tenth of its production to the support of the common schools of the town. At its tithe of one barrel a day, this royalty was worth $5,000 to $6,000 a year.

The industry's growth was explosive. Total production in 1859 was 2,000 barrels with one producing well. A year later, of 240 wells drilled, 201 were producing, and total annual production exceeded 500,000 barrels. By 1865, more than $300 million was invested in the oil industry, and there was production of sorts all across the country. By 1869, the tenth anniversary of the industry, total production had exceeded 42 million barrels, with more than 5,000 wells drilled and, for the most part, quickly abandoned. From 1859 to 1874, the oil industry in the United States produced more than 90 percent of all the world's oil and until 1883 never produced less than 80 percent.

The oil industry came of age early during the Civil War, providing turpentine substitutes, lubricants for machinery, and lamps for the army. Because trains and ships ran on coal or wood, petroleum was not the factor it became in later wars, though even then it was regarded as a strategic target. A promising field in what is now West Virginia, near Washington's burning spring, fell to 3,000 Confederate cavalry troops, who destroyed the wells and burned 300,000 barrels of oil in storage. Worse yet, water infiltrated the field and ruined it. Not until 30 years later did West Virginia regain significant production.

While wildcatters, speculators, and drillers scattered across the landscape to seek new fields away from the clutter of Pennsylvania, other segments of the infant industry rapidly took shape. A rotary drill with a diamond bit was tried as early as 1860, and explosives were used for fracturing that same year.

From the glut of the early 1860s, when prices fell to ten cents a barrel, demand rose rapidly to match supply, and prices increased enough by 1865 to make stripper wells profitable. Transportation remained a major bottleneck, however. Then, as now, oil wells often came in far from the market, which meant moving the oil to market. This was frequently more costly than producing it.

In the beginning, the only way to move the oil was in barrels, six to a horse-drawn wagon, hauled to the nearest railhead. The railroads then carried the oil to its destinations on flatcars. By 1862, more than 6,000 teams of horses were hauling oil from Oil Creek to one of three railroads. The bill for the short haul was twice what the railroad charged to haul the oil the rest of the way to New York. At $20 a barrel, producers could afford it, but as more oil came on the market and the price broke, they had to find a cheaper way. And they did.

Producers on the Allegheny River floated their product. Those on Oil Creek, a rivulet, had to be more resourceful. They appointed a superintendent who assessed the producers for funds to build a series of dams along the creek. Oil-loaded scows, called "gipers," gathered behind the dams. When the water level in the dam ponds got high enough, the dams were broken all at once, creating a surge that floated the armada of scows to the Allegheny. One hitch interrupted the smooth flow of scows. Gipers often collided and sometimes burst into flame.

Next the producers put down plank roads to conquer the mud. Then new rail links crisscrossed every part of the oil country. The first tank cars consisted of two 240-barrel wooden tanks on a flatbed car. Metal cylindrical tanks, little changed to this day, followed the wooden tanks.

Pipelines were tried, but, since they were made of wood and often unburied, they were highly vulnerable to vandalism, particularly by teamsters who were angry at

16

The Roberts' Torpedo revolutionized oil production by restoring wells clogged with paraffin or sediment. When the Torpedo was exploded, the blockage was removed and sometimes new fissures opened.

the prospect of being replaced by pipelines. Samuel Van Syckel built a successful five-mile line in 1865 that cut the cost of transit to the railhead from $3 to $1 a barrel, but the line soon succumbed to financial difficulties. It passed to William H. Abbott and Henry Harley, who already had a line. These two created a pipeline system and bought out the charter of the Western Transportation Company, the only entity empowered by the Pennsylvania legislature to transport oil by pipe to the railroad. They called their new system the Allegheny Transportation Company. About that time, in 1865, the Pennsylvania Railroad formed the Empire Transportation Company expressly to generate oil business, and in 1866, the Standard Oil Pipeline opened for business between Pithole and West Pithole.

Neither teamsters nor railroaders were happy with the pipelines, and both groups fought them tenaciously, and not always fairly. In 1862, the teamsters' pressure defeated a pending pipeline-enabling bill in the Pennsylvania legislature. Not until 1872 did the legislature finally

Detonating the charge was a highly risky business. Many shooters were killed, either in hauling or in handling the shot.

17

The Abbott and Harley Pipe Line Terminal at Shaffer Farm, which included storage capacity and railroad loading facilities, foreshadowed the trend toward consolidation that marked successful ventures in the oil business.

On April 9, 1872, teamsters learned that a train was loading oil for Standard Oil of Cleveland. An excited crowd gathered and prevented the train from pulling out.

18

The Columbia Conduit Company, barred from laying pipe under the Pennsylvania Railroad, used tank wagons to haul oil across the tracks.

grant the right of eminent domain to pipeline companies, and then only over the violent objections of the railroad interests. Producers benefited enormously.

By 1866, there were 58 refineries in Pittsburgh, 30 in Cleveland, and many more on the East Coast, all producing more or less standard grades of lamp oil and other products, specified by color. The composition of a product varied widely from one producer to another, or from one batch to another, and one never knew when he lit a lamp whether it would smoke or explode. Order, of a sort, was soon to come.

In 1863, John D. Rockefeller, a sharp, unemotional young man from Cleveland, left his job as clerk and bookkeeper in a commission house and joined the petroleum refining firm of Andrews, Clark and Company. Two years later, the firm became Rockefeller and Andrews, and two years after that, it merged with four other refiners — William Rockefeller & Company, Rockefeller & Andrews, Rockefeller & Company, and S. V. Harkness & H. M. Flagler — "to unite skill and capital in order to carry on a business of some magnitude and importance." This was the understatement of the century, for thus was created the nucleus of the Standard Oil Company.

One of Rockefeller's objectives was to refine a consistent, dependable product that would neither smoke nor explode but burn cleanly batch after batch, a "standard" oil. In 1868, Rockefeller formally organized the Standard Oil Company of Pennsylvania, the first to bear that name. In 1870, the Standard Oil Company was officially organized in Ohio and absorbed all of the Rockefeller interests.

Rockefeller and his associates meant to bring order to the industry by dominating it, which they set out to do with great success. Rockefeller felt that the road to dominance lay not in producing oil but in refining, transporting, and marketing it. His resources and acumen enabled him to undercut his rivals on transportation costs through secret deals with the railroads. Later, he snapped up the pipelines at purchase prices depressed by conditions he had created. Operators rarely knew when they were dealing with a Rockefeller company since many cleverly hid their alliances. Late to learn and slow to organize, many fell and were gobbled up or driven out of

John D. Rockefeller as he appeared in 1880 when he was indicted for conspiracy to monopolize trade and transportation in the oil industry.

business. Rockefeller justified his ruthless strategy by his self-proclaimed mission to bring order out of chaos.

One of the more interesting Rockefeller gambits was the South Improvement Company, created in 1872 by the Pennsylvania legislature "to construct and operate any work or works, public or private, designated to include, increase, facilitate or develop trade, travel, or transportation of freight, livestock, passengers or any traffic by land or water, from or to any part of the United States." The legislature authorized 2,000 shares, 900 of which were held by the Rockefeller interests.

The South Improvement Company contracted with the railroads for rebates on all petroleum and products shipped, rebates available only to members of the South Improvement Company, that is, Rockefeller associates. This set off an oil war in which the usually divided producers embargoed all crude shipments and forced the railroads to cancel the contracts. The Pennsylvania legislature then revoked the charter of the South Improvement Company and further granted the right of eminent domain to pipelines, breaking the railroads' hold on oil transport — but not Rockefeller's.

The setback hardly fazed Rockefeller, who soon formed the Pennsylvania Refiners Association, later the Central Refiners Association, with himself as president. There followed the "Treaty of Titusville," by which refiners agreed to buy crude only from the Petroleum Association at a price set by the latter. But as Standard's hegemony increased, the scattered independent producers soon lost their preeminence, and crude oil prices fell back into line. Rockefeller said, "We proved the producers' and refiners' associations were 'ropes of sand.'"

Standard Oil was not the first major oil company. Others had started to grow earlier, and many followed. In 1860, the oil exporting company of Warden, Frew and Company began operations in Philadelphia and, in 1866, merged with Peter Wright and Sons to become the Atlantic Petroleum Storage Company. In 1870, this company acquired a small refinery near Philadelphia and incorporated as the Atlantic Refining Company. Its independence was short-lived. It became an affiliate of Standard Oil in 1874.

In 1866, the Vacuum Oil Company was organized in Rochester, New York. It later became an affiliate of Standard of New York. In 1875, Continental Oil Company was founded as the Continental Oil and Transportation Company. Carter Oil Company came along in 1877 and fell to Standard Oil in 1894. Tidewater Oil Company got its start in 1878 as a pipeline company out to break the Standard transportation monopoly. Pacific Coast Oil Company began in 1879 as part of the Standard Oil Trust. In 1881, Kendall Refining Company set up in Bradford, Pennsylvania, to make lamp and cylinder oil. In 1887, Ohio Oil Company, later Marathon, was formed as part of Standard. In 1889, Sun Oil Company came along. Not all survived. John Wilkes Booth's aptly named Dramatic Oil Company began producing only after its organizer's death, and it faded quickly.

By 1879, Standard Oil, meaning Rockefeller, controlled 90 to 95 percent of the refining capacity in the United States. As the industry grew and followed the railroads across the country, Rockefeller began buying oil fields as well as refineries and pipelines. He put Standard Oil well on the way to becoming an integrated oil company on today's model.

To control this massive empire, Rockefeller in 1882 set up the Standard Oil Trust of New Jersey. The trust owned and exercised rights of ownership in the shares of individual companies that made up the trust. Eventually, 37 different firms incorporated or chartered under the laws of the various states in which they operated. In 1899, the trust was dissolved and converted into a holding company, the Standard Oil Company of New Jersey.

Despite his belief that he was providing a needed service, Rockefeller became an object of public fear and political hatred. Standard Oil provoked antitrust legislation at the national and state levels, with states sometimes faster to move than the federal government. Indeed, an antitrust action in Ohio drove Rockefeller to set up Standard Oil in New Jersey. Meanwhile, Texas law and regional chauvinism teamed up to keep Standard Oil from participating in the pivotal events taking shape in that state. Because of this, Rockefeller missed out on Spindletop and the other discoveries that occurred there during this period. Among companies that grew out of these discoveries were Texas Fuel Company (later, Texas Company) in 1901; J. M. Guffey Petroleum Company (later, Gulf Refining Company), organized with Mellon money from Pittsburgh in 1901; Gulf Oil Corporation in 1907; Humble Oil Company in 1911; and Houston Oil Company, later sold to Atlantic Refining Company.

As the new century arrived, the attack on Standard Oil gathered force. In 1906, the U.S. attorney general brought suit under the Sherman Anti-Trust Act against Standard and its 33 subsidiaries. In 1908, Missouri revoked the company's franchise to operate in that state. In 1909, the courts enjoined the holding company, Standard Oil of New Jersey, from exercising right of ownership in its 37 affiliates. Finally, in 1911, the Supreme Court ordered the dissolution of the Standard Oil group and forced the holding company to divest itself of all shares held in subsidiary companies.

Each component returned to the state in which it operated. For many years, the shares remained in the hands of the same group of men who controlled the original trust, including Rockefeller; but gradually own-

ROCKEFELLER: "I GUESS WE WILL KEEP ALL OF THEM"
—Morris in Spokane *Spokesman-Review*

JOHN D. IS GOING TO PUBLISH HIS AUTOBIOGRAPHY
TOLD IN SIX CHAPTERS

CHAP. I — I GRABBED EVERYTHING IN SIGHT

CHAP. II — I GRABBED EVERYTHING IN SIGHT

CHAP. III — I GRABBED EVERYTHING IN SIGHT

CHAP. IV — I GRABBED EVERYTHING IN SIGHT

CHAP. V — I GRABBED EVERYTHING IN SIGHT

CHAP. VI — I HAVE IT ALL — LET'S GO OUT AND PLAY GOLF

Popular opinion, expressed in cartoons like these, opposed Rockefeller's policies.

ership spread among millions of shareholders and true disintegration of the empire was complete. The holding company was dissolved, but the name lingered on in the many Standard Oil companies incorporated in the various states, including Standard Oil Company of New Jersey (now, Exxon).

Dissolution of the Standard Oil Trust set the pattern of the oil industry as it is known today in the United States. Some names have changed. Mergers and acquisitions have converted the larger companies into fully integrated firms that compete with one another at home and abroad. Production, refining, transportation, and marketing became functions of virtually all companies in the business as automobiles began rolling off Henry Ford's assembly lines and the oil appetite of a world at war became insatiable. There was no slackening of demand as the boom of the 1920s followed.

By 1900, the United States had produced its first billion barrels of oil, and 8,000 automobiles (4,000 of which were sold that year alone) were being driven on its roads. In 1901, the great Spindletop field came in, soon followed by new and bigger fields along the Gulf Coast and in Oklahoma. Howard Hughes, Sr., designed the tricone drilling bit in 1908, and the first service station opened in Detroit in 1911, followed three years later by the first modern service station in the East, built by Atlantic Refining Company. This station accented the growing importance of the automobile and consumer marketing to the oil industry. Gasoline became an increasingly important product, supplanting lamp oil.

Consumption expanded rapidly, but so did production as new find followed new find throughout the country. Montana, New Mexico, the Texas Panhandle, the Permian Basin, Michigan, Arkansas, Kansas, several important California fields and, again, Oklahoma added to the flood that kept prices depressed and increased the pressure for production control. Tulsa called itself the "Oil Capital of the World" and retained that title up until the East Texas field was discovered in the early 1930s. The Oklahoma City field literally drenched the capital building in a deluge of crude when the discovery well came in. In successive decades, U.S. production doubled and redoubled and redoubled again — 209 million barrels in

22

Howard Hughes, Sr., invented a bit with rotary cutters that was faster and cleaner than the fishtail it replaced.

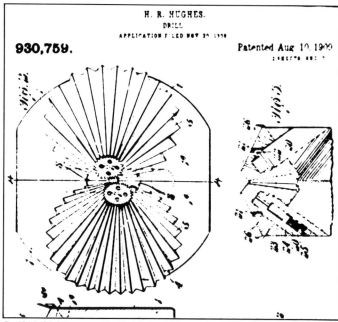

By 1914, this bit was used in 11 states and 14 countries. While bits have changed since then, the essential design has remained the same.

This service station at 40th and Walnut in Philadelphia was one of the first stations built by the Atlantic Refining Company.

1910, 443 million in 1920, 898 million in 1930.

By the end of the 1920s, it seemed that all of the easy oil had been found and that only little fields like the shoestrings of Michigan and Kansas remained. There was still plenty of oil production. New production from the old Bradford field of Pennsylvania, discovered in 1871, broke the market for eastern crude in 1930. Yet the world clearly seemed to be running out. The first man to say as much was a Pennsylvania state geologist who had predicted in 1874 that the country would run out of oil in four years.

Meanwhile, automobile production was outpacing the oil industry's ability to produce gasoline. Henry Ford's adaptation of the assembly line for automobile manufacture lowered the price of the Model T, and the American automobile industry, which had produced 4,000 cars in 1900, manufactured 187,000 in 1910. The more people bought "Tin Lizzies," the more the demand for gasoline increased. The growth of the automobile industry stimulated growth in the petroleum industry, but new sources were not found immediately.

Many prospectors were driven by the dream of the big strike. One of these, Marion Columbus ("Dad") Joiner, an itinerant wildcatter said to have found the first oil in Louisiana's Caddo field in 1905, believed that intuition worked as well as geological study in locating places to drill for oil. In 1927, instructed by his intuition, Joiner leased 10,000 acres of Daisy Bradford's ranch in East Texas, an area written off both by oilmen and geologists. Dad Joiner had a wooden derrick, a dilapidated rotary drill, and two mismatched boilers nicknamed "Big Joe" and "Little Joe". He managed to sell off enough leases to raise money to drill 3,000 feet, but ill luck dogged him. He abandoned Daisy Bradford No. 1 at 1,098 feet after six months, moved his rig 100 feet and spudded Daisy Bradford No. 2, which he abandoned at 2,518 feet 11 months later.

By January 1930, Dad Joiner, by now 70 years old, wanted to spud Daisy Bradford No. 3. First, however, some changes had to be made. He persuaded Ed Laster, a

24

Dad Joiner (front row, left) congratulates geologist A.D. Lloyd at the East Texas discovery well, Daisy Bradford No. 3. Oilmen and geologists tried to discourage Joiner, but he never gave up on the idea that he would find oil in East Texas. His persistence paid off, but he never reaped financial success from his discovery.

skilled and determined driller, to take charge of the drilling operation. Then he had to shop around for more money to keep things going. He did not have much left to sell, but he let eight more acres go for $125 and sold stock in his venture just to pay for labor and supplies. Drilling resumed. On September 3, 1930, there was an oil show. During the following month, the well was shut down while Joiner scrounged for casing to complete the well. He found some, and the crew put the last of it in the well on the night of October 2.

Late in the day of October 5, working by the light of lanterns, Laster heard a muted roar deep in the hole. "Put out the fires!" he shouted. The crew barely had time to douse the flames of the lanterns when oil and water shot over the crown block and fell like a rainstorm on the 4,000 faithful who were watching. The well came in at 300 barrels a day. This show was respectable, not spectacular, but enough to get others interested.

Two weeks later, another well was spudded ten miles away. It came in on December 28 at 15,000 barrels per day. A month later, still more wells came in, one at 17,000 barrels per day. Clearly, Dad Joiner had stumbled onto something. His intuition paid off.

Indeed it had. At first people thought three separate oil fields had been discovered in East Texas. But soon they realized that Dad Joiner had barely nicked the edge of the field. At 6 billion barrels, it was the biggest field in the United States, bigger than the next 20 largest fields put together.

In all, more than 25,000 wells were drilled in the East Texas field, but little good that did Dad Joiner. He suffered the fate of many for whom the search is everything, the achievement anticlimax. He had sold it all to bring it in, and when he did, there was little left for him. Fast movers like H. L. Hunt and Harry Sinclair made the fortunes in East Texas. They enjoyed the benefits of Dad Joiner's intuition.

Like Titusville and Spindletop, East Texas marked another major juncture for the oil industry. It was not a case, as Spindletop seemed to have been, of Providence answering a need. Despite the initial sense of success at finding such a huge field, East Texas production thoroughly wrecked the market with a flood of oil. The price of oil plummeted. The posted price dropped to 15 cents a barrel within a few months. It had been $1.10 when the discovery well was completed. By August 1931, the price had dropped to ten cents a barrel, and some sales went for as little as a penny a barrel. The field was headed toward production of one million barrels a day. Something had to be done.

The situation precipitated the adoption of the nation's first prorationing laws, which oilmen conceived as market protection, but which ultimately served as a primary tool for conservation. Unknowingly, the East Texas operators were among the earliest environmentalists.

CHAPTER 3 –
U.S. SINCE 1931

Dad Joiner profited little from his enterprise and persistence in East Texas. He had, as they say about prize fighters, sold at least 110 percent of himself to finance his miraculous wildcat. Though he signed over his last 10,000 acres of leases for $1,250,000 to the legendary H. L. Hunt, he saw little of that money. What money he had left he spent on more wildcats and lawyers. Joiner spent much of his remaining 17 years in court, suing and being sued, and eventually he died broke.

Joiner said he had no regrets. He was content to have proved the existence of oil where none was thought to be and to have discovered the greatest field ever found up to that time. His fate was familiar to oilmen by that time: Drake, Higgins, Lucas, Guffey, and others had by their vision and persistence changed the world, yet died with little to show for it.

The profit often went to those who came afterward, those with the business and financial acumen to finance, transport, manufacture, and market the oil that others found. Those who elbowed aside the pioneers went on to build the companies and fortunes. Few of today's great oil companies can trace their beginnings to an original founder who single-handedly drilled an important discovery well in a new field. Virtually all trace their corporate beginnings to businessmen and financiers who were able to exploit and build on the initial discoveries of others.

The great East Texas field wrenched the oil industry and, indeed, the legal fabric of the nation as has no other oil development before or since. The Depression had slowed demand, and this flood of oil put enormous stress on the nation's storage, transportation, refining, and marketing facilities. At one point, East Texas was producing a million barrels of crude every day, fully one-third of all U.S. production. Prices naturally tumbled.

Older, less prolific fields throughout the country found it uneconomic to produce. They shut down, intensifying the effects of the depression. Some 600 other producing fields were also forced to close down by the East Texas glut, which made a bad economic situation worse in

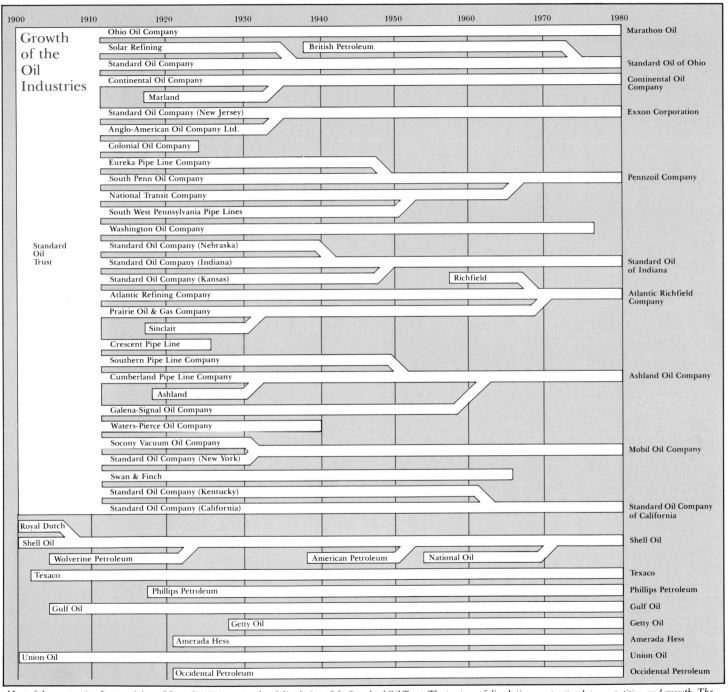

Growth of the Oil Industries

1900	1910	1920	1930	1940	1950	1960	1970	1980	

Ohio Oil Company → Marathon Oil

Solar Refining → British Petroleum

Standard Oil Company → Standard Oil of Ohio

Continental Oil Company → Continental Oil Company

Marland

Standard Oil Company (New Jersey) → Exxon Corporation

Anglo-American Oil Company Ltd.

Colonial Oil Company

Eureka Pipe Line Company

South Penn Oil Company → Pennzoil Company

National Transit Company

South West Pennsylvania Pipe Lines

Washington Oil Company

Standard Oil Company (Nebraska)

Standard Oil Company (Indiana) → Standard Oil of Indiana

Standard Oil Company (Kansas)

Richfield

Atlantic Refining Company → Atlantic Richfield Company

Prairie Oil & Gas Company

Sinclair

Crescent Pipe Line

Southern Pipe Line Company

Cumberland Pipe Line Company → Ashland Oil Company

Ashland

Galena-Signal Oil Company

Waters-Pierce Oil Company

Socony Vacuum Oil Company → Mobil Oil Company

Standard Oil Company (New York)

Swan & Finch

Standard Oil Company (Kentucky)

Standard Oil Company (California) → Standard Oil Company of California

Standard Oil Trust (left margin label)

Royal Dutch → Shell Oil

Shell Oil

Wolverine Petroleum → American Petroleum → National Oil

Texaco → Texaco

Phillips Petroleum → Phillips Petroleum

Gulf Oil → Gulf Oil

Getty Oil → Getty Oil

Amerada Hess → Amerada Hess

Union Oil → Union Oil

Occidental Petroleum → Occidental Petroleum

27

Most of the companies shown originated from the 1911 court-ordered dissolution of the Standard Oil Trust. The purpose of dissolution was to stimulate competition and growth. The plan succeeded, since most of these companies now have a net worth much greater than the old Standard Oil's $666 million.

many parts of the oil country, notably in Pennsylvania.

For a while production was unrestrained as thousands of independent oil producers in the East Texas field operated as independently as possible. Each one insisted on his God-given and constitutional right to take as much oil as possible as fast as possible. To attempt to control production was to infringe upon the rights of private property. The independents remembered the Standard Oil monopoly. They regarded any attempts at control as conspiracies by the moneyed interests and major oil companies to sweep them up or put them out of business.

Many oil people long had realized that unrestrained production could kill a field before its time, increase eventual production costs, and, in the end, leave more oil in the ground than was taken out. Producers realized the need for conservation as early as 1865, when Pennsylvania oilmen at Tarrville agreed to case and pack their wells to preserve the field. Many states passed conservation laws before 1900, but it took East Texas and the Depression to bring the need for conservation into focus.

At the root of the problem were two complicating legal concepts. One was the law of capture, which held that oil belonged to the person who found and produced it, however and whenever the oil got there. Thus, while sound reservoir engineering regarded a pool or field as a unit where what affects one part affects all, the law of capture encouraged producers to think only of their own drill holes and to try to extract as much as they could, even if it meant draining a neighbor's well, as it often did. It was like one soda with many straws. The child who sucks the fastest gets the most.

Second was the concept of land ownership. In most countries, landowners have claim only to what is on the surface, and the government owns the mineral rights underneath. In the United States, landowners claim the surface rights, all the minerals beneath, and the air above. They may deal or not as they please, and with whomever they please. Where there are many owners, as in East Texas, to enforce a uniform approach to an oil pool is very difficult.

East Texas, therefore, made urgent the need for establishing some rules of conservation or proration to

28

Who Gets What Share?

Mineral rights belong to the landowner (left), but the courts upheld the rule of capture, by which a bird belonged to the man on whose land it fell, not to the hunter; and a man who dug a well got the water.

This confusion beset the new oil industry. The oil belonged to the property owner, but the rule of capture said that it belonged to the man who drilled the well. Producers sank wells fast to offset their neighbors.

The courts unraveled the snarl by reconciling the rule of capture, property rights, and the state's police powers. This code encourages true conservation. Each property owner gets his share, and the pool produces more oil.

prevent wasting a valuable asset. Unfortunately, the loudest and strongest voices in favor of proration belonged to the major oil producers, upon whom the independents largely relied for transporting, storing, and marketing their oil. Suspicions were aroused not only among the independent producers, but also in the public mind and in the courts. Any moves to restrict production were equated with price fixing instead of resource conservation.

A long and sometimes violent battle ensued, during which the governors shut down the fields in Oklahoma and Texas and placed the area under martial law. Federal and state governments participated in regulating the oil industry. Gradually, the concept of field unitization took hold.

In East Texas, bootleggers ran not liquor but "hot oil," which was produced in violation of state proration regulations. Gun battles between operators and enforcement agents were as common as bribery. Illegal refineries, which rivaled the illegal stills and breweries of the concurrent Prohibition Era, flourished. In the absence of effective state enforcement, major oil companies tried to exercise some control through pricing and reduced production from their own properties. But nothing worked. Actual production bore no relation to the allowables set by the Texas Railroad Commission, the agency responsible since 1917 for refereeing the state's petroleum industry.

Everyone's goal was "dollar oil." But that price could not hold when hot oil went for ten cents or less. Posted prices had little effect, even though the major companies controlled transportation and could pick and choose from whom to buy oil. Posting a low price did little to stem the flood. Posting a higher price opened the valves wider. Only on the eve of World War II did oil reach the dollar-a-barrel level.

In the end, the oil industry came under federal regulation through the National Industrial Recovery Act (NIRA), subject to the not very friendly oversight of Harold Ickes, secretary of the interior and administrator of the oil industry provisions of the National Recovery Administration (NRA). The oil states themselves had encouraged federal intervention through their own inabil-

ity to control the oil trade. The Supreme Court tossed out the NRA in 1935, but at about the same time, Congress passed the Connally "Hot-Oil" Act, which prohibited the movement in interstate or foreign commerce of oil or products manufactured or withdrawn from storage in violation of a state law or regulation. In effect, illegal production became a federal offense.

Though a little late, the states were getting themselves organized. Earlier in 1935, a congressional committee headed by William Purington Cole, Jr., of Maryland recommended no federal legislation but urged the states to adopt an interstate oil compact. This they did. By the end of the year, the Interstate Oil Compact to Preserve Oil and Gas was passed by Congress and signed by the president. This established the Interstate Oil Compact Commission. Most states have since become members, though the moving spirits then were Oklahoma, California, Texas, and New Mexico.

The East Texas field finally came under control with the Oil Compact in place, a federal law with teeth threatening the hot-oil bootleggers, and the Texas Railroad Commission in the strong hands of General Ernest O. Thompson. The hot-oil wars abated, and the last great oil boom, with all the trimmings of shantytown and rampant individual enterprise, was over. Exciting times promised to follow, and wildcatters would still be around. The concepts of conservation, proration, and pool management were now firmly established, though as late as 1962, courts still ruled that oil belongs to the person who finds it. The law of capture has not been repealed.

After that, oilmen developed fields with an eye to their ultimate productivity, with wells spaced accordingly and output paced to reasonable market demand. While wildcatters continued to discover new fields by guess and by hunch, there were no more Dad Joiners or oil fields like East Texas in the United States. Instead, there were the well-financed independents and the established oil companies who had the resources to hire geological crews and develop and apply new technology and equipment, as well as the logistical backup to maintain expensive exploration crews in remote areas.

Those turbulent years forged the oil industry and its relations with government in the shape still generally known today. The major corporate names existed or came into being in the 1930s. Richfield Oil, for example, was incorporated in California in 1937. Twenty six large companies, with Standard Oil of New Jersey far out in front, together owned two-thirds of the capital structure of the industry and controlled 60 percent of the drilling, 90 percent of the pipelines, 70 percent of the refinery operations, and 80 percent of the marketing. In the ensuing years, mergers and consolidations reduced the number of names, but strong independents and aggressive competition have also reduced the dominance of the largest companies.

The 1930s was a time of technological innovation. The first commercial catalytic polymerization unit to make 100-octane aviation gasoline began operating in 1935. In 1934, a research chemist at DuPont Laboratories spun a new fiber called polymer 66 which, as nylon, was to have great implications for the petroleum industry. Practical researchers in many areas produced a flood of new developments to help find new oil, drill and log wells, and monitor the wells and reservoirs. The results were deeper wells, production from below 10,000 feet, multiple completions in the same well, offset wells, and the first ventures into offshore exploration, drilling, and production, albeit in shallow waters.

Many exotic and mundane developments, now commonplace, came during the 1930s. Among these occurrences was the Chicksan Company's invention of the metal swivel joint that made all-metal lines flexible. Phillips Petroleum Company designed the first pipeline for simultaneous movement of different products over long distances. For the first time, a refinery used natural gas to make ammonia. Atlantic Refining Company developed the first all-welded seagoing vessel. Cameron Iron Works patented a pressure-operated blowout preventer that soon had enormous value. The first polymerization process that used heat and pressure alone now appeared. Gun perforating, a method to increase production from a pay zone, was first used practically. The first acid treatment of a well increased production tenfold. Galveston Bay saw the first offshore geophysical surveys by boat. The first submersible drilling barge was used on a lake in Louisiana. Repressurizing a field through reinjection of

the gas produced with the oil was first demonstrated. Recording instruments were developed for continuous measurements in wells and during drilling. Photographic recording was first used. A scraper, or "pig," was first used to clean a large-diameter pipeline while it was still in use. The first well was drilled in the open waters off the Gulf of Mexico, and the discovery well of the Creole field, a mile off Cameron Parish, Louisiana, was spudded. In 1937, the Sun Company built the first commercial catalytic cracking plant at Marcus Hook, Pennsylvania. It took four years to build and cost $11 million.

Also during this period, Standard Oil chemists invented synthetic butyl rubber. Gamma rays were first used for well logging. Standard Oil of New Jersey invented the fluid catalytic cracker, which revolutionized the production of gasoline. Gulf Research Corporation invented the airborne magnetometer, which sought out oil-bearing formations from aircraft. Developed in 1939, this tool was used to trace enemy submarines during World War II before the oil industry could enjoy its benefits.

Oil companies themselves made some of these innovations, but most of this constant and growing stream of technology came from the oil service industry, which was at least as old as the oil industry but saw its greatest development in the 1930s and afterwards.

After all, Drake got his pipe and machinery from local suppliers who were already providing equipment to brine-well drillers. It did not take these firms long to see the new opportunities. Advertisements soon plastered the oil country: Keystone Iron Works of Conneautville and Titusville, Pennsylvania, "Oil tools and machinery, very best materials"; Barr & Johnson of Erie, "Oil Pipe Made to Order"; and Liddell & March of Erie, "Engines, drilling tools and pipe for oil wells." Many fortunes were made and many services rendered by men who turned their ingenuity not to finding, refining, or marketing oil but to providing the necessary tools.

Many names are still familiar: Halliburton for cementing and well treating; Hughes for the tri-cone bit; Dowell for well treatment; Dresser for derricks and other drilling equipment; O'Bannon for pump rods; Schlumberger for electrical well-logging; Baroid for sophisti-

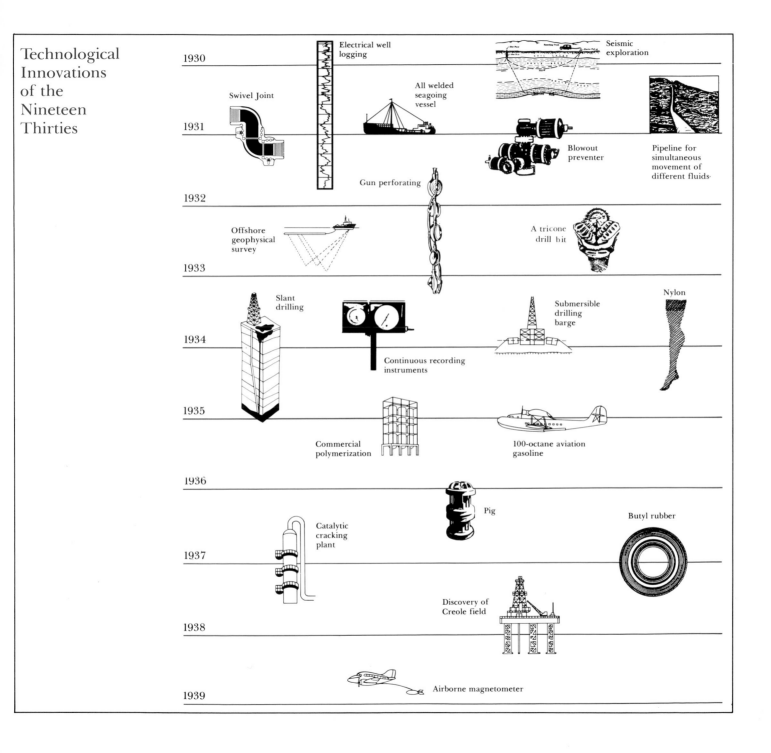

Technological
Innovations
of the
Nineteen
Thirties

1930

1931

1932

1933

1934

1935

1936

1937

1938

1939

Electrical well
logging

Seismic
exploration

Swivel Joint

All welded
seagoing
vessel

Blowout
preventer

Pipeline for
simultaneous
movement of
different fluids·

Gun perforating

Offshore
geophysical
survey

A tricone
drill bit

Slant
drilling

Submersible
drilling
barge

Nylon

Continuous recording
instruments

Commercial
polymerization

100-octane aviation
gasoline

Pig

Butyl rubber

Catalytic
cracking
plant

Discovery of
Creole field

Airborne magnetometer

31

cated drilling muds; Reed Roller Bit for core drills, reamers, tool joints, and rock bits; A. O. Smith Corporation for high-yield casing; Brown Oil Tools, Inc.; Bowen Company of Texas; McCullough Tool Company; Unit Rig and Equipment Company; Wagner Morehouse, Inc.; and Houdry Process.

War came to Europe again in 1939. Though the United States would not be involved for another two years, the American petroleum industry was affected early. The Bureau of Mines began a monthly survey of aviation gasoline (avgas) to cover stocks, production, and demand, just in case. The bureau later extended the survey to the expected aviation-gas content in crude oils.

Early in 1941, President Roosevelt declared a state of unlimited emergency and set up the Office of Petroleum Coordinator for National Defense, with Harold Ickes as coordinator. Ickes immediately took control of all production in California and appointed 78 industry leaders to the Petroleum Industry Council for National Defense. After the Japanese attack on Pearl Harbor, the council became the Petroleum Industry War Council with responsibility for mobilizing oil-industry resources to deal with the war emergency.

As petroleum administrator for war, Ickes divided the United States into five Petroleum Administrative Districts (PAD), which still exist, with regional and district committees for production, transportation, refining, and supplies. I served on one of these committees as an independent in PAD 3, the Texas Gulf Coast area.

The earliest crises dealt with German U-boats, which were sinking tankers almost as fast as the United States could send them to sea. In those days, the coasts of Florida and the Hatteras banks were blackened with oil and flotsam. The skies were greasy with the smoke of ships burning within sight of golfers on the shore. Millions of tons of ships and oil and thousands of men went to the bottom of the sea in those months when the U-boats seemed to rule the oceans. German submarines lay in packs off Miami Beach, safe from a nation whose antisubmarine technology was sadly outdated.

While there was little the oil industry could do to prevent sinkings, it was uniquely equipped to solve the manufacturing crises. The United States and its allies

32

Harold Ickes (foreground) and oil-industry leaders at Phoenixville, Pa. dedicated the world's largest pipeline during World War II, one of many examples of the manner in which Ickes united the forces of government and industry.

From February 1 to the end of May 1942, 50 tankers, such as Standard Oil's R. P. Resor *pictured here, were sunk off the East Coast of the United States.*

desperately needed 100-octane aviation gasoline. Thanks to the Bureau of Mines survey, the United States knew how much avgas it had, how much could be made, and how much potential there was in the nation's crude. The industry lost no time gearing up its refineries in response to the government's request to boost avgas output. Within 24 months, production increased to meet the needs of war.

Yet another crisis was rubber. The United States had no natural rubber at all, and the Japanese advance into the rubber-producing areas of Asia cut off virtually all supplies. Fortunately, American research chemists had developed several types of synthetic rubbers, all based on petroleum. The government undertook a massive synthetic-rubber program.

In brief, the government, with the help of industry, designed a standard plant to manufacture a rubber based on styrene and butadiene, two chemicals derived from petroleum. Then a number of these huge, nearly identical plants were built around the country. Industrial firms,

some of them oil companies, some chemical companies, and some joint efforts, operated the plants and produced millions of tons of rubber for the wheels of war.

After the war, the government sold all these plants to private business. They formed the nuclei of what became great petrochemical manufacturing complexes. Many of these installations still exist.

The end of the war saw a remarkably fast and surprisingly easy transition and return to normal market conditions. When the Petroleum Industry War Council was dissolved at the end of 1945, the oil industry had already returned to peacetime competition. Gasoline rationing ended almost at once, and ample supplies of high-quality fuel immediately became available to the public.

The plastics and petrochemicals industries increased in importance. Many new compounds, laboratory curiosities before the war, found the technology and the uses to make mass production feasible. After the war, technology, demand, and capacity brought the great explosion of synthetic goods and expansion of the petrochemical industry that continues today.

The organization of the domestic petroleum industry did not change. The integrated major companies, generally considered the top 15 or 20 in the industry, and the large integrated independents all had much the same organizations as they had before the war. Basically, the companies had four major divisions: (1) production, including exploration, development and drilling, and activities related to production; (2) transportation and storage (basically, pipeline and marine); (3) refining; and (4) marketing. This structure is still used today.

Until long after the war, the United States remained the major petroleum producer in the world and a net exporter. In the 1950s, however, U.S. production was challenged when the Middle East fields began producing great quantities for world markets, largely under the control of the seven major international oil companies, dubbed the Seven Sisters. Five of these companies are based in the United States: Gulf, Texaco, Mobil, Socal, and Exxon. Royal Dutch/Shell is British-Dutch, with a 67 percent-owned subsidiary in the United States, Shell Oil, itself the fourteenth largest company in the country. The

33

seventh company, British Petroleum, has also become an important factor in the United States through Sohio, its subsidiary, and the latter's holdings in Alaska.

The United States, with its millions of cars, was the logical market for all the inexpensive and profitable foreign crude. The battle of the pump followed. New gas stations sprouted everywhere as the companies tried to convert their low-cost crude into cash at the corner gas station.

During this period, the very large companies with large foreign holdings saw great growth and prosperity. They enjoyed a market strength they never lost, despite vicissitudes not long in coming.

At some point in the mid-1960s, the Middle East passed the United States as the world's largest producer. About then, the United States began to depend heavily on imported crude to supply half of the oil used for domestic consumption. More ominously, while domestic crude production peaked and began to decrease in 1970, the discovery and addition of new domestic reserves fell even faster. In the Middle East, reserves increased much faster than production. What the doomsayers had been foretelling for a century looked real. The United States could run out of oil.

In the past, when doomsayers predicted that the United States was about to run out of oil, it always seemed that oilmen found new sources to relieve the pressure. The new oil, however, usually was more expensive to find, produce, and transport than the old and had to be sought in inhospitable places. In 1968, the Atlantic Richfield Company (ARCO) drilled the discovery well on the North Slope of Alaska and brought in the biggest field in North America ever, at least twice the size of the East Texas field. With one stroke, the proved reserves of the United States doubled.

The Prudhoe Bay field was no easier to find than any other. Geologists hoped to find oil in the barren tundra sloping down from the Brooks Range to the Arctic Sea. But it took drilling many dry holes, under severe conditions, before Prudhoe Bay State No. 1 came in at about 8,600 feet, testing some 2,400 barrels a day. Another well, seven miles away, confirmed the find. By then, British Petroleum, the first drilling company, had given up. But

it soon came back and is now a major operator.

Finding Alaskan Arctic oil was one thing. Getting it to the market was another. For a decade, none went to market. Alaska certainly could not use a million or two barrels of oil daily, and there was no apparent way to move production to the "Lower Forty-eight." Oilmen brought much ingenuity and imagination to the problem, including schemes that used ice-breaking tankers and submarines that traversed the North Pole. As it turned out, the only practical solution was a pipeline.

The Trans Alaska Pipeline System (TAPS) was one of the greatest engineering achievements of the twentieth century. It was conceived as an economically feasible, though very expensive, way to get the oil from Prudhoe Bay to the ice-free port of Valdez, 800 miles south. There, tankers were loaded to carry the oil to markets in the rest of the United States at a price competitive with other sources.

Engineering design problems associated with the climate and terrain, plus a multitude of environmental demands, rules, regulations, litigation, and legislation, delayed construction of the line for years. The initial cost estimate of $900 million skyrocketed to nearly $9 billion. Oil finally began flowing in June 1977, some two years after construction actually began. Today, TAPS supplies nearly 10 percent of total U.S. crude demand.

In 1973, the United States learned how dependent it was on foreign oil and, more important, that this source of energy is finite. It will run out. In that year, the Organization of Petroleum Exporting Countries (OPEC), which represented 35 percent of world production and one-third of U.S. consumption, first declared an embargo on oil shipments, then increased the price of crude many times. As a result, the American petroleum industry became inextricably tied to international developments. It could no longer rely entirely on its own resources to meet its energy needs.

Through a cartel that John D. Rockefeller might envy, competition in the international area was throttled. While there was enough oil in the world to depress markets and drive down prices, as in the days of East Texas, most production and reserves were so firmly under cartel control that no nongovernment entity could produce and

Or buy a Volkswagen.

In the period after the 1973 oil embargo, the United States learned that it had to conserve its use of oil.

market enough oil to affect the price against the cartel's wishes. Only sharply reduced world demand and usage, brought on by high prices and recession, could, and in the early 1980s did, force the cartel to reduce prices. Price controls, which the United States imposed on the industry during the 1970s, unfortunately aided the OPEC countries by its so-called entitlement program. This actually subsidized imports of oil at the expense of domestic production.

Controlled at home, nationalized abroad, the oil industry in the United States languished in the late 1970s. Then, in January 1981, the federal government decided to decontrol oil. Given new incentive, the industry embarked on the highest level of activity ever witnessed. Drilling and exploration soared to an all-time peak. Despite an economic downturn that produced a temporary surplus of oil and doldrums in the industry, opportunities over the next 20 years appear to be unlimited. Careers in the petroleum industry offer great prospects for young people today.

I have covered 40 years in this chapter, the lusty youth and vigorous middle age of our industry. A person born in the 1930s and living a normal life span into the next century will likely have seen the United States consume 87 percent of all its oil. In another 50 years, petroleum promises to be a residual industry.

America's energy future, into the 1990s, lies with costly new technology and new sources, much of which will be the outgrowth of existing petroleum technology.

35

"HE'S CONVERTING HIS ZIPPO TO COAL."

Rising oil prices forced consumers to seek alternate sources of energy.

CHAPTER 4 –
INTERNATIONAL TO 1939

Just as Titusville, Pennsylvania, sprouted a forest of derricks in the mad boom following the Drake well, so drilling sprang up quickly whenever a strike occurred, or seemed imminent. Canada soon became a major producer from fields in Ontario north of the Pennsylvania oil patch. Derricks mushroomed in Rumania, Poland, Burma, Russia — wherever there was a smell of oil.

Preeminent among these new oil patches was the Russian city of Baku. The sea surrounding it was so redolent of petroleum spirits that it was routinely set afire for the amusement of visitors. The area had been the petroleum center of the Eastern Hemisphere and a focal point of commercial, political and religious strife for centuries. For all its ancient provenance and its mystic, mythical ties to religion and the supernatural, Baku soon became a lusty, brawling oil boomtown. It had everything connected with boomtowns, including a few local variations.

Always politically unstable, the region had been inhabited by Persians, Russians, Armenians, Turks, Christians, and Moslems of various sects, as well as the fire worshipers. Oil brought neither peace nor stability. The multilingual, multinational, multicultural mix of the people there had a flavor all its own. Baku produced more rich men than the entire U.S. industry. Yet the workers of the field, among them the Georgian Iosif Dzhugashvili (later Joseph Stalin), were wretchedly exploited and, unlike their U.S. counterparts, had little chance to enrich themselves.

The Tartar, Armenian, Russian, and Persian millionaires who came there lived well but precariously. They built enormous palaces; had ten servants for each family member, as well as escorts of armed retainers, imported Italian architects, and Russian harems; and built underground vaults with steel doors behind which to retire when one of the frequent riots swept the Peninsula.

During the revolutionary uprisings of 1905, the worst of these riots occurred. Moslems and their Russian Orthodox enemies slaughtered one another by the thousands. Prominent people spent weeks in underground shelters, sometimes to no avail. A Tartar mob besieged one Armenian millionaire in his mansion. The mob finally broke in and killed his family. Before his capture, however, the millionaire picked off 40 of the attackers with his rifle. Oil was truly a high-risk business in that setting.

It was a time of unfettered free enterprise in the Russian oil fields. Derrick piled upon derrick. The earth spouted oil in quantities not seen again until Spindletop. The average Baku well produced 280 barrels a day, and production rates of several thousand barrels were not unusual. Gushers were allowed to run wild. If the producer had enough land, he caught the spewing oil in earth ponds, but if he had no storage space, the oil simply ran away and was taken by whoever could catch it. Thus, the Droojba gusher, though it produced 40,000 barrels a day, bankrupted its Armenian owners because they had no means of storing it. Others captured the oil and profited while the owners had to pay for cleanup.

Caspian gushers produced immense quantities of sand along with the crude. Oilspouting sand volcanoes 350 to 400 feet high were common. Sand sometimes completely buried buildings and facilities .

Chaotic as conditions were, Baku was not at first free and open to all comers. When the Russians took over the area from the Persians in 1806, they had made oil concessions a monopoly of the crown. The czar in turn farmed out the monopoly to a man named Meerzoeff, who built the first real refinery in Baku in 1861. Meerzoeff had neither the resources nor the will to develop the properties, and so their immense potential was not realized for many decades. By that time, the United States had firmly established itself as the leading petroleum producer and exporter and had left Russia far behind.

Charles Thomas Marvin, an English journalist who visited Baku frequently, wrote in 1884: "Experience in all countries, in all ages, has shown that nothing is more fatal to the development of an industry than for the state to render it a closed monopoly." The petroleum industry at Baku was no exception to the general rule. The protective system of the Russians, following on centuries of free trade under the Persians, stunted the growth of the petroleum trade. The industry grew, but its development

was nothing like it would have been had there been no government restriction. In addition, Marvin noted that "the fatal monopoly check upon foreign and native enterprise was not removed until 1872," that is to say, 12 years after American oil had already secured a foothold in the European market.

Indeed, the American industry supported itself during the Civil War and earned foreign exchange for armaments by exporting oil to Europe. John D. Rockefeller opened an office in New York in 1867 to handle his export trade. Soon Rockefeller's Standard Oil could boast that its product was "carried wherever a wheel can roll or a camel's hoof be planted." By 1872, Standard was well on its way to absorbing all domestic competition. None of significance had arisen abroad.

When Russia canceled the Meerzoeff monopoly, development at Baku began with a roar. In 1872 alone, oilmen sank 415 new wells. Foreign investment and expertise succeeded in getting oil out to market despite continuous social chaos. The Nobel brothers, Ludwig,

In the late 1880s, gushers were common in Baku. Like Spindletop, the area soon was invaded by investors. The plots were divided again and again until they were so small that people said that "you couldn't swing a cat by the tail between the derricks."

Robert, and Alfred, the last the inventor of dynamite, invested heavily in Baku. They spent millions on modern refineries, pipelines, tank cars, and ships. The Nobels were the first to commission the building of a ship specifically to carry oil, in 1877. By 1879, under private and competitive development, the Baku wells were producing a million barrels a year.

In the early 1880s, the French Rothschilds entered the field and built their Caspian and Black Sea Company into the biggest exporter of kerosene in the European market. In 1883, they completed building a railroad which went 550 miles from the relatively isolated Baku to Batum, on the Black Sea. From there, oil was shipped by water to European markets. In 1905, an eight-inch pipeline was built.

Europe demanded more of the heavy industrial oils than kerosene, which for the Baku refiners became a low-priced by-product that vastly undercut Standard's principal product. Standard's products were still found wherever wheels could roll, but they were meeting increasingly stiff competition.

Rockefeller, however, soon established his own network of national subsidiaries in various countries, complete with refineries and marketing organizations, and acquired his own tankers to enable him to ship oil to market cheaply. Supported by efficient management and accounting, still the model of the American petroleum industry, Standard was soon able to compete with Russian oil around the world.

Thus, the stage was set for the great competition for oil supplies and world markets that soon began. The old players — Russia as the world's leading producer and exporter, the Nobels and Rothschilds as the chief entrepreneurs — made way for new casts of characters. One cast brought together the Far East and the old Ottoman Empire. The second cast acted out its play in Persia and then the Middle East.

The first cast was more akin to Rockefeller than to Drake, Lucas, and Joiner. These men were hardheaded, cold-eyed entrepreneurs, businessmen, financiers, diplomats, and wheeler-dealers with names like Samuel, Kessler, Deterding, Gulbenkian, and Churchill. Their enterprises turned closely on national policies. Their companies were tied more tightly to their home governments than any in the U.S. industry would ever be.

Marcus Samuel, born in London in 1853, came of a family that dealt in curios and gifts, which included exotic shells from the Orient. In the 1880s, the family expanded into trading general merchandise and established a trading post in Hong Kong. That led to interests in Japanese coal, which, in turn, led to concerns in lamp oil. As the lamp-oil business grew, Samuel acquired tank ships and sold oil from Baku and the United States around the world. With Rothschild's help, he eventually challenged Standard Oil in the Far East. By persuading the controllers of the Suez Canal to pass his ships while barring Standard's on "safety" grounds, he was able to ship his oil much more cheaply than Standard could.

For some years, relying on his contract with the Rothschilds for distribution of their Baku products, Samuel was content to sell other people's oil. Then, in 1897, he formed the Shell Transport and Trading Company to produce and refine oil as well as market it. He got

Marcus Samuel, founder of Shell Oil, was so impressed by the oil tankers developed by Ludwig Nobel that he tried to gain an edge over Standard Oil by sailing his own fleet of Nobel-style tankers through the Suez Canal.

J.B. August Kessler built the Royal Dutch Petroleum Company into a world power. He died from overwork at age 47.

When Kessler died, Henri Deterding was elected to his position. Deterding's hopes for a merger of Royal Dutch and Shell were realized in 1907.

a concession in the Dutch East Indies, now Indonesia, which gave Shell a source of crude near its major markets.

The first field in the region had been discovered in northern Sumatra in 1880. The Royal Dutch Petroleum Company, an integrated firm with wells, pipeline, and refinery, operated these properties under Managing Director J. B. A. Kessler. But Kessler lacked transportation, markets, and development capital and could not sell all of his product. In 1896, he hired an ambitious young Dutchman, Henri Deterding, who had been working as an accountant in the Far East for the Netherlands Trading Society. Deterding's first task was to help beat off Standard Oil's attempted takeover of Royal Dutch, then to find markets for Royal Dutch's product in the face of determined Standard opposition and price cutting. Deterding fought tenaciously and built a fleet of tankers and depots throughout the Far East that enabled Royal Dutch to beat Standard on price and delivery. But Royal Dutch's crude position was endangered when its Indone-

sian wells began to produce salt water instead of oil.

Kessler, through connections, solved the crisis by buying a cargo of Baku crude at favorable prices. In the process, Deterding learned the value of alliances, particularly in the face of Standard Oil competition. He saw that Shell and Royal Dutch were on parallel courses, both operating in the Far East, both competing with Standard and one another, yet neither strong enough to take on Standard alone. To him, consolidation made a great deal of sense, but the staunchly nationalistic Kessler refused to consider any dilution of the Dutch character of Royal Dutch Petroleum.

When Kessler died in 1900, Deterding immediately began to woo Samuel toward a marriage of Shell and Royal Dutch. At first, Samuel was cool, but Standard's constant pressure eventually convinced him that if Shell was to survive, he had to find a way of stopping the worldwide price cutting that only Standard could afford.

In 1903, Samuel and Deterding agreed to organize a marketing venture with the Rothschilds. They created the Asiatic Petroleum Company, with Samuel as chairman and Deterding as managing director. This was the first step toward the eventual merger of the two companies. It was also the first of a pooling of resources, now commonplace in the petroleum industry.

The final consolidation came in 1907, when Samuel, backed to the wall by relentless price wars in Europe, had to sell six tankers to German interests and sought the merger with Deterding himself. Shell's financial position by then allowed it to have only a 40 percent interest in the Royal Dutch/Shell group, and Royal Dutch retained 60 percent, an apportionment that remains today. The uneven split humiliated Samuel and gave the group a problem. The British government regarded the merged company as a foreign corporation, even though Deterding promptly moved his headquarters to London. The merger gave Royal Dutch/Shell the financial power and corporate organization to take on Standard Oil on all fronts.

Simultaneous developments in the Middle East had long-lasting consequences for the oil industry. Considering the immense reserves of the Middle East, it seems incredible that the search for oil there was at first long and fruitless. This classic story of failure and persistence finally led to spectacular triumph. The players here were men of high-level international finance and politics, many of whom never saw an oil barrel, let alone visited the lands that produced oil. Their empires were based on the daring, imagination and persistence of wildcatters cast in the American mold — men like the Australian mining engineer William Knox D'Arcy and New Zealander Frank Holmes.

As early as 1872, Baron Paul Julius von Reuter, better known for his news agency, was intrigued by the historical reports of fire temples in Persia and concluded that there must be oil there. He obtained concessions to explore the mineral resources of northern Persia, but Russia considered the area within its sphere of influence and obliged Persia to cancel the concessions. In 1889, the persistent Reuter obtained more concessions south of the "Russian" province and drilled three wells, all dusters.

In 1900, William Knox D'Arcy, spurred by the same reports and aided by a geologist, began his search in Per-

At the turn of the century, William D'Arcy won a 60-year concession in Persia that became the basis for British Petroleum.

sia. He spent the fortune he had made in gold in Australia. After two years of traversing the inhospitable desert, he found nothing. Then he turned to the remote Qasr-i-Shirin district. But first, he needed a concession. He negotiated directly with the shah and "walked off with the country." For $100,000 in cash, another $100,000 in stock, and a 16 percent royalty, his First Exploitation Company obtained the exclusive right to drill in 500,000 square miles of Persian territory for the next 60 years. After two more years of baksheesh and bandits, as well as heat, disease, bad food and water, and immense logistical problems, one of D'Arcy's test wells showed a little oil, then went dry. And D'Arcy was broke.

London bankers who had been supportive when he was rich had no time for D'Arcy when he was broke. But the British Admiralty, looking to national defense, introduced D'Arcy to Burmah Petroleum people who had a lot of money and markets but not enough crude. D'Arcy assigned his shares in the First Exploitation Company to Burmah Oil for the money to keep drilling, under Burmah's watchful and soon despairing eye. Years went by.

In 1908, at a place called Masjid-i-Sulaiman, D'Arcy got a cable from Burmah Oil ordering him to stop operations. Money and patience had run out. But the wily D'Arcy decided to keep drilling until the confirming letter arrived in another four weeks. Two weeks later, on May 26, the well produced a monstrous gusher. It had taken D'Arcy eight years of relentless effort, his personal fortune, and a great deal more, operating under the harshest conditions to find what nature hid so well but now bestowed with such abundance.

To exploit D'Arcy's concession, Burmah Oil in 1909 organized the Anglo-Persian Oil Company, whose shares were also sold to the public. D'Arcy got his own money back as well as stock in Burmah Oil, but he lost control of his company. In 1913, the British government took a majority interest in Anglo-Persian through the initiative and manipulations of First Lord of the Admiralty Winston Churchill who disliked Royal Dutch/Shell and, wanting to secure a reliable oil supply for the navy, was most eager to keep APOC under British control.

Considerations of defense forced the British government to look everywhere for oil to fuel a war it knew was coming. Britain viewed growing German influence in the rotting Ottoman Empire with alarm.

In the late 1890s, Germany tried to gain concessions and build a railroad from Berlin to Bhagdad, which, to the British, meant Berlin to Bombay. Germany had acquired vast concessions from the sultan for a right-of-way and 20 kilometers on either side. The sultan, corrupt but no fool, believed that the Germans had something more in mind than a railroad. He commissioned a young Armenian from a kerosene-refining family to conduct a study of the land through which the German concessions ran. Then only 21, Calouste Sarkis Gulbenkian, who had gained attention by publishing a learned paper on oil in a French journal, reported that there appeared to be great oil potential in these lands, which included virtually the entire Middle East, except Persia. The sultan promptly had great tracts of land removed from its owners and

41

Reynolds (left), Williams, and Crush, members of D'Arcy's Persian exploration party, enjoyed a spot of "tea" near Masjid-i-Sulaiman in 1908.

placed on the Civil List, which is to say, he put them in his own pocket.

Beginning in 1912, Gulbenkian tried to organize a group to develop these lands. In 1914, the group met in London, where the first agreement was signed. At this meeting, Anglo-Persian Oil, the British government-backed company, now British Petroleum (BP), got 50 percent. Royal Dutch/Shell and the German Deutsche Bank got 25 percent each. For engineering the deal, Gulbenkian got 2½ percent each from Royal Dutch/Shell and Anglo-Persian. Eventually, he was known as Mr. Five Percent and became one of the wealthiest individuals in the world. But war intervened first, and the agreement, which formed the basis of the Turkish Petroleum Company (TPC), later the Iraq Petroleum Company, was worthless for years.

At war's end, the German interest in Turkish Petroleum was assigned to France. The Americans wanted in but were rebuffed by Anglo-Persian and Royal Dutch/Shell. But after many years of negotiations piloted by Gulbenkian with French support, plus increasingly tense diplomatic exchanges, an American consortium called the Near East Development Company and consisting of Standard Oil of New Jersey, Gulf Oil, Standard Oil of New York, Atlantic Refining Company, and the Mexican Oil Company obtained a 23¾ percent share in Turkish Petroleum, equal with the other three participants, Anglo-Persian, Royal Dutch/Shell and Compagnie Française des Petroles. Gulbenkian still retained his 5 percent.

The agreement required each participant to undertake all activities in the territory only through TPC and forbade any independent operations outside TPC's framework. Exploration and development began at once and soon brought results.

In June 1927, an immense gusher producing 95,000 barrels a day came in near Kirkuk, in Iraq. It was evident that there were enormous resources in the area. This whetted the American appetite for more production. Each American company had a much smaller share in TPC than did the three European participants. When the Americans claimed they could operate in the area but outside the limits of the TPC agreement, such as in Kuwait, their partners objected. Yet, in setting up the

Calouste Gulbenkian received his oil training in Baku where he was recognized quickly as an expert in the industry and became widely respected as a businessman.

consortium agreement, the participants had not defined the territorial limits with respect to independent operations. Such a definition became imperative.

Members met in London in 1928 and spent days haggling over the boundaries of the 1914 Ottoman Empire, which all concurred were the limits of the agreement. Finally, Gulbenkian came to the rescue. He called for a map, took a crayon and drew a red line around the Middle East, excluding Iran and Kuwait, and said: "That was the Ottoman Empire which I knew in 1914. . . . If anybody knows better, carry on." This became known as the Red Line Agreement.

Kuwait's exclusion from the agreement was the then unseen Achilles heel in the majors' control of world oil. Through this chink in the armor of protective agreements, independents like J. Paul Getty eventually gained access to Middle East oil.

While major companies and governments carried on their interminable discussions over who would exploit the Turkish Petroleum Company's properties, Frank

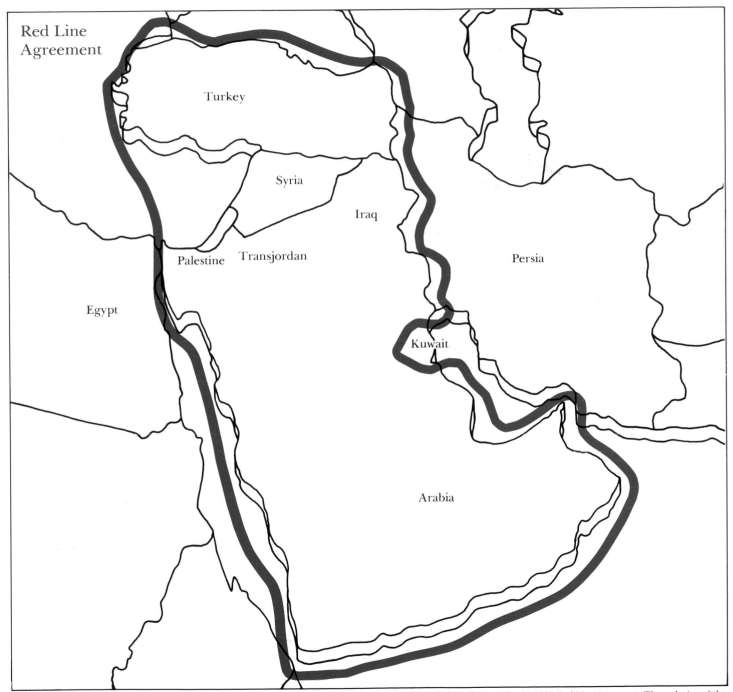

Red Line
Agreement

Turkey

Syria

Iraq

Persia

Palestine

Transjordan

Egypt

Kuwait

Arabia

This map shows the line drawn by Gulbenkian proscribing the pre-World War I Ottoman Empire and establishing the basis for the 1928 Red Line Agreement. The exclusion of the then British Protectorate Kuwait provided the loophole through which the first U.S. oil companies gained a foothold in the Middle East after World War II.

Holmes, a self-educated New Zealand mining engineer, began negotiations with the kings and sheiks of the Arabian Peninsula along the Persian Gulf, an area that major company geologists had written off, much as they had eastern Texas. Holmes, against the opposition of Anglo-Persian, succeeded in gaining concessions from King Ibn Saud of Saudi Arabia and the sheik of Bahrain.

Needing support to exploit the concessions, Holmes approached Shell, Burmah Oil, Anglo-Persian. When they showed no interest, he turned to the Americans. Gulf responded but was squelched by its partners in Turkish Petroleum. Finally, Standard Oil of California (Socal, now Chevron), a newcomer to the international scene, took up the concessions in Saudi Arabia and Bahrain.

In the early 1930s, after years of profitless exploration and drilling, these areas became prolific producers, and the American companies at last had their own firm foothold in the Middle East.

But Socal had a problem. New to the international arena, it had no distribution and marketing network. The major national companies and Standard Oil of New Jersey, irritated that they had been outflanked, were not inclined to help. The Texas Company had the distribution and marketing abilities. What it lacked was production in the area. In 1936, the two formed Caltex to produce, distribute, and market oil from their Arabian properties. This organization formed the basis of the Arabian American Oil Company (Aramco).

Frank Holmes got the usual pioneer's desserts. His syndicate sold its claim on a fractional royalty in the Kuwait concession to Gulf for a few thousand pounds. If he had he held onto it, as Gulbenkian held on to his, it would have been worth millions.

The Middle East differs from all other oil-producing regions culturally, politically, economically, geologically, and in its enormous reserves of crude. Once termed the cradle of civilization, it may now be the cradle of civilization's future, for good or ill. Its character was shaped largely between the two world wars by British and French interests seeking oil reserves and military advantages in defense of their empires. Despite the involvement of U.S. firms there, the region did not become vital to American interests until after World War II.

44

In 1922, Major Frank Holmes (standing) met with Abdul Aziz, Sultan of the House of Sa'ud, and Percy Cox, British High Commissioner.

On May 29, 1933, Abdullah Suleiman, Saudi representative, and Lloyd Hamilton, Socal representative, signed the agreement that became the foundation for Aramco.

The Middle East was an oilman's dream at a time when a 300-to-400-barrel-a-day well in most of the world was considered a bonanza. Wells there capable of producing 10,000 barrels a day were often plugged. Today, many wells still produce 40,000 to 50,000 barrels a day with natural pressures that also force the oil 50 miles or more through pipelines and directly into seagoing tankers.

Of course, there was plenty of oil elsewhere in the world; in fact, there was so much in the Western Hemisphere that American companies had little need to develop their Middle East holdings until the late 1930s.

Peru had an integrated oil industry of sorts when Francisco Pizarro landed there in 1527. Indians collected oil in trenches, boiled it to the desired consistency, and used it to waterproof containers. A Pennsylvania driller had a pumping well there in 1867, and in 1896 the country produced as much as 47,500 barrels. Peru remains a significant producer, though well down the world list.

Nearly all Latin-American countries have some petroleum production, though few started as early as Peru, and only two, Mexico and Venezuela, are currently significant on the world scale.

Mexico's beginnings may be as old as Peru's. Oil seepages have long been found on the east coast. The Aztecs used pitch and asphalt in their cities far in the interior as well. In 1876, an American sea captain found oil at Tuxpan, south of Tampico on the Gulf of Mexico. He refined it and sold kerosene for a time.

Despite small discoveries during the next 20 years, Mexico's annual production by 1900 was only about 10,000 barrels. That was enough, however, to pique the interest of Edward L. Doheny, a mild-mannered but successful California wildcatter, and his partner, Charles Canfield. The two former mining prospectors had parlayed a seven-barrel-a-day hand-dug pit hole into several important California fields and were looking for new worlds to explore.

They went to Mexico, which in those days was as unstable as the Ottoman Empire and as wide open as the Baku oil fields. But once again, in spite of an abundance of seeps and some small production, professional opinion held that oil would never be found in quantity in Mexico. Cecil Rhodes, the African empire builder, had spent half

Edward Doheny and Charles Canfield successfully explored for oil in Mexico. One of their first gushers was Juan Casiano.

45

Everett DeGolyer (seated), a young geologist fresh out of college, worked for Mexican Eagle. His innovative methods helped discover the Potrero del Llano No. 4.

a million dollars in an unsuccessful bid to find oil there.

Doheny and Canfield had better luck. Chopping their way through the swampy jungle near Tampico, they found oil bubbling out of a spring. They went on a land-buying spree and sewed up 450,000 acres, which they were able to do without attracting too much attention because everyone was rushing to Spindletop. But oil spring or no, they soon spent $3 million in extensive exploration and drilling, methodically developing their property. Their modest success was enough to attract an English engineer, Weetman Pearson (later Lord Cowdray), who staked out a concession next to the Doheny-Canfield properties. Pearson brought in a gusher that burned for 40 days. He went back to England for financing and formed the Mexican Eagle Oil Company.

In 1910, both Mexican Eagle and Doheny brought in spectacular gushers; one of them, Mexican Eagle's No. 4, Potrero del Llano, flowed 110,000 barrels a day. Gusher after gusher came in, and the area became known as the Golden Lane. With the British there, Standard could not be far behind, and Doheny soon became Standard's production arm in Mexico.

The oil industry does not operate in a political vacuum. Oil is intensely dependent on the good will of governments, both at home and abroad. Companies are seldom allowed to confine themselves to the business of getting oil out of the ground and into the machinery of society.

Mexico was a testing place and proving ground for the national aspirations and international politics now associated with oil. Revolution came hard on the heels of the oil boom. Dictators came and went. Private armies and bands of brigands roamed the land, extorting, burning and destroying. Parent governments of foreign oil firms became involved, and their wounded sensitivities led to testy diplomatic exchanges. The Mexican public became increasingly irritated over foreign exploitation of their resources and the behavior of foreigners on their soil.

The industry came under increasing pressure. The law that gave the landowner exclusive right to subsurface minerals was changed in 1917. The new Mexican constitution claimed all mineral rights for the state and gave

46

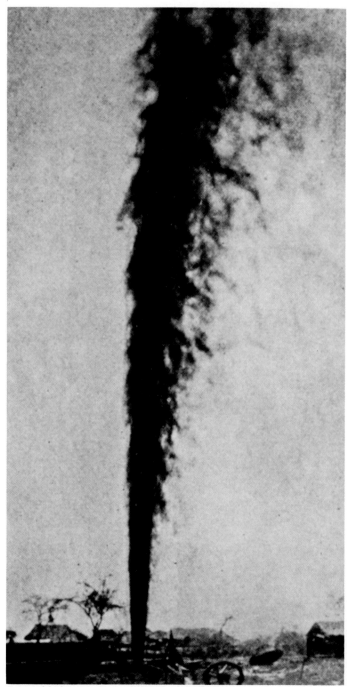

Potrero del Llano No. 4 came in spewing 110,000 barrels per day. It eventually produced 130,000,000 barrels, the greatest quantity any single well in the world has ever produced.

Mexican nationals the right to exploit them. Turmoil continued as new and more restrictive legislation was passed. Although by 1918, Mexico was the world's second-largest producer, and production peaked in 1921 at 193 million barrels, output steadily declined thereafter. Part of the decline was natural, as the fields played out. Even the Golden Lane began producing salt water. But just as much was institutional. Conditions did not encourage new exploration and production.

In 1938, the final blow came in Mexico. This first massive crack indicated that the basic foundation of the government oil structure was unsound, though it would be another 20 years before the edifice crumbled. On March 18, 1938, President Lázaro Cárdenas of Mexico issued a decree that expropriated all foreign oil properties, mostly American, British, and Dutch (Royal Dutch/ Shell had bought out Mexican Eagle in 1919). In June, a new law set up Petroleos Mexicanos (PEMEX), a government corporation, to handle the country's oil business.

The reaction of foreign companies and their governments was predictably negative. As a result, Mexico found no foreign markets for her oil and got little outside help in developing her resources. Production fell. PEMEX became an outcast in the oil world. Later, companies as well as governments wished they had been more clairvoyant and less hostile. What grew out of these events were suspicions that still exist between the United States and Mexico.

Venezuela, the other major Western Hemisphere producer, gave oil exploration concessions as early as 1866, though it did not become a producer for many years. There was little doubt that Venezuela had oil, and its proximity to the Panama Canal made it attractive for serving Asian markets. But it presented immense technical problems in production and transportation. A lot of money and expertise was required to solve these problems. Henri Deterding and Royal Dutch/Shell provided both.

Though others had worked earlier concessions, Shell first developed the Lake Maracaibo field and, in 1910, brought out the first production through a ten-mile long, eight-inch pipeline, the first of many to lace the region. Shell, through its Venezuelan oil concessions, in which the ubiquitous Gulbenkian had a hand, proved the prime mover in Venezuela, getting at one time one-sixth of its total crude from the country. But others soon followed. In 1920, Standard Oil of Indiana organized the Creole Petroleum Corporation to handle its Venezuelan concessions. Standard Oil of New Jersey organized a Venezuelan company in 1921 and in 1928 bought out Creole, making it the rival of Shell. By 1928, more than 35 American companies had concessions in Venezuela and controlled more than 50 percent of the country's output, although only a dozen or so survived the Great Depression.

In the late 1930s, company relations with Venezuela went through a crisis. In 1938, a new military government threatened nationalization if the companies did not accede to demands for higher taxes and royalties. Shell, which had lost its Mexican production and was backed by an English government anticipating war, willingly compromised in return for stability and a new 40-year concession. Though Standard Oil was truculent, the U.S. government also saw the need for assured oil supplies in the event of war. In time, all parties came to friendly agreement, and the other companies, like Shell, were happy to settle for stability. As in Mexico, a pattern had been set.

On the eve of war, the future Allies utterly controlled the world's oil.

CHAPTER 5 –
INTERNATIONAL SINCE 1939

48

Oil, or the lack of it, determined the outcome of World War II, as it had World War I. In short, the Allies had it; the Axis did not.

The Allies had unlimited oil reserves in developed fields as well as enormous manufacturing capacity far from the fields of war. Once they had beat back the German submarine attack, their armies swept across Europe and Japanese-held beaches in the Pacific. They were propelled on a tide of Western Hemisphere oil.

Allied forces safely stood astride the approaches to the huge reserves of the Middle East and blunted every thrust by Axis armies to reach Baku, Iraq, Iran, and the oil states of the Arabian Peninsula. Rommel ran out of fuel in the North African desert as sleek Italian cruisers carrying deck cargoes of gasoline cans tried vainly to supply him across the Mediterranean Sea. The ships were easy prey to Allied aircraft.

In the Pacific, Japan had gained control of the substantial oil production of Indonesia and Burma, but at the end of a very long supply line. While the Japanese surface fleet proved a worthy match for that of the Allies, American submarines virtually cleared the sea of Japanese merchant shipping, including tankers.

Despite adequate supplies of new equipment, the German and Japanese war machines slowly ground down for lack of oil. In Germany, the war spurred a huge synthetic-fuel effort from which the world may yet benefit, but it was not enough. Even with plentiful supplies of coal and the hang-the-cost attitude brought on by conditions of war, the Axis could not match the flood of cheap crude and the range of high-quality products available to the Allies.

The war nearly established the U.S. government in the oil business. Texaco and Socal, which held the Aramco concession in Saudi Arabia at the start of the war, believed that they would be unable to retain it unless the United States helped subsidize the Saudi government at the same level the British had. At that time, the chief source of Saudi income was not oil but the flow of pilgrims to Mecca. The war had stopped that trade.

Trooz, Belgium, was one of the main World War II fuel stations. Gasoline was shipped there by train and then pumped from railroad cars into empty cans.

By war's end, shortage of fuel forced the Nazis to use horses to transport their wounded.

The oil companies persuaded President Roosevelt and Harold Ickes to extend Lend-Lease aid to Saudi Arabia. The companies' concern piqued the government's interest in these vast new reserves at a time when the United States was again worrying about its own dwindling sources. Planners felt that the United States urgently needed additional reserves outside the country.

Texaco and Socal wanted government support rather than government involvement. But the United States favored the Anglo-Iranian Oil Company model, in which the government held controlling interests. Under Ickes's urging, the president authorized formation of the Petroleum Reserve Corporation, which was to acquire 100 percent of the Saudi Arabian concessions. Not surprisingly, Texaco and Socal resisted, agreeing to sell only one-third interest.

About this time, a government-sponsored geological expedition, headed by the eminent geologist Everett Lee DeGolyer, visited the Persian Gulf area. DeGolyer's prescient report stated: "The center of gravity of world oil production is shifting from the Gulf-Caribbean areas to the Middle East, to the Persian Gulf area, and is likely to continue to shift until it is firmly established in that area." Though it would be two decades before that shift was complete, a more accurate forecast would be difficult to find.

The war ended the German threat to the Middle East, and the oil companies quickly lost interest in the notion of government support and government participation. The Petroleum Reserve Corporation also faded. It was a shotgun affair demanded by war and not the sort of relationship that either the U.S. government or U.S. business was likely to consider in normal times.

But Harold Ickes did not give up easily. He next proposed a 1,000-mile pipeline to carry oil from Saudi Arabia to the Mediterranean. He wanted the government to build and own the line in return for 20 percent of the concessions, which would be held for naval petroleum reserves. Socal and Texaco agreed, but the rest of the industry solidly opposed it. So did the British, who saw their influence threatened.

Eventually, Texaco and Socal decided to build the line themselves. In 1945, they began construction of the Trans Arabian Pipeline, or Tapline, as it came to be called. For its time, the project was as daring as the Trans Alaska Pipeline System was later. It traversed scorching deserts and barren mountains far from sources of supply. But its greatest obstacles, ultimately, were politics and war.

The first Arab-Israeli War, in 1949, interrupted pipeline construction. Later that year, Syria and Lebanon at last permitted the line to cross their territories. Four years and $200 million after construction began, the first oil from Arabia was loaded into tankers for Europe at Sidon, in Lebanon. But the line was always a hostage to Middle East politics and passions, the frequent target of terrorists and guerrillas, and subject to the whims of governments currently in power. For all that, economic considerations finally shut it down. When tanker rates fell so low in 1975 that it was cheaper to ship Arabian oil by sea, Tapline became an empty steel string, 30 inches in diameter, which stretched a thousand miles across the land. Its silent pumping stations filled with drifting sand. In 1980, the Iraq-Iranian war created sufficient political pressures in the Middle East to see the line back in temporary operation, enabling Iraq to ship oil while avoiding the Persian Gulf.

At the end of World War II, just two companies, Texaco and Socal, sat on the world's largest oil reserves. This worried their competitors. Aramco could flood world markets with cheap oil and undercut the other companies, British and American, that were still restricted from independent operations by the Red Line Agreement of 1928. Socal and Texaco were not parties to the agreement and thus not bound by it to share the oil with the remaining Seven Sisters.

Standard Oil of New Jersey and Mobil, the American participants in the Iraq Petroleum Company, tried to break out of the Red Line Agreement even before the war. But Gulbenkian and the French interest would not consent. After the war, Standard and Mobil renewed their efforts to break out, claiming the French and Gulbenkian had forfeited their shares because both had been under enemy occupation during the war.

This claim set off a marathon of negotiations and horse trading which led to a potentially earthshaking lawsuit. Compagnie Française des Petroles (CFP) and Gul-

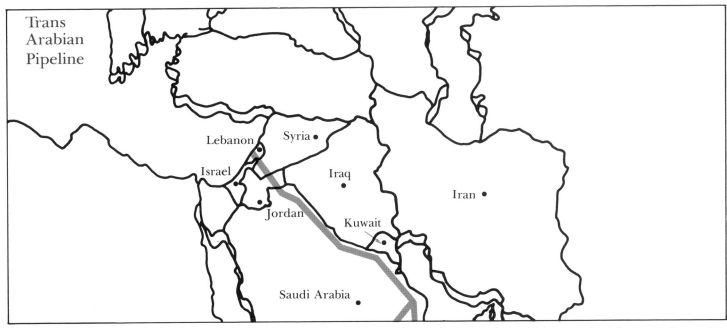

Trans
Arabian
Pipeline

Built to eliminate the long trip around the Arabian Peninsula and then through the Suez Canal, the Trans Arabian Pipeline provided access on the Mediterranean. Construction began in eastern Saudi Arabia and went west to a port in Lebanon.

benkian refused to give in without a dogged fight. As proposals and counter proposals were made and rejected, governments became increasingly involved. Smaller American companies watched with alarm. They feared that the Saudi Arabian oil reserves might become the private preserve of a very few large business and government interests. The shadow of the old Standard Oil monopoly seemed about to rise from the desert sand.

Eventually, the Americans pacified the French with an agreement that would help expand CFP production in its Iraq concession and also provide new pipelines to the Mediterranean. The agreement effectively kept the French out of Saudi Arabia. A United States Senate report 25 years later stated that the French never forgave the Americans for that.

Calouste Gulbenkian was not so easily satisfied. As Gulbenkian's son, Nubar, put it: "That [Red Line] Agreement could have precluded the Americans from taking an interest in Saudi Arabia, and so, quite bluntly and ruthlessly, but with the best of big business good manners, they set about trying to pull down the edifice conceived by my father.

"They took the attitude that as it was they who had won the war. . . the American oil companies should have greater freedom in the Middle East. . . . The Americans are a generous people, but they are also people with a strong business sense and, as a nation, are generally able to defend their commercial interests on ostensibly moral grounds." In this case, the moral grounds were the anti-trust implications of the Red Line Agreement.

Gulbenkian finally signed the new agreement at 2:00 A.M. on November 2, 1948, thus ending the Red Line Agreement. The new agreement left the Americans free to develop Saudi Arabia on their own and assured much more rapid development of Iraqi production of free oil. The delay was caused by Gulbenkian's fear that the Americans would expand Saudi Arabian production at the expense of Iraq's, thus reducing his income.

The settlement with Gulbenkian allowed a new Aramco to come into being. Standard Oil of New Jersey

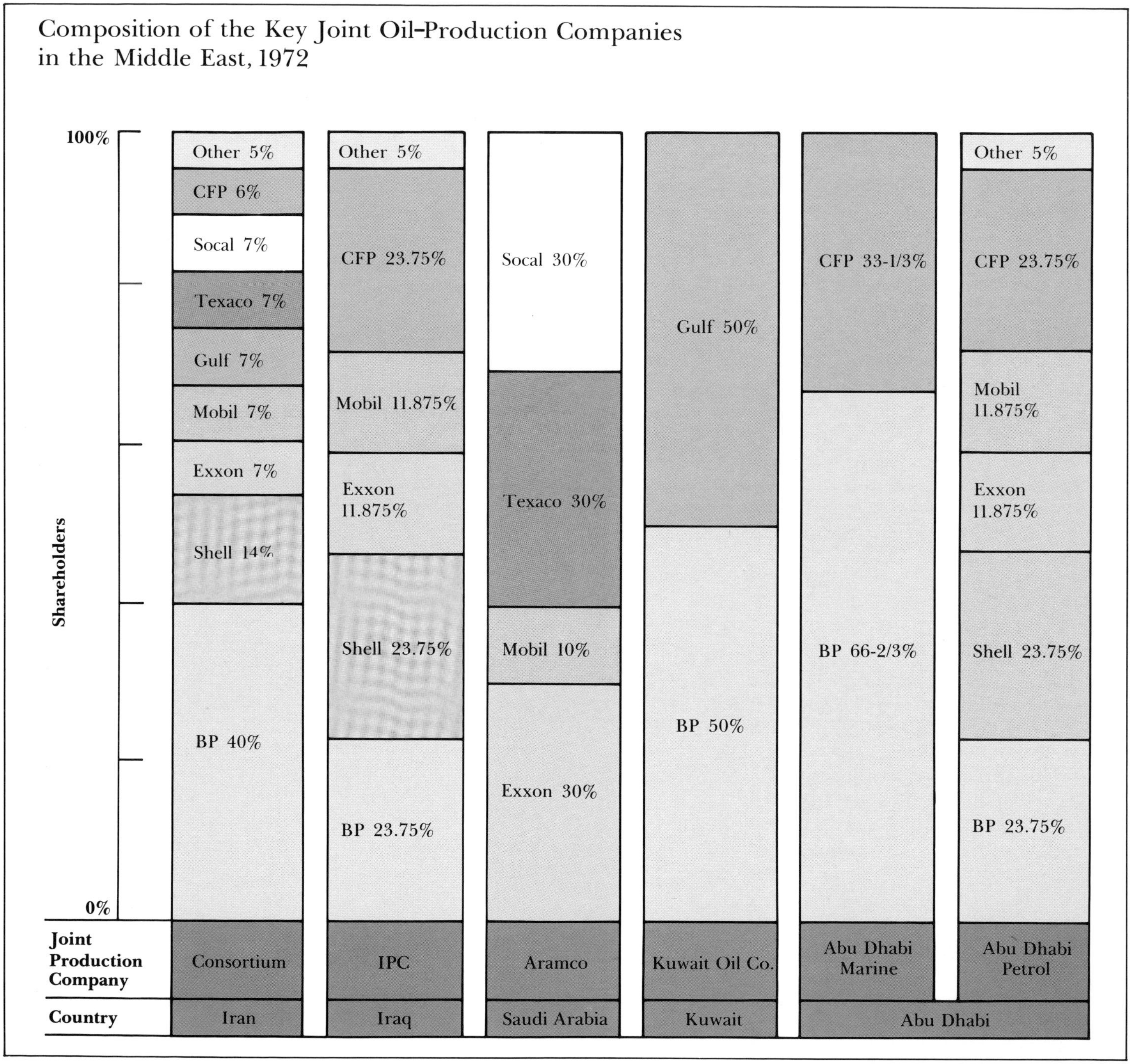

Composition of the Key Joint Oil-Production Companies
in the Middle East, 1972

Shareholders

100%

Consortium	IPC	Aramco	Kuwait Oil Co.	Abu Dhabi Marine	Abu Dhabi Petrol

Consortium (Iran)
- Other 5%
- CFP 6%
- Socal 7%
- Texaco 7%
- Gulf 7%
- Mobil 7%
- Exxon 7%
- Shell 14%
- BP 40%

IPC (Iraq)
- Other 5%
- CFP 23.75%
- Mobil 11.875%
- Exxon 11.875%
- Shell 23.75%
- BP 23.75%

Aramco (Saudi Arabia)
- Socal 30%
- Texaco 30%
- Mobil 10%
- Exxon 30%

Kuwait Oil Co. (Kuwait)
- Gulf 50%
- BP 50%

Abu Dhabi Marine (Abu Dhabi)
- CFP 33-1/3%
- BP 66-2/3%

Abu Dhabi Petrol (Abu Dhabi)
- Other 5%
- CFP 23.75%
- Mobil 11.875%
- Exxon 11.875%
- Shell 23.75%
- BP 23.75%

0%

| Joint Production Company | Consortium | IPC | Aramco | Kuwait Oil Co. | Abu Dhabi Marine | Abu Dhabi Petrol |
| Country | Iran | Iraq | Saudi Arabia | Kuwait | Abu Dhabi | |

52

and Mobil (then Socony Vacuum) joined the original partners, Texaco and Socal. Mobil, in a huge miscalculation, took only 10 percent of the new arrangement, while Standard took 30 percent and became an equal partner with the other two.

Three years after the war, the same companies and company-government combinations that dominated the industry in 1939 had resumed their dominance. The international politicians achieved a semblance of stability. They seemed to have put things back the way they were before the war. But the solution could not and would not hold. Nationalistic groups resented the presence of foreign oil companies and began to press for greater control over their own irreplaceable resources. In the Middle East, the creation of the state of Israel in 1948 upset the complex arrangement there.

One of the first breaks with the prewar order occurred in Venezuela. In 1945, a new government came to power. A new oil minister, Perez Alfonso, pushed through a law, effective in 1948, that made the government a 50 percent partner in all oil operations within the country. The kingdoms and sheikdoms of the Middle East soon demanded no less.

Gulbenkian himself had said: "The governments and people of the Middle East must eventually have an increasingly larger share of their own wealth, and the days of the capitalist entrepreneurs are coming, no less inevitably, to their end." He also said that because the local authorities were not yet technically capable of producing and marketing their oil, it would be some time before the oil companies would face final eviction. As an Iranian minister had told him years before, "Gradually, you will become well-paid contractors for the government, but the bulk of the profits will come to us."

While the Seven Sisters were reluctant to increase the payout to 50 percent, others were quite tempted by half of an enormous pie. While Aramco rebuffed Saudi demands to grant terms similar to those conceded to Venezuela, individual entrepreneurs stood at the palace gates willing to yield those terms and better.

First among these was J. Paul Getty, an independent who made his first million by the time he was 23 and bought the Tidewater Oil Company in 1934. In 1948 and 1949, when the Neutral Zone between Kuwait and Saudi Arabia was opened to bids, Getty bid on the Saudi side, offering spectacular terms for those days: a down payment of $9.5 million and a much more generous royalty than the majors paid. On the Kuwait side, the American Independent Oil Company, Inc. (Aminoil), a ten-company consortium headed by Phillips Petroleum Company, agreed to equally munificent terms for its concession.

But the unwritten rules of oil exploration had not changed. Not until 1953 and the expenditure of many millions of dollars did either operator find oil in the Neutral Zone. Eventually, proved reserves in this small area came to 11 billion barrels and made Getty reputedly the wealthiest individual in the world. Getty and Aminoil showed what could be done. True, the game for the lone wildcatter with a poor-boy rig no longer existed, but others besides the Big Seven (or eight, counting CFP) could land a concession and find oil. A new rush was on.

In 1945, only about six U.S. companies outside the top five had any foreign concessions. Among these were Atlantic Refining, Richfield Oil, and Sinclair Oil. A mere handful of firms dominated by Royal Dutch/Shell and Anglo-Iranian also had foreign concessions. In 1953, this situation changed drastically. Twenty-eight U.S. firms entered the international scene. Among these, Amerada, Continental, and Ohio (now Marathon) were jointly working a concession in Dophar, at the tip of the Arabian Peninsula, with the usual lack of initial success. In Venezuela, Atlantic, Phillips, and Sinclair produced in a small way alongside such giants as Shell and Standard Oil of New Jersey.

Between 1953 and 1972, however, some 350 oil companies and 50 government companies either entered the international business or increased their activity. Americans formed the majority of the private companies. Roughly, the principal groups included 15 large U.S. oil companies; 20 medium-sized U.S. oil companies; 10 large U.S. natural gas, chemical, and steel companies; 25 foreign private firms; and 15 foreign government companies. The presence of nonoil companies made clear that the international oil business was far from an exclusive club, though the entry fee was still high.

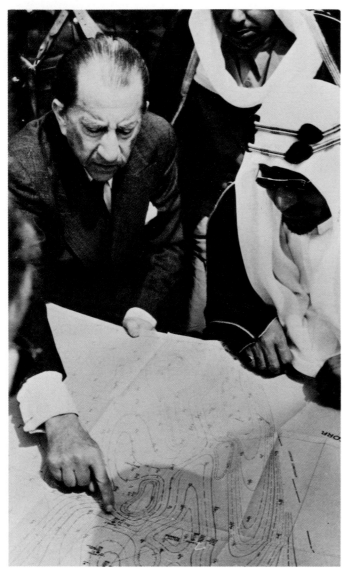

John Paul Getty, unlike Aramco members, was not reluctant to share oil profits 50-50 with the Middle East sheiks.

54

The majors, of course, equally pressed for exploitation of their concessions and acquisition of new ones. What was significant in the advent of the newcomers, particularly in the older concession areas such as the Middle East, was the forced surrender of parts of the majors' concessions because of bidding by the newcomers. Thus occurred a steady erosion of the favorable terms under which the majors had initially produced their crude.

It would soon get worse for all. By the early 1950s, despite ups and downs from the earliest days of the concession, the British, through the Anglo-Iranian Oil Company, were still very much in charge of oil production in Iran. This situation caused much irritation among the Iranians, whatever the social and material benefits the company had brought to the nation. The idea began spreading that Iran would not merely change this or that detail of the agreements, as it had in the past, but would overthrow the company and shatter the bases on which it stood. In 1950, Anglo-Iranian depended on Iran for 76 percent of its crude and had no intention of giving up its concession. That bottom-line view left the company insensitive to mounting winds of change.

First the Iranian government demanded a share in the management of the company. The company responded with increased royalties, but no shares. Iran rejected that. The company offered a better deal, 50-50 plus cash advances along the lines agreed to in Venezuela and Saudi Arabia, but no shares. By then, the Iranians refused even to negotiate.

In 1951, prompted by the oil minister, Mohammed Mossadegh, Iran nationalized the oil industry, the first such event since the Mexican takeover in 1938. Mossadegh became prime minister. The British were not pleased, and the international oil industry was greatly alarmed; if nationalization could happen in Iran, it could happen anywhere.

As Iran took over the physical assets of the company, it found no market for the oil. Since it had no marketing or shipping organizations of its own, the country depended wholly on the world petroleum industry to lift its product and sell it. The world industry not only refused to do this but threatened litigation against anyone buying or selling this illegal oil. The effect was enormous.

Iranian production fell from 660,000 barrels a day in 1950 to 20,000 in 1952. Abadan, the world's largest refinery, was shut down. The Iranian oil industry lay practically idle for three years, earning nothing, while production elsewhere increased correspondingly to meet world demand. Anglo-Iranian itself suffered little, since it drew on crude reserves outside Iran.

The crisis culminated in 1953, when Mossadegh, overreaching himself at home, took over the army and caused the young shah to flee. Riots followed. The army took control. Mossadegh fell, and the shah returned with Anglo-Iranian in his wake. However, Anglo-Iranian was no longer alone. In exchange for the U.S. aid that had helped topple Mossadegh, Anglo-Iranian now had to share the field with the five U.S. majors, Royal Dutch/Shell, and CFP in a consortium known as the Iranian Oil Participants. The British had lost their monopoly in Iran.

The consortium was set up in 1954, the year Anglo-Iranian changed its name to British Petroleum (BP). As contractor to the government-owned National Iranian Oil Company (NIOC), the consortium operated the facilities, exported the oil, and marketed it. But NIOC owned the facilities and the oil. Although Mossadegh was gone, the seeds of national ownership he sowed took root and soon spread like rampant ivy throughout the oil-producing countries.

The immediate benefits to Iran were hardly different from what they would have been if the country had accepted Anglo-Iranian's terms three years before. The nation's petroleum industry was still wholly dependent on foreign companies for production, refining, transportation, and marketing. Iran still received approximately half of the profits.

But the implications for the years ahead were incalculable. The companies, as predicted, had become contractors. In time, NIOC would become more self-sufficient and less dependent on the consortium, which faced an increasingly precarious future.

There were other consequences. At the insistence of

55

Ayatollah Khomeini, guarded by turbaned and robed Moslem security officers, returns to Iran after years in exile. This event signaled the end of the old order for the U.S. oil companies that had invested in Iran.

the U.S. antitrusters, the Iricon Group was included in the consortium. In 1955, it was given 5 percent. Initially consisting of Aminoil (itself a ten-company consortium), Richfield, Standard of Ohio, Getty, Signal, Atlantic, Hancock, Tidewater, and San Jacinto, Iricon gave independent American oil companies small but important access to cheap foreign crude and whetted their appetites for more.

In the years following, the Iranian government, through NIOC, continued to make deals with other companies and governments that were increasingly beneficial to Iran. The first, with ENI, the Italian government oil company, gave Iran 75 percent, a source of much distress to the traditional oil industry. Some deals eventually gave Iran 90 percent. By 1975, Iran had made 20 separate agreements with 34 foreign companies of nine different nationalities. Only two of these companies were majors.

In 1978, political events in Iran heated up again, much as Gulbenkian had predicted. The population became increasingly nationalistic, especially since it did not enjoy the benefits of increased oil production and higher oil prices. Workers in all areas, and especially in oil production, went on strike to protest the shah's regime. Oil production and export came to a halt. The strikes, opposition to the shah, and the suspension of oil shipments took the United States by surprise. For a while, rationing was discussed. Even after appointing a civilian government, the shah was forced into exile.

The shah's departure paved the way for the return of the Ayatollah Khomeini, the leader of an anti-Western fundamentalist Moslem group. Although Khomeini at first promised to resume the oil production and exports to the West, which had been disrupted by revolutionary events, relations between the United States and Iran completely deteriorated after the ailing shah was allowed to enter the United States for medical treatment. Student revolutionists stormed the United States Embassy, took 60 hostages, and vowed not to release them until the shah returned to Iran to face trial. Over a year of negotiations commenced. The shah's death brought little improvement in this situation. American plans to rescue the hostages also failed and increased Iran's distrust of the West.

Events in Iran went from bad to worse. Iran became

engaged in a continuous, debilitating war with Iraq, which interrupted the flow of oil from the Middle East far more than the revolutionary situation did. The refinery at Abadan was lost and won several times and was finally forced to shut down as a result of damages suffered during the constant battles. Iraq resumed use of the old Trans Arabian Pipeline temporarily, until it had to suspend oil exports altogether because it had no oil to send.

The impact of these events was not as momentous for the United States as might have been expected, except in terms of price. Iran's production of goods and services fell 12 percent in the months following the revolution, and the war with Iraq stopped both Iranian and Iraqi oil shipments. U.S. oil supplies were not affected significantly however, because Kuwait canceled proposed production cuts and Saudi Arabia increased production. Nevertheless, the fear of shortages alarmed buyers. This drove up the price of oil, despite the success of conservation measures such as the 55 MPH highway speed limit. Ever since the first oil crisis of 1973, the United States had been trying to develop an internal program toward energy self-sufficiency, away from dependence on OPEC oil. By 1980, strengthened by the development of Alaskan oil and the promise of decontrol of domestic oil prices, this program had come a long way. In fact, by 1982, an oil glut was developing.

Meanwhile, a new phase was beginning in the Sahara Desert. The newly independent states of the northern tier saw little oil exploration before the war. Immediately afterward, the majors rushed in, staked concessions and literally walked into a minefield. The vast Sahara battleground was strewn with land mines that the oil companies had to clear at a cost of many millions of dollars before they could even begin exploration.

The North African countries, along with most other countries having postwar oil potential, retained close control of their resources. They granted smaller concessions for shorter periods to more bidders and imposed stringent rules that required exploitation or surrender. Many newcomers took concessions in Nigeria, Morocco, Algeria, and especially Libya.

In Libya, the government of King Idris granted many short-term and relatively small concessions to a

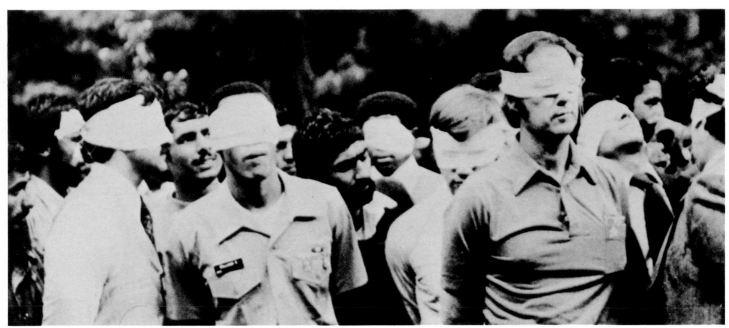

Student revolutionaries parade the American hostages through the grounds of the United States Embassy in Teheran.

large number of bidders, both majors and independents, rather than grant large concessions to one or a few major companies. He was quite forthright in his intentions: "I didn't want my country to be in the hands of one oil company." As a result, Libya soon led all other countries in the proportion of oil produced by independents. In this lay further trouble for the industry as a whole.

In 1969, a coup deposed the monarchy. The Revolutionary Command Council under Colonel Muammar Qaddafi told the companies that Libyan oil was priced too low in relation to its quality as well as transportation and production costs. He demanded a price increase of 40 cents a barrel. The companies offered a 5-cent increase in the posted price, since the true market price was far below what was actually posted.

Libya then ordered Occidental Petroleum, the major independent operator in the country and the one most dependent on Libyan crude, to reduce its production from 800,000 barrels per day to 500,000 and then to 440,000. Other operators faced similar restrictions, but none was hit as hard as Occidental, and none was as vulnerable. The company surrendered and in 1970 agreed to a 30-cent increase, which rose to 40 cents in five years. Under pressure, the other companies soon followed.

Companies feared that what Libya gained other countries would soon demand. When those demands were granted, Libya would then insist on retaining its advantage. Demands would spiral upwards. To prevent that, the companies decided to present a united front in dealing with the producer countries. The resultant Libyan Producer Agreement stated that the companies could deal only with issues that all producing countries had agreed upon, an attempt to head off leapfrogging and whipsawing of the companies between Libya and the Persian Gulf producers. The companies further agreed to support each other in the event one or another was cut off from supplies in a reprisal move, a type of support Occidental had sought but did not get.

Unwittingly, the companies handed the producing countries, until then rather loosely organized in the

Muammar Qaddafi of Libya recognized the power of oil very quickly and was one of the first to use it as a weapon.

Organization of Petroleum Exporting Countries (OPEC), the tool that cemented them into a powerful cartel and deprived the companies of their last freedom of action. Gradually, the international petroleum industry changed so that the companies needed the governments more than the governments needed the companies.

OPEC was formed in 1960 by Iran, Iraq, Kuwait, Saudia Arabia and Venezuela. Later, Qatar, Indonesia, Libya, the United Arab Emirates, Nigeria, Algeria, Ecuador, and Gabon joined. In its first ten years, when the international oil industry was strong relative to the governments and the national oil companies, OPEC had little influence and effect. But when, in 1970, Libya succeeded in increasing its revenues from the companies and the other countries followed, OPEC began to gather increasing strength.

Until 1973, the governments and the countries together negotiated changes in existing agreements. Then came the Arab-Israeli War of 1973, after which the Arab countries used the oil weapon to full effect for the first time. OPEC embargoed oil shipments to all countries friendly to Israel and, also for the first time, unilaterally increased the price of crude. On October 16, 1973, the price of Middle Eastern crude overnight went from an average of $3.01 a barrel to $5.12.

In December 1973, OPEC again unilaterally raised the posted price to $11.28. The companies and the consuming countries, rich and poor alike, had no defense against this enormous economic shock. Oil revenues of OPEC nations jumped in one year from $20 billion to $100 billion.

The next price milestone occurred in 1979, when Iran stopped shipping its crude, owing to revolution and war. Oil buyers around the world panicked. The price of oil doubled from $17 to $34 a barrel, and sometimes even higher on the spot market. However, in 1982 this situation started to change dramatically. U.S. and European demand for OPEC oil declined so significantly that an oil glut began to develop on the world market. Prices dropped until March 1983, when OPEC set its new price at $29, although market forces have since overwhelmed OPEC's efforts to hold the line.

Great new deposits of crude oil have opened up in subsequent years. The North Sea has become a major producing area, which provided temporary energy sufficiency to England and reduced Europe's dependence on Middle East supplies. Paradoxically, North Sea oil would not have been economically feasible without OPEC's price increases. Alaska and other Arctic sources also promise ample supplementary but high-cost supplies. And Mexico has rejoined the ranks of top oil producers with the discovery of large new onshore and offshore reserves.

National oil companies on the model of Mexico's PEMEX have become the rule in OPEC and non-OPEC countries alike. The United States is the only nation today without a national government-owned oil company. OPEC pronouncements in 1970 and 1971 clearly stated the intent of OPEC nations to follow Mexico's lead and seek participation with the companies operating in their states — in other words, nationalization, partial or total. The process began with Algeria in 1971 and ended with Iran in 1979.

Libya quickly followed stated intent with action. By 1973, the nation had acquired 100 percent of three major operators, British Petroleum, Shell, and a Texaco-Socal venture, and had taken 50 percent of another four. In 1978, when the 40-year concessions negotiated in 1938 expired, Venezuela nationalized the production properties of the majors then operating and reduced their roles to service and contracting companies. Nigeria nationalized British Petroleum holdings there. And Iran rejected the consortium.

By 1979, most countries had increased their participation in the operating companies and had gone to full nationalization. In many, national oil companies operate alongside the private companies and generally employ the international companies as contractors. Aramco continues to operate in Saudi Arabia, even though increasingly as an arm of the Saudi government and an adjunct to Petromin, the Saudi national oil company.

Throughout the world, most governments take between 80 and 90 percent of the net financial benefits from oil and gas. Even the United States, with its vast array of taxes, including the windfall profit tax enacted in 1980,

receives between 80 and 90 percent of most oil production. Taxes in one form or another are now much the largest factor in the price a consumer pays for petroleum products. This holds true in both producing and consuming countries, as the latter have discovered that product taxes can and have become major sources of revenue.

Today, in the 1980s, I can see that the shape of the international petroleum industry has changed completely in the more than 40 years I have known it. I expect that it will continue to change, though along lines now discernible. The many hundreds of companies in the business hold their international reserves of crude largely at the sufferance of sovereign governments that can change the terms overnight, and have. They compete successfully with national oil companies, not so much because of their advantages as superior marketing organizations as because of their superior technical and financial resources. Know-how is important, but none of these advantages will last forever.

In Iran, NIOC alone employs thousands of native

59

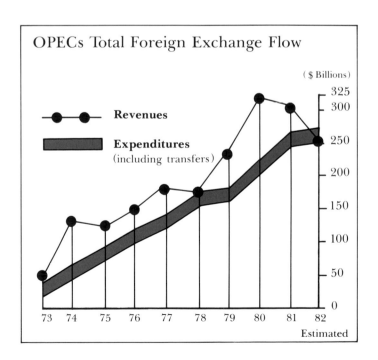

OPECs Total Foreign Exchange Flow

OPEC Loses Its Domination as World Market Shrinks

Oil Production
In millions of barrels a day

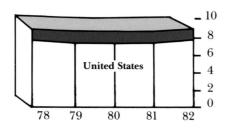

United States

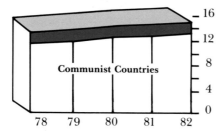

Communist Countries

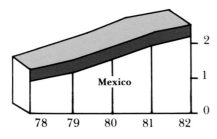

Mexico

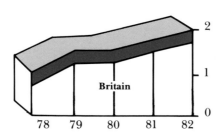

Britain

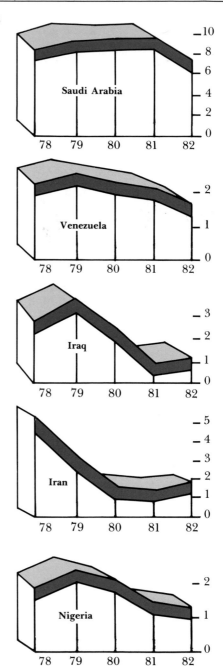

Saudi Arabia

Venezuela

Iraq

Iran

Nigeria

technicians and engineers who learned the petroleum business while working for major international companies on offshore rigs in the Gulf of Mexico or the North Sea. NIOC, Petromin, Pemex, Maraven, Pertamina of Indonesia, and many other national companies, plus the Soviets, developed within their own countries the technology they need to run their own affairs without foreign help. They developed refineries to produce their products, acquired shipping capacity, built pipelines, and even established the marketing organizations to sell their product abroad. For instance, the Soviets have built a pipeline to transport natural gas from Siberia to Western Europe. Business schools in the United States have trained a whole new generation of young foreign businessmen who are the equal of any in the world today. More and more, oil will become a government-to-government business, particularly in the less-developed world.

No national company yet equals the private international petroleum industry for technology, transportation and marketing, and particularly efficiency. Given the nature of government organizations in any place and in any age, I seriously doubt that any government company will ever rival private companies in terms of efficiency. However, no international company can stand against the sovereign power of even the smallest country. National ownership invariably brings into play a range of political considerations that results in gross overemployment and indecision. PEMEX, for example, employs many more workers than the level that prevails where private industry dominates.

But the government companies are integrating downstream as they force private industry to retreat from its upstream international operations. Perhaps, as Gulbenkian once observed, it will be some time before the final eviction. I am convinced that most of the world's producing countries now recognize the need to have national and private industry working side by side. Some nations that have totally nationalized their petroleum industries are trying quietly to bring private-equity ownership back. India is an example. Departing from its go-it-alone attitude of past years, it recently began offering contracts to foreign companies to hasten oil exploration and development. Even China's as yet undisclosed petroleum policies are based on foreign private investment and know-how. The trend toward extreme nationalism appears to have abated as most countries fortunate enough to have oil realize that the combination of government ownership and private development is the only way to have their cake and eat it too.

CHAPTER 6 —
WHERE OIL IS FOUND

Unfortunately, oil is seldom where the people and the needs are. For instance, although there are several million barrels of petroleum on Manhattan Island on any given day, none of it occurs there naturally. All that oil is stored in fuel tanks of various kinds, from those of motorcycles and passenger cars to those of electric utilities and apartment houses.

Manhattan is not unique. The same is true of most of the world's largest cities — London, Rome, Istanbul, Moscow, Rio de Janeiro. Each stores enormous amounts of petroleum, yet none is located near a natural oil source that could come close to meeting its daily needs.

A few cities do sit on top of large oil deposits — Oklahoma City and Los Angeles are two — or are located in the middle of prolific producing areas. But these are exceptions.

Of some 200 sovereign nations in the world, only 66 are petroleum producers, and of these, a mere ten account for 52 percent of world petroleum reserves and produce 25 percent of world energy generated from all sources. Yet these ten — Algeria, Colombia, Indonesia, Iran, Iraq, Kuwait, Libya, Nigeria, Saudi Arabia, and Venezuela — have only 6 percent of the total population and 10 percent of the world's land area and consume only 3 percent of all energy from all sources. The world's oil reserves total 545 billion barrels, and Saudi Arabia alone has more than 50 percent of it.

By contrast, the United States, with 5 percent of the world's population and 7 percent of its land mass, has less than 6 percent of the world's petroleum reserves but consumes 30 percent of the world's energy, while producing less than 23 percent from all sources, including coal and natural gas.

In a perfect world, that imbalance would not be bad. Oil transportation is cheap enough. Before World War II, when companies and countries exercising colonial suzerainty controlled the sources of oil, it mattered little that oil was often produced far from its markets. "Great Powers" determined who should be allowed to have oil, a matter deemed of no concern to the countries

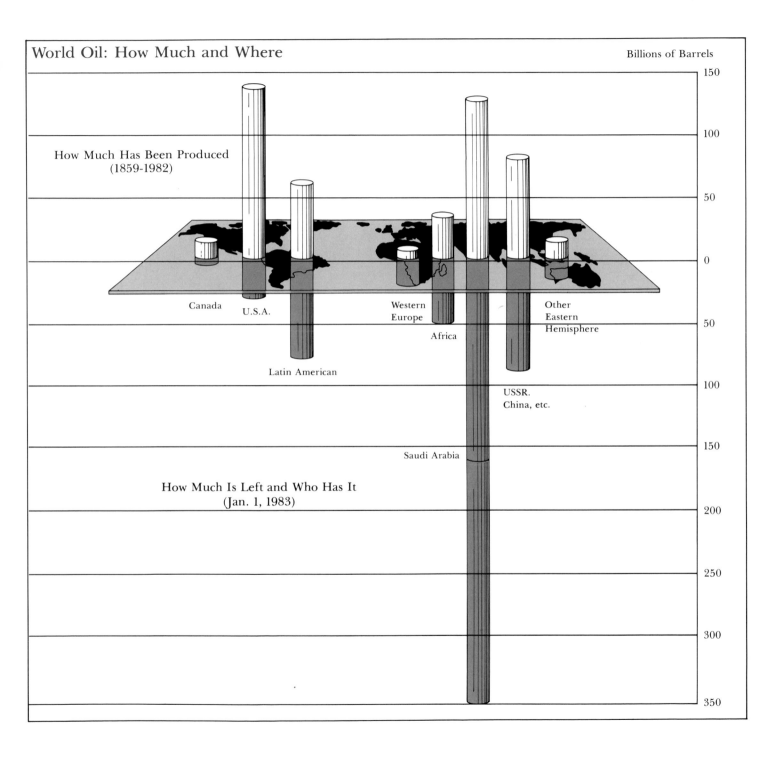

World Oil: How Much and Where

Billions of Barrels

How Much Has Been Produced
(1859-1982)

Canada

U.S.A.

Western
Europe

Africa

Other
Eastern
Hemisphere

Latin American

USSR.
China, etc.

Saudi Arabia

How Much Is Left and Who Has It
(Jan. 1, 1983)

150

100

50

0

50

100

150

200

250

300

350

63

that actually owned it. In those circumstances, geography was less important than the relations between adjoining nonoil-owning nations in Europe.

Dorsey Hager's *Fundamentals of the Petroleum Industry*, published in 1939, illustrates the true position of oil-owning nations in those days: "The most recent addition to the oil-producing nations is Iraq in Mesopotamia where France, England and the United States are jointly interested in developing the oil resources. . . . English, Dutch and American capital have developed the Dutch East Indies, although English-Dutch capital owns the major portion of the oil production. English and Dutch capital control Burma and Persia; English capital controls Egypt; French capital controls Algiers. In Poland and Rumania, English, Dutch and American capital control. . . . In South America, United States, English and Dutch capital predominate."

But Hager also looked ahead to the world to come, in which world politics would have more to do with the free flow of oil than would world geography: "Russia has nationalized oil and is developing its own resources. . . . . The Argentine government has nationalized its oil fields and controls them." Mexico had also just nationalized its industry, but that development was such a shock to the world industry that apparently Hager could not bear to mention it. He merely reported, "Mexico was developed by American, English and Dutch capital."

While the political location of oil, and U.S. access to it, has changed drastically in 40 years, its geographical location has not. Of the many new discoveries since 1939, few were made outside of known or potential areas. By 1939, oilmen and geologists had very good ideas about where oil could and could not be found.

Even Alaska was considered a possible source more than 40 years ago. By then, oil had been discovered in the Athabasca area of Canada and was refined and shipped by plane to remote mines in that isolated region. In a prescient comment, Hager said, "Any marked development there will not become active for many years, not until the price is high enough to repay exploitation."

Since the world's resources are not infinite, while the world's demand for energy appears to be, scarcity eventually drives up the price of oil. In a normal supply-and-demand market, cheaply exploited sources are used first, as was the case in the continental United States and in the world before OPEC. Cheap oil delays the development of more expensive sources and even causes older fields that are becoming expensive to be shut down.

Today, the price of oil reflects geopolitics far more than it does either geography or geology. The ownership of oil is fragmented and largely nationalized, no longer under the control of private companies that operate in a highly competitive market. Recognizing that oil is a depleting and nonrenewable resource, its new national owners seek to extract the maximum financial benefit from it while they can, thus administering the world price at levels that bear little relation to production costs. Conversely, these high prices make feasible the development of high-cost deposits that in a free market probably would not have been developed for many more years, if ever. With much of the world's oil under political control, other high-cost sources are being developed for national-security reasons. In the long run, the development of these diversified and alternate sources of energy not only may be beneficial but may well drive the price of energy down.

Total world oil consumption today is about 20 billion barrels per year. Total annual productive capacity of all oil fields is about 22 billion barrels, and total proved reserves more than 545 billion barrels. In addition to proved recoverable reserves, there also exist probable, possible, speculative, and even ultimate reserves.

Proved reserves are defined by the American Petroleum Institute as "that portion of a reservoir actually delineated by drilling and the adjoining portions that reasonably can be expected to be productive based on geologic and engineering data."

Each of the other designations is considerably less precise than the one before and likely to be overstated or understated, depending on the point of view of the stater. Mexico appears to be a good example here. One estimate of the world's ultimate recoverable reserves is 1.7 trillion barrels, of which 925 billion barrels — nearly twice the proved reserves — is yet to be discovered. Many, including myself, believe that these figures are highly optimistic and misleading in regard to the economic reality of recovery.

The recoverable reserve is the actual quantity of oil likely to be retrieved from the reservoir, a factor determined by economics as well as geology and technology. It is generally a conservative figure based on proved and probable reserves.

Probable reserves are an estimate of oil yet to be found in an otherwise tested and producing area but not delineated. Possible and speculative figures are usually mathematical exercises based on wishful thinking, statistical analysis of the past, or, occasionally, cerebral evaluation of as yet untested geological features.

Geologists have continually predicted that the world would run out of oil. The end always seemed about 20 years away. But year after year, proved reserves have increased faster than consumption and, on a global basis, are still doing so, although the margin is rapidly narrowing. However, in the United States, despite increased exploration and drilling, proved reserves have not kept pace with production. Including Alaska and offshore fields, U.S. reserves since 1970 have increased by only three billion barrels per year, while production has totaled about four billion per year. Recent discoveries, such as Chevron's Point Arguello field, on the edge of the Santa Maria Basin in southern California, have eased the concern that the United States is running out of oil.

Reserve figures relate to use, not availability. Oil planners believe that reserves should stand at 10 or 20 times the amount of oil consumed and that, as oil use increases, reserves should grow proportionately. Budget planners would not run a money budget this way, but that is how the world energy budget has been run for the last 50 years and more. The ratio of reserves to production was five and a half times greater in 1960 than 30 years before. The ratio is now declining, however, as would be expected of a nonrenewable resource, and stands at about 30 times world producton.

As a proportion of world-published recoverable reserves, U.S. reserves have declined from more than 35-percent in 1945 to only 5 percent now, while Middle East reserves have gone from 31 percent to more than 60 percent.

The United States is not now and never has been a have-not nation. It is still the third-largest producer in the world, behind the Soviet Union and Saudi Arabia. Other factors, however, reveal the imbalance more clearly.

The United States has a population of 225 million, a gross national product of about $3 trillion, and an annual energy consumption of nearly 36.7 million barrels per day (bpd) oil equivalent. Petroleum satisfies less than half the U.S. total energy requirement, and only half of that petroleum is domestic. The rest is imported.

Saudi Arabia, on the other hand, is the second-largest producer at upwards of five million bpd. It has ample capacity to increase output, as well as the world's largest reserves of over 163 million barrels. Its estimated population is about seven million, and its energy use is less than 1 percent of that of the United States.

The Soviet Union, the world's largest oil producer at around 11 million bpd, is hard pressed to increase output and more and more resembles the United States as its energy situation develops. Russia's proved reserves of some 60 billion barrels, though twice those of the United States, are not spectacular against a population of 332 million, a gross national product exceeding $1 trillion, and an energy consumption about half that of the United States, although rapidly increasing. The country exports petroleum to the East European bloc, and the growing energy demands of that entire group of nations make them dependent on the Middle East, too.

Although petroleum is found all over the world, it is located at two major poles. One is centered on the Gulf of Mexico and the Caribbean; the other, on the Persian Gulf. The latter swings by far the greater weight. Though each pole forms a great geologic region which contains huge oil deposits that extend over thousands of square miles, neither is politically homogeneous. Each is divided into numerous nations whose boundaries are defined not by geology but by surface features and political accidents that have little to do with natural resources.

The United States is the major oil producer on the North American continent. It has seven distinct oil areas: Northeastern and East-Central States; Mid-Continent; Louisiana and the Southeastern States; Rocky Mountains; Texas; California; and Alaska. Oil discoveries started in Pennsylvania, then moved south and west.

The Rocky Mountain area includes a large part of

the eastern slopes of the Rockies, and most of this area's current production comes from Wyoming. Geologists hope for future discoveries in the Overthrust Belt, the geologically complex deep sediments along the flanks of the Rocky Mountains that extend into Canada.

Texas, of course, is one of the world's major producing areas. Indeed, the state has given its accent, technology, and vocabulary to the entire industry, from Alaska to Siberia and the Persian Gulf. Despite the decline in U.S. production and the relative importance of the United States as an oil nation, Texas output alone figures as 35 percent of U.S. production and is greater than that of both Libya and Venezuela. Thanks to state prorationing laws aimed at conserving the reservoirs, actual production is carefully regulated, and is less than it would be otherwise. Texas has about 175,000 functioning wells, and about 7,000 to 8,000 wells are drilled there every year.

Louisiana, second to Texas in production, has seen its output increase rapidly in recent years, thanks to offshore discoveries. Nearly 95 percent of the state's production originates along the Gulf Coast or offshore since the petroleum deposits here are based on a great river delta.

Mississippi has become an important producer, though its output of 50 million barrels per year is minuscule next to that of its Gulf neighbors. Alabama also has some production, as does Florida, but neither is significant.

California is the third-largest producing area in the United States. It has many large fields; some 40,000 wells have been drilled, but it matches neither Texas nor Louisiana in number and size. Most of the oil comes from onshore pools, though the state has extensive offshore production, particularly in the Santa Barbara Channel. Until recently, offshore resources have not been as fully explored or developed as those in the Gulf of Mexico. Some estimates envision as much as 100 billion barrels of oil in the California continental shelf, only 10 percent of which has so far been discovered. The potential there is rich and waits to be developed.

The newest and the largest single proved petroleum deposit in the United States is in Alaska. While the search

66

The Delacrois A-1, an exploratory well in southern Louisiana, confirmed the presence of oil and gas there. These operations help increase U.S. oil independence.

The Baldwin Hills field in Los Angeles was prolific in the 1930s. Drilling in California dates back to 1864, when Benjamin Silliman inspected the area and wrote back to Pennsylvania for a rig.

for Alaskan oil began as far back as 1898, nothing of consequence was found until 1957. Since then, at least ten sedimentary basins that could contain oil have been identified, and important discoveries have been made in the Cook Inlet area and especially on the North Slope. South Alaska has substantial proved reserves, but the North Slope's Prudhoe Bay field, which contains more than ten billion barrels, dwarfs them. Despite the high costs of production and transportation, Alaska is expected to become an increasingly important source of domestic oil, accounting for some 25 percent of total U.S. production.

Estimates of future discoveries are hazardous. With some three million wells drilled, the United States is by far the most thoroughly explored area in the world. For a while, oil geologists felt that the odds were that there would be no more East Texas or Prudhoe Bay fields, but discoveries in Alaska and in the Santa Maria Basin have defied this expectation. Baltimore Canyon, off the New Jersey shore, and the Georges Bank, off New England, are other offshore areas that look promising, although exploration has so far produced no commercial oil. Until the wells are drilled, firm predictions cannot be made.

Elsewhere in North America, Canada continues as an important oil-producing country. As far back as 1861, oil was drilled in Ontario from formations that are contiguous with those in the Northeastern Trend of the United States. After discoveries in Alberta in 1947, the country really entered the petroleum scene.

Today, western Canada has significant production and bright prospects. Alberta has deposits in the foothills of the Rockies, as well as potential in the Overthrust Belt. One substantial deposit in the central plains alone exceeds a billion barrels. The largest deposit is the Athabasca Tar Sands, 13,000 square miles of oil-soaked sandstone that contains at least 625 billion barrels of heavy, sulfur-laden crude oil. The sands are expensive to work and, so far, have mainly been the target of experimental processes. Eventually, however, the same high prices that make Alaskan, offshore, and Arctic sources feasible will make the tar sands a viable source of petroleum supply.

Eastern Canada is rather poor in oil, though there is much hope for offshore areas. Large sections of northern

This North Slope rig is built from special steel that withstands the severe Alaskan conditions, including temperatures that go down to −65 degrees Fahrenheit.

67

and Arctic Canada appear promising, though the high costs involved have so far prevented adequate exploration.

Mexico, the third North American producer, was once second only to the United States. In the early 1920s, Mexico produced 25 percent of total world production from the Golden Lane along the Gulf Coast. The geologic nature of these deposits allowed their rapid exhaustion, and production quickly declined. But with new discoveries in recent years, Mexico is once again a major factor in world oil, with reserves estimated at more than 40 billion barrels, though this amount is open to discussion depending on who is making the evaluation. Since many of Mexico's foreign loans are secured by oil holdings, no exact evaluation is in sight.

All Mexican deposits lie either along the Gulf Coast or offshore. The most recent discoveries are on the Yucatán Peninsula, and offshore, where wells with tremendous productive capacity are regularly brought in. Production is being restrained, however, in order to sustain reservoirs and hold income to levels the country can safely absorb without excessive inflation.

Central American countries south of the Yucatan Peninsula produce little or no oil. Small as their economies are, these nations consume many times as much as they produce. The same is true for the Caribbean Islands, including the largest, Cuba, which, despite some promising geological indications, produces only a few hundred barrels from a few shallow wells.

The one exception is the island of Trinidad, geologically an extension of Venezuela. A Caribbean nation of less than 2,000 square miles with fewer than 1.5 million people, Trinidad and Tobago produces four times as much energy as it consumes and has 671 million barrels of reserves. When oil was first produced there in 1866, Trinidad was the largest producer in the British Commonwealth. In recent years, onshore production from some 3,500 wells has been falling, but offshore output has been increasing rapidly. An interesting feature of the island is Pitch Lake, a 137-acre pool of bitumen that has been mined for many years and is apparently replenished from an underground source.

Venezuela, the major petroleum country in the Caribbean — indeed, in all of Latin America — accounts for some 75 percent of the oil produced in South America. It exported its first barrel of oil to Spain in 1539 and has had an oil industry of sorts ever since. Though the first modern concession was granted in 1866, the first really commercial field was brought in near Lake Maracaibo only in 1914. After that, many additional rich deposits were found and developed all along the coast and some distance inland. Despite nature's bounty — the country's ultimate producible reserves are estimated at some 70 billion barrels — Venezuelan oil is high in cost, high in sulfur and relatively heavy. The 13,000 or so wells in the country average only 300 bpd, and 80 percent of them are pumped. Nonetheless, Venezuelan oil is extremely important to the Western Hemisphere.

Venezuela has another, and so far largely untapped, heavy-oil resource in the Orinoco Tos Belt, where recoverable reserves are estimated at some two trillion barrels. But like the Canadian tar sands and the Trinidad pitch, it is costly to produce in usable form.

Colombia has reserves of less than one billion barrels. Current production runs about 1.5 million but has been as high as 2.26 million. Ecuador shares some of the same formations, but more bountifully. Reserves are estimated at about 1.3 billion barrels against a daily production of about 1.9 million barrels. There is also hope for significant offshore development in the Pacific.

The earliest commercial producer in South America was Peru, which in 1900 yielded 265,000 barrels from deposits on the coastal plain. This occurred some 14 years before Venezuelan oil even became a factor. The oldest field, discovered in 1869, is still producing from some 7,700 wells. Additional reserves have been found in the interior. Estimated reserves are less than one billion barrels against current production of 30 million barrels per year.

Chile is a latecomer to the oil world. The first discovery there occurred in 1946 on the island of Tierra del Fuego. Economically more advanced than its neighbors, and with greater energy consumption per capita, Chile has great need of oil but does not have nearly enough of it. Production of about ten million barrels a year from reserves of 210 million barrels meets only half or less of the nation's growing energy needs.

Argentina shares Tierra del Fuego and its oil deposits with Chile. The hope of oil nearly caused the two nations to go to war over a tiny neighboring island whose possession has been in dispute.

Oil was discovered in Argentina in 1910, but production did not reach significant levels until the 1950s, when the country's output increased dramatically. Production is now about 152 million barrels per year from reserves of 2.6 billion barrels. A relatively developed country, Argentina needs all the oil it can produce and could probably meet its current needs from domestic production if it had an aggressive program of exploration and development of promising offshore areas.

Bolivia shares some oil formations with Argentina and, in addition, has other deposits of valuable low-sulfur oil. A lightly populated and underdeveloped country, Bolivia produces around 14 million barrels annually from reserves of 138 million barrels, far exceeding its domestic needs. As a result, the country is a net exporter

68

of oil and gas. It sends oil over the Trans Andean Pipeline to Chile and to the West Coast of the United States. In addition, it sends gas through a pipeline to Argentina.

The most populous, most rapidly industrializing, most energy-using nation in South America is Brazil. In relation to its petroleum needs, Brazil is the most deficient producing country in South America. From its vast area of 3,286,470 square miles, only 10 percent smaller than the United States, Brazil's petroleum production is about 162,000 bpd. Energy production from all sources satisfies only a third of Brazil's needs, and the disparity is growing. Nonetheless, exploration continues, and promising discoveries have been made offshore, even though vast areas of the interior remain unexplored.

Although Brazil has great reserves of oil shale, current reserves, pitiful for such an ambitious giant, are only about 800 million barrels. Not surprisingly, with its great agricultural resources, Brazil is a leader in developing plant sources for energy, such as alcohol from sugar.

In the Persian Gulf area, supply far exceeds demand for most countries. Armed with this fact, as well as a similarity in level of development, culture, language, tradition, religion, and general world outlook, these nations act in concert and thus wield a weight in world energy affairs greater than any of them would as a single nation.

The leading producer among these nations is Saudi Arabia, where exploration began only in 1936, and where the first discovery was made only in 1938. The country's productive capacity, currently estimated at around 14-million bpd and expandable, dwarfs its actual output, which the government adjusts up or down for international policy reasons. Recent production has varied between eight and ten million bpd.

Although Saudi Arabia cooperates with and is an active member of OPEC, its capacity is such that it alone can unilaterally flood or starve world petroleum markets and thus control the price of oil.

Bordering Saudi Arabia on the Arabian Peninsula are Kuwait, Bahrain, Qatar, United Arab Emirates (Abu

Lake Maracaibo, Venezuela, was the first really commercial field in South America. The derricks and drilling platforms are mounted on caisson piles set in the shallow lake bottom.

Dhabi, Dubai, Sharjah), and Oman, as well as the Neutral Zone between Kuwait and Saudi Arabia. All are desert lands with small native populations who were, until recently, largely nomadic, underdeveloped, and needing little energy. The combined petroleum reserves of these peripheral countries exceed 107 billion barrels, onshore and offshore in the Persian Gulf. Their combined populations are estimated at less than three million.

In the same geologic trend as Saudi Arabia is Iraq, the great Mesopotamian sedimentary basin where oil has been found and used since ancient times. Iraq contains one of the world's largest single oil fields at Kirkuk with original recoverable reserves estimated to be at least 15 billion barrels. The country's total reserves are about 35 billion barrels, and current production is 2.3 million bpd.

Another great petroleum country of the region is non-Arab Iran, believed to be the site of the oldest commercial petroleum works in the Middle East. The first modern discovery was made there in 1908. At 48 billion barrels, Iran's reserves are the third largest in the area, after Saudi Arabia and Kuwait. But with 36 million people, the country is far more populous than its neighbors and has had ambitious development programs that demanded heavy funding over a short period. Iran's production for domestic use and export has been second only to Saudi Arabia's. Political upheavals beginning in the late 1970s, as well as the war with Iraq, severely cut production, internal consumption, and export.

Syria became an oil country in 1968 and now has reserves of about three billion barrels and modest production of 200,000 bpd. Turkey, the most populous nation in the region, with over 40 million people and large energy needs, has very little oil — only 88 million barrels of reserves and 52,000-bpd production, far short of domestic needs.

Many African countries have become oil producers since World War II. Algeria and Libya are more akin to the Middle East in topography and world outlook and are often justifiably linked with the latter area, while other areas of the continent also have significant petroleum production.

Algeria has a number of prolific oil fields, first dis-covered in the Sahara Desert in 1959. The country's reserves are almost ten billion barrels and production is about one million bpd. With 18 million people and a land that is 90 percent desert, Algeria can use its oil. Libya, its neighbor, has fewer than three million people but more than twice as much oil. In deposits similar to Algeria's, Libya has reserves exceeding 23 billion barrels and a daily production of two million barrels, virtually all for export.

In Egypt, important oil fields were found on both sides of the Gulf of Suez as far back as 1908. More recently, oil has been found west of the Nile. The country's reserves are three billion barrels, a respectable figure if the land were not so populous and the demands for energy were not large and increasing.

Tunisia and Morocco are not as well favored. With six million and 18 million people, respectively, their petroleum resources are in inverse proportion. Tunisia has reserves of two billion barrels and daily production of 80,000 barrels, while Morocco produces less than 100,000 barrels a year.

South of the Sahara, almost all petroleum has been found on the west coast, particularly around coastal and offshore Nigeria, where commercial oil was first dis-covered in 1955, though the area had been explored as early as 1908. The country's desirable low-sulfur paraffinic crude comes from several fields with overall reserves of at least 12 billion barrels. Here, a great river, the Niger, laid down its enormous deltaic deposits. The geology is similar to that of the Lousiana Gulf Coast, another great river delta.

Other oil-producing countries in Africa are Gabon and Angola, with relatively small outputs and reserves. Zaire, People's Republic of the Congo, and Cameroon also have some oil. Oil users hope that more will be found elsewhere on the huge continent, which has not been adequately explored.

The continent of Europe has several oil-producing areas, though none of great consequence, with the exception of the North Sea. Western Europe, excluding North Sea production, provides only 3 percent of its oil needs. More than 90 percent of Europe's oil comes from the Middle East and Africa.

Such oil as there is in Europe, again excluding the North Sea, is found in Austria, Germany, and Italy, with lesser amounts in France and Holland. Spain, England, and Denmark have small onshore production, while Sweden, Norway, and Switzerland have virtually none.

In Germany, geology along the Baltic Sea is similar to that of the Texas Gulf Coast, but not nearly as bountiful. Nonetheless, the country is the largest producer in Western Europe. Scattered in several basins, West Germany has reserves of 321 million barrels and a production of 110,000 bpd, not one-fifth of the nation's needs. Interestingly, most German oil is located in West Germany; the East has barely 22 million barrels in reserve.

Austria is the second European producer, with reserves of 168 million barrels and daily production of 40,000 barrels. The rest of Europe is even less significant.

The North Sea, however, is a major resource that falls mainly in the offshore areas of England and Norway. Thanks to these deposits, England is on the verge of becoming self-sufficient in energy, while Norway can be-

come an oil exporter. Norway's recent discovery in the North Sea may equal what the Siberian gas pipeline was expected to supply Western Europe over the next 25 years. Total North Sea reserves may well exceed 35 billion barrels. The oil is extremely expensive, however. Its exploitation is made feasible today only by the high price of OPEC oil. It may never be economical to work some North Sea deposits that elsewhere would be giant fields. The search for oil will continue along the shores of Europe, including the Mediterranean area.

Rumania is the largest European petroleum producer after the Soviet Union. It has had commercial output since 1860. The famous Ploesti oil fields, targets of Allied air raids during World War II, once supplied 90 percent of Rumanian output but now account for only 30 percent. The balance comes from a number of newer fields, many discovered since the 1950s. The nation's reserves are about 1.3 billion barrels, and output is 300,000 bpd. The country is energy-sufficient.

The Soviet Union is the most important oil country

71

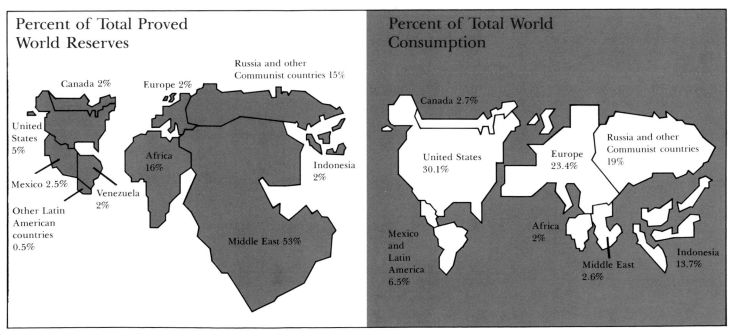

Countries of the world drawn in proportion to the amount of their oil reserves (left) and the amount of oil they consume (right) appear the inverse of each other. The United States, for instance, is very large measured by consumption, but small by reserves.

in Eastern Europe and, indeed, in the world, outside the Middle East. Oil history begins with the great fields at Baku. These fields are still highly productive and have been joined by a great many more all across the Soviet Union, including those in the coldest reaches of the Siberian Arctic.

Soviet deposits are relatively new and have not been developed as extensively as have similar deposits in the United States. The nation's total reserves are estimated at more than 60 billion barrels, with production at about 12 million bpd. The USSR's growing energy needs are catching up to its capacity.

Among the other Eastern European nations, only Yugoslavia, Hungary, and Albania have production or reserves of any size. Observers do not expect that they will be able to develop new resources fast enough to keep up with demand, let alone for export.

The continent of Asia is much larger than the Middle East, with an immensely larger population, containing as it does such populous lands as India, with 660 million people; China, with 897 million people; and Indonesia, with a population of 150 million. Asia has oil, but not in proportion to its area, its population, or its needs.

Little is known of China's petroleum potential, though minor fields were discovered there as early as 1897, and records indicate the use of petroleum products in 211 B.C. The country's geology is complex and not well explored. To date, indicated reserves are approximately 15 billion barrels and production exceeds one million bpd. Current production from all sources exceeds demand. In September, 1982, ARCO was granted a concession for exploration and development there.

With offshore exploration and onshore development anticipated soon, many expect that China will become a major factor in world oil by the end of the century. China probably will not export its oil, since its energy needs are likely to increase at least as fast as supply.

Historically, the major petroleum source in Asia has been Indonesia, formerly the Dutch East Indies. The geology that gave rise to the islands proved particularly suited to hydrocarbon formation and resulted in extensive deposits of high-quality, low-sulfur petroleum. Development began about 1890. All of the major is-

lands — Sumatra, Java, Borneo, West Irian — are substantial producers, with Sumatra accounting for 40 percent of the total. Reserves exceed 12 billion barrels, the largest in Asia outside China, and production runs about 1.5 million bpd.

Malaysia, with Brunei, occupies the smaller part of the island of Borneo. Early production in this area was onshore but is now largely exhausted. Malaysia's still extensive output of 200,000 bpd comes mainly from recent offshore discoveries. Reserves are fairly significant, close to 5 billion barrels.

Elsewhere in Asia, once preeminent Burma has gone into a long decline brought on by depleted fields and political lethargy. The country has reserves of only 66 million barrels and produces 24,000 bpd.

Up and coming may be India, which needs all the energy it can find to fuel its ambitious development programs and to raise the standard of living for its 660 million people. Oil has been found in the extreme northeast, in Assam, and in the west, extending offshore into the

Drilling in the Ardjuna field in Indonesia is a tricky business. Because the shallow waters there make it difficult to construct onshore production facilities, all the oil is produced, processed, stored, and exported from offshore facilities like these.

The Glomar Java Sea, *a drill ship under contract to ARCO China, Inc., was the first to find natural gas deposits in the South China Sea. Unfortunately, while working on its third exploration well on October 25, 1983, it sank as a result of tropical storm Lex.*

Arabian Sea. Intensive exploration is being carried out in promising formations throughout the subcontinent. Present reserves are an inadequate 2.9 billion barrels and production a comparative trickle of 173,000 bpd.

Japan needs energy to fuel its superheated economy but has little of its own. Although oil was found there as early as 1892, intensive exploration has produced little. The geology is unfavorable. As a consumer of petroleum products, Japan is third behind the United States and Russia, and must import some 2 billion barrels per year. Its own reserves total about 29 million barrels, little more than a five-day supply, assuming that it could be produced.

Pakistan has reserves of 87 million barrels. Offshore exploration in the Indus Delta appears promising but, so far, is unproductive. Bangladesh has not been that lucky, though there is interest offshore in the Bay of Bengal.

Elsewhere in Asia, the Philippines have seen much exploration but little oil. Small deposits have been found in Thailand, Cambodia, and Vietnam.

Given their size and advanced states of development, Australia and New Zealand are deficient in petroleum. Because of their position on the globe, they were largely bypassed by the great geologic events that directly and indirectly helped form great deposits of oil elsewhere. Nonetheless, Australia has had production since 1914. A one billion-barrel field allows the country to produce more than half its energy needs from domestic petroleum. Total reserves are about 2.5 billion barrels and production is less than 500,000 bpd.

Everyday, almost everywhere in the world, new discoveries, political changes, and increases in population keep these figures in a state of constant flux. Countries that once had little need for oil are now developing, building their industrial base, and increasing their demand for oil. Older industrialized nations are also increasing their demand for oil, as well as looking for alternate sources of energy. Given these constant shifts, virtually the only certainty is that the map of the world's oil supplies always will continue to change.

CHAPTER 7 –
THE MAKING OF OIL

Some years ago, the picture of a green dinosaur, a brontosaurus, was used as the symbol for the Sinclair Oil Company. The dinosaur was supposed to connect the modern gas station with the age of reptiles. Scientists then theorized that reptiles left behind the raw materials that eventually became the oil of today.

The dinosaur no longer symbolizes the origins of oil. Scientists now believe that very few dinosaurs laid down their lives for the modern motorist. They now know that the formation of petroleum is the result of a much more complicated process that involved trillions of one-celled animals and plants, the sun, and maybe even dinosaurs.

Petroleum is, in fact, almost an inadvertent by-product of earthbound life, the remains of innocent bystanders caught between the relentless grindstones of geologic events in the lithosphere, on the one hand, and the climatic changes of the atmosphere on the other.

Life persists in the biosphere, an exceedingly thin layer between the atmosphere and the lithosphere. Where mountains wear down and disappear, some residues of life also disappear, only to emerge again as oil.

Carbon, oxygen, hydrogen, calcium, phosphorus, sulfur, iron, and smaller amounts of other elements — the stuff of lifeless air, rock, and water — combine with the help of the sun's energy to form the complex carbon-based molecules that make up all living things, lichens and sequoias, ants and elephants, dinosaurs and men, diatoms and clams.

The carbon atom makes all life possible. It forms the basis of petroleum, lignite, coal, peat, tar, pitch, graphite, diamonds, and synthetic rubber. A common element with unusual qualities, carbon may play a role throughout the universe similar to the role that it plays on earth. Carbon-based molecules have been found in meteorites and in the spectra of distant constellations. There may indeed be carbon-based life elsewhere in the universe, though scientists cannot yet prove it.

Carbon is a peculiar element. While it does not readily combine with others, it will, under the proper conditions, react with virtually every other element in nature.

Abundant sea life, which left its record in fossils like these, helped form petroleum on the ocean subfloor.

appropriate number of linkages. These attachments represent a property called "valence," which governs the way atoms link to other atoms.

Imagine the electrons, or linkages, on some atoms as grasping claws — they actively seek out and seize the attachments on other atoms. Such elements are reactive. Elements such as helium and neon figuratively fold their hands and will not join others. Still others are neither accepting nor grasping but may play either role. These include boron, silicon, phosphorus,and carbon.

Carbon, although often found as a pure element, is fairly easily persuaded to hook up with other atoms. Once formed, carbon-based elements tend to be stable; carbon's linkages hold tight.

On this versatile element nature has built all of life — bone and shell, leaf and stem, heart, blood, and brain. When combined with hydrogen in a ratio that approaches four hydrogen atoms for each carbon atom, thus forming compounds known as "hydrocarbons," this element becomes the essence of all natural petroleum,

In its pure form it has several guises, depending on the conditions to which it has been subjected. High temperature and extremely high pressures yield diamond, the hardest substance known. High temperature alone can produce graphite, a soft substance that is an excellent lubricant and the heart of the so-called lead pencil. Yet both materials are pure carbon, differing only in molecular-crystal structure. The diamond molecule is three-dimensional, a pyramid with a carbon atom at each corner. The graphite crystal is flat and six-sided with a carbon atom at each of six corners. Diamond crystals are locked together; graphite crystals slide over one another.

There is no way to describe to a nonchemist in literal, everyday terms the workings of an atom and its relation to other atoms. The best that can be done is to invent a metaphor. The nucleus of a carbon atom, for example, is surrounded by four electrons. One way to think of the carbon nucleus is to imagine it as a little ball with four attachments or linkages (electrons) on it. Imagine the nuclei of other elements as little balls also, each with its

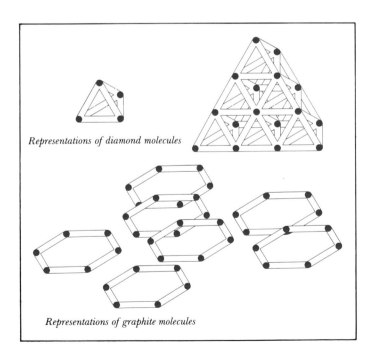

Representations of diamond molecules

Representations of graphite molecules

the gaseous, liquid, or solid mixtures of numerous hydrocarbons and hydrocarbon compounds that occur naturally in the earth.

The range and complexity of naturally occurring hydrocarbons is extremely large, and the variation in composition from one deposit to another is enormous. All crude petroleums differ in the fractions of the various hydrocarbons they contain. The specific chemical compounds can and do have molecules of different shapes and vary in size or weight from those based on the single-carbon atom or, in petroleum shorthand, C_1, to those based on large and complex molecules of 80 and more carbon atoms ($C_{80}+$). At its simplest, the one-carbon compound has four hydrogen atoms hooked to the carbon to give the compound CH_4, or methane, a gas.

Methane is consistently simple. A C_1 hydrocarbon is always CH_4. Not so for molecules of C_2 and more, called higher hydrocarbons. If there are not enough other atoms to go around, the extra carbon valences can become carbon-to-carbon links or bonds. In a two-carbon hydrocarbon (C_2) then, two of its eight hooks can be linked carbon-to-carbon and the other six bonded to hydrogen to make C_2H_6, or ethane, also a gas. If not enough hydrogen is available to use up or saturate all available carbon bonds, those left over also become carbon-to-carbon bonds, C_2H_4, or ethylene. With still less hydrogen comes C_2H_2, or acetylene. Such hydrogen-short hydrocarbons are called "unsaturated compounds" because there is still room in the molecule to add more atoms. Large C_3 and higher molecules with more than one double or triple bond are called "polyunsaturated."

Think of unsaturated molecules as unfulfilled. They are much more reactive than contented saturated molecules and thus used widely as the building blocks of chemistry. A saturated hydrocarbon can be made to react by knocking off one or more of its hydrogens, thereby freeing a carbon linkage or valence to connect to something else, either another carbon or a different element altogether.

The first four carbon molecules, C_1-C_4, with the addition of hydrogen, form hydrocarbon gases: methane, ethane, propane, and butane. Larger molecules, C_5-C_7, cover the range of light gasoline liquids; C_8-C_{11} are naphthas; C_{12}-C_{19}, kerosene and gas oil; C_{20}-C_{27} lubricating oils; and above C_{28}, heavy fuels, greases, waxes, asphalts, bitumens, and materials as hard as stone at normal temperatures. Accompanying the gaseous compounds may be various amounts of nitrogen, carbon dioxide, hydrogen sulfide, and, occasionally, helium. U.S. gas wells are the only ones in the world that produce helium in commercial quantities.

Gas and oil wells seldom produce pure usable hydrocarbons. What flows from a well is an unpredictable mixture of chemicals. The major product of natural gas wells is usually methane, but other gases are present as well. A well in Colorado produces 90 percent carbon dioxide, CO_2. One in Montana yields 57 percent CO_2. The Santa Maria field in California once produced enough CO_2 to justify setting up a dry-ice plant on the site. Dry ice is frozen CO_2. Hydrogen sulfide, H_2S, a foul-smelling and extremely poisonous compound, also exists in large quantities.

Liquid hydrocarbons from natural-petroleum wells may have nitrogen, oxygen, and sulfur in quantities ranging from traces to significant amounts, as well as traces of such metals as vanadium. These impurities may be found alone, as elements, or chemically linked to the hydrocarbon molecules. The problem of extracting the pure hydrocarbons the chemist wants is complicated further by the degree of saturation and structure of the hydrocarbons themselves.

In any particular crude certain hydrocarbon compounds predominate and characterize the mixture. But whatever the specific compounds, all crudes are variations on the hydrocarbon theme, CH_2, with 84 to 87 percent carbon by weight, 11 to 14 percent hydrogen, 0.06 to 2 percent sulfur to 2 percent nitrogen, and 0.1 to 2 percent oxygen.

Physically, crudes may be black, heavy, and thick, like tar, or brown, green to nearly clear straw color, with low viscosity. Most crudes are lighter than water, running from 79 to 95 percent of an equivalent amount of water. But some are heavier than water. Chemists call specific gravity the weight of a substance relative to the weight of the same volume of water. Oilmen use their own arbitrary scale to rate oil: the American Petroleum Institute (API)

Hydrocarbons		Saturated	Unsaturated	Isomers
Gases	C_1 Methane			
	C_2 Ethane	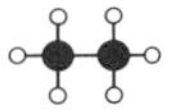	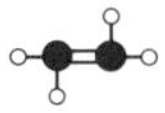	
	C_3 Propane	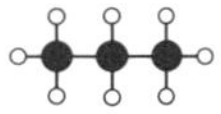	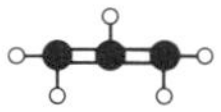	
	C_4 Butane	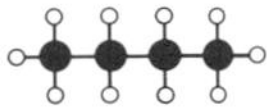	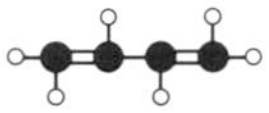	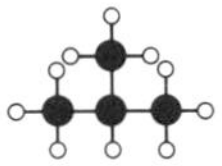
Liquids	C_5 to C_{10}	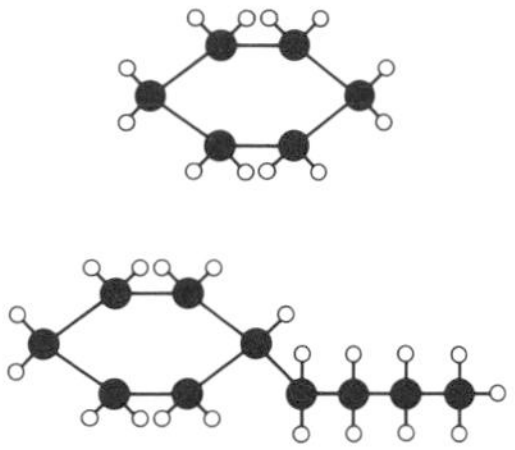	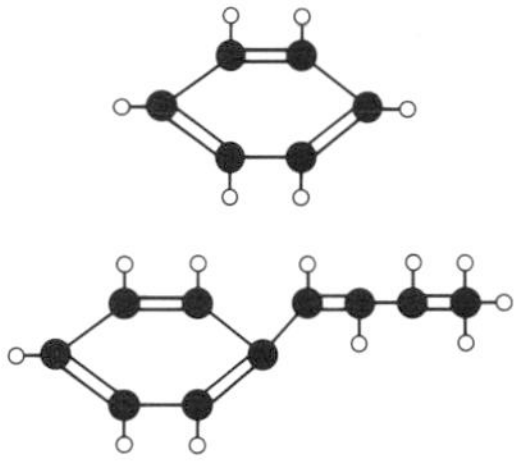	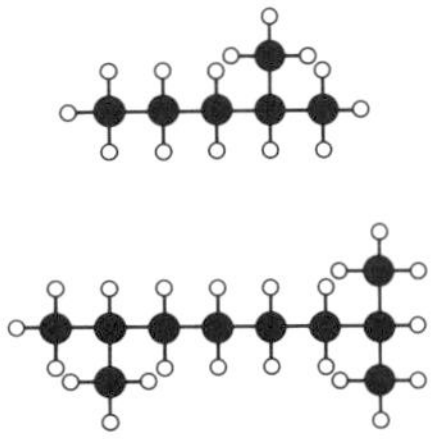
Solids	C_{11} to C_{50}	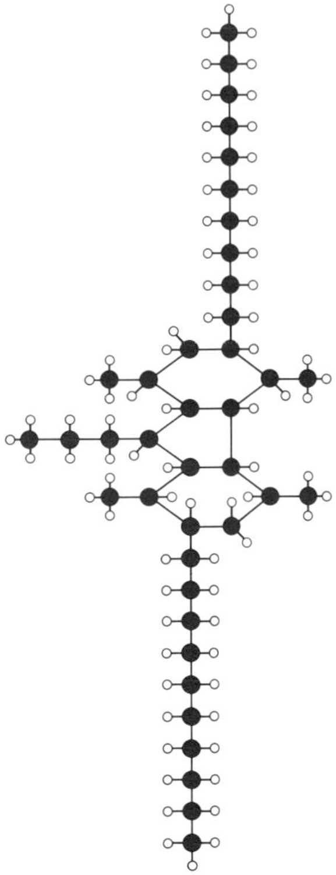	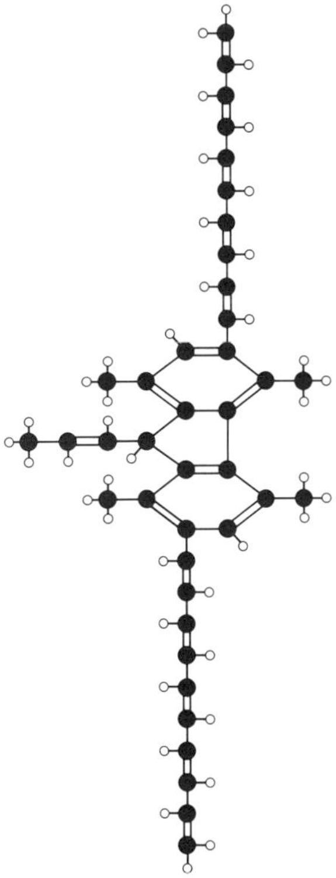	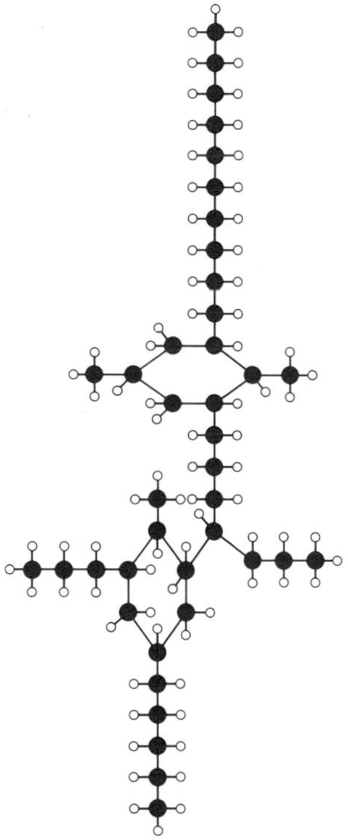

Carbon atom

Hydrogen atom

gravity. Expressed in degrees, API gravity divides the number 141.5 by the actual specific gravity of the oil at 60 degrees Fahrenheit and subtracts 131.5 from the resulting number. Paradoxically, the higher the API gravity, the lighter the crude. The high-gravity crudes are generally more valuable. Oils of 10 degrees API gravity or less are heavier than water and will sink. Heavy crudes, fuel oils, and asphalt are in this category.

In general, since carbon is 12 times heavier than hydrogen, the more carbon-to-carbon binding, the heavier the crude oil. Oilmen classify crudes as light, containing high proportions of lower hydrocarbons; heavy, or deficient in lower hydrocarbons; paraffinic, heavy with unsaturated hydrocarbons; or aromatic, loaded with ring-shaped and reactive unsaturated hydrocarbons. They also label crudes as sweet or sour, depending on the sulfur content.

Refiners prefer light, paraffinic, sweet crudes of high API gravity, since they contain a large proportion of gasoline components. These crudes are less corrosive, burn cleaner, and require less processing to yield valuable products. They command a premium price. Aromatic sour crudes are sometimes so difficult to process that they are barely worth exploiting, even in today's market. Aromatic crudes are also called asphaltic because their higher weight constituents are asphalts and tars. The higher weight constituents of paraffinic crudes are waxes.

How a particular crude got to be what it is and where it is depends on the biological material from which it was originally formed, as well as its geological history. Most current theories date the origin of petroleum and its related fuels to the final breakdown of animals and plants by bacteria that occurred tens of millions of years ago.

In that long-ago era, sun-powered life processes converted inorganic elements into organic compounds. These include oxygen-containing compounds such as cellulose, starches, and sugars; nitrogen compounds such as protein and amino acids; fats and oils; and even calcium compounds that comprise bone and shell, although plants of all types are the principal group.

All living things die and cease to be sun-driven factories converting the inorganic to the organic. Micros-

copic life-forms, bacteria, start the process of decay. These bacteria-driven processes produce gases, mostly methane, still called "swamp gas" or "sewer gas."

All biologic materials participate in this cycle. A dead animal or plant becomes food for bacteria. Bacteria convert them into forms suitable for plant nourishment. These new plants, in turn, become food for new animals that become food for other animals. So it has gone for some billions of years.

Virtually all petroleum is found in what are clearly marine sediments collected beneath the seas and oceans. How did they get there? The relationship between chemistry and geology is most apparent here.

Most biological material recycles locally. But much organic residue is washed down and blown out from mountainside and forest floor into streams, rivers, lakes, and seas. There, eventually, it mixes with sand and silt and the calcium carbonate shells of microscopic animals and sinks to the bottom. Without oxygen, biologic activity soon ceases, and the organic-rich sediments collect.

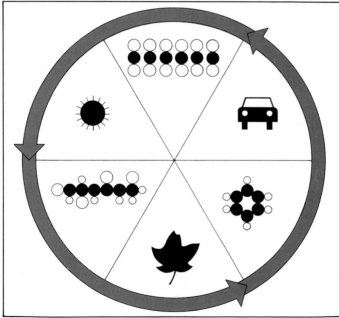

The carbon cycle is the basis for all life on earth. Decaying organisms become hydrocarbons and are used by CO_2-emitting autos. This CO_2, photosynthesized by the sun, turns into carbohydrates and eventually food for plant and animal growth.

78

Over time, changes in the ecology alter the mix of sediments. In some cases, water becomes too salty to support any life, and biologic decay halts early, leaving sediments exceptionally rich in organic content. Some conditions favor biologic activity, so that previously sterile but organic-rich sediments undergo biogenic decay that goes to completion and leaves no organic residue.

Climatic changes or changes in heat flow from the earth and sun (which apparently happened some 250 million years ago) tremendously influence the quantity and distribution of biologic material. Such catastrophic events also influence the identity of the organisms but not their chemistry; thus the extinction of some species — such as the dinosaurs — and the rise of others do not affect the origin of oil. All species are grist for the earth's oil mill.

Earth's organic life-originated material is a minute proportion of the planet's total mass, even of that small percent contained in the lithosphere, biosphere, and atmosphere. With so small a store to draw on, even organic-rich sediments from ancient times seldom exceed 1.5 percent organic matter. In all probability, all existing petroleum was generated in rocks that had as little or even less organic content.

Today sediments average 2.5 percent organic material by weight, varying from 1 percent to 7 percent near shores. In some lakes, though, the organic content in bottom oozes is as high as 40 percent. Some Black Sea locations have 35 percent, while deep ocean sediments have less than 1 percent. Ancient sediments, dating from geologic times, average 1.5 percent organic matter, with a range of 0.5 percent to 5 percent in or near existing oil fields. Thus, if ancient sediments were originally similar to modern ones, they have lost up to 40 percent of their organic material during their burial.

Scientists today believe that virtually all petroleum originated in marine sediments that collected beneath seas and oceans. A very few were aeolian deposits, onetime wind-blown sand dunes. The exceptions are so few that they suggest some geologic accident that caused the oil to migrate from its place of sedimentary origin in a marine environment to a place in an igneous rock, where it could not have originated.

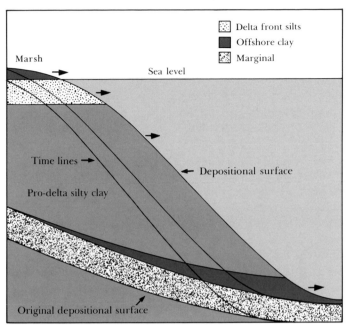

This cross section of a delta shows how the clays migrate over each other toward the sea. Newer deposits lie on top of older deposits.

Whether the base material was terrestrial or marine, animal or vegetable, also seems a distinction without a difference. Petroleum contains trace amounts of complex compounds based on both chlorophyll and hemoglobin that obviously came from plants and animals. In any case, by the time the vast bulk of organic material has been swept from its place of origin by rivers and streams and attacked by numerous bacteria, its final passage, before settling with a sediment, is likely to be through some type of single-celled animals or plants such as bacteria or algae. Such life-forms represent by far the greatest weight in aquatic environments.

In the Caspian Sea, for example, are an estimated 400 million tons of bacteria, fully 75 percent of all life in that inland sea. Other microforms of life constitute more than 24 percent of the total. Fish, currently the economic mainstay of that area, total a mere 900,000 tons, or 0.16 percent. Scientists have little reason to believe that the biologic content of ancient seas differed much from that of the seas today.

Scientists have strong evidence that petroleum originates from the final breakdown products of life that settle to the bottom with sediment, and not directly from leaves, twigs, and the bodies of animals. The inclusion of occasional dinosaur fossils in marine strata or sediment is the result of the accident of entombment, for none apparently lived in salt water.

The sediments range in size from boulders, which rumble along the bottom of rushing streams and stop when the current slackens, to the finest of silts, which remain in suspension as long as there is the least movement in the water, or in the air, where they are carried thousands of miles on the wind or air currents.

Large-particle sediments, such as boulders and stones, carry little organic material. The usual sediments, however, are not gravel but sands of varying coarseness and ever-finer silts. In a moving stream, the sands deposit first, settling near the shore, where they are subject to wave action. Silts travel farther, sometimes a thousand miles to sea. The organic content of the sediment is inver-

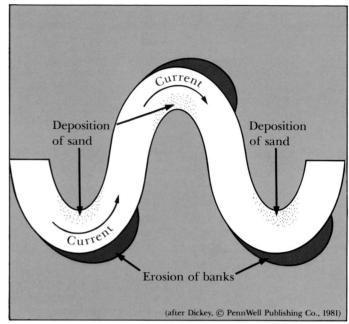

(after Dickey, © PennWell Publishing Co., 1981)

A river collects the flood-plain deposits, dropping the sand on the inside of the next bend but carrying the clay and silt farther downstream.

80

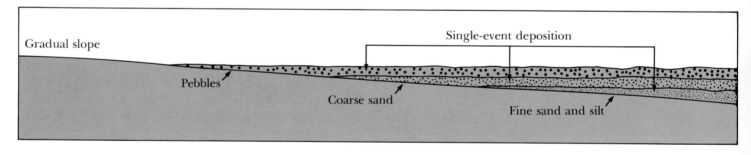

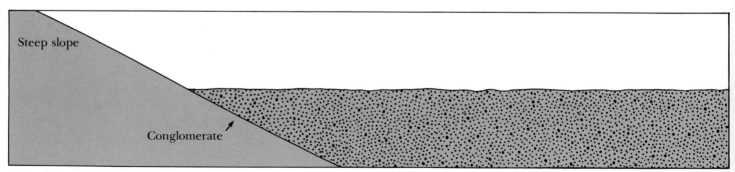

The sorting of sediments is a function of the means by which and the distance over which the material is transported. Fine-grained matter travels farther and faster than do coarser grains. The longer the process operates, the greater the separation between the sizes (upper). If the matter goes a short way and is deposited rapidly, a totally unsorted sediment remains (lower).

sely proportional to its coarseness and directly proportional to the relative rates of deposition.

Because the sand settles first, it has the least time to capture organic material to take to the bottom, though settling time is only one factor. Silts and clays carry two to four times as much organic material as sand. The tiny particles of fine silts are more likely to capture significant amounts of microscopic material than are the larger particles of sand. Distance and time in suspension are also factors. The longer the organic material remains in a zone conducive to bacterial action or oxidation, the less of it remains entrapped with the sediment.

Thus in river deltas, major locales of sedimentation, the production of organic material in the delta plain may be tens of grams per square meter per day, but only a tenth of this production takes place at the delta front and on the continental shelf, and a hundredth of it beyond the shelf.

New sediments soon cover old ones and their entrapped organic constituents. The accumulating weight compresses the underlying material, squeezes out the contained water, and greatly reduces porosity and permeability. *Porosity* refers to the number and size of the spaces between each particle, while *permeability* refers to the degree of connection between those spaces. A solid substance may be porous but impermeable if the pores do not connect to form a passage for the movement of fluids.

During this period of compression sediments also subside, with corresponding increases in temperature. The earth's interior temperature increases with depth, averaging 1 to 1.6 degrees Fahrenheit per 100 feet, though with great variation from place to place.

Under the influence of pressure, heat, and perhaps low-level radioactivity and chemical reaction with minerals, the contained organics undergo a transformation. Originally cellulose, starches, sugars, proteins, lignins, waxes, and fatty oils, the biological material becomes kerogen. This is not a particular organic compound but a mixture of high-molecular-weight fatty acids and their salts. Kerogen contains a higher ratio of hydrogen to

81

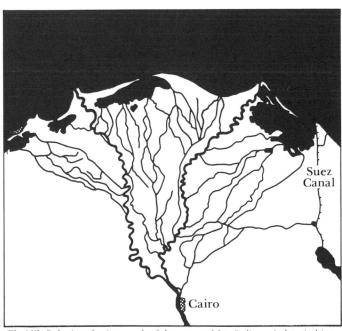

The Nile Delta is a classic example of the arcuate delta. Sediment is deposited in a broad arc at the mouth of the river. The river breaks this blockage and creates new channels, which themselves become blocked, and the process repeats itself.

In the bird's-foot delta, exemplified by the Mississippi Delta, deposition occurs along the sides of the channel until levees are built up. Eventually, the river breaks through, creating deep channels that resemble a bird's foot.

82

Millions of years ago, kerogen, an accumulation of dying plant and animal material, mixed with sediments and became trapped in rocks like these in Colorado. For years scientists have tried to find means of extracting the oil from oil shale (modern kerogen), but so far, production costs have been prohibitive.

carbon and less oxygen than the initial organic material. The material collects in the silts and clays rather than in the sands and tends to cement the fine particles into an impervious mass.

Scientists believe that kerogen is the precursor of petroleum. It is the organic constituent of oil shales. These shales are sediments whose development into petroleum-source rocks was prematurely arrested, sediments that did not stay buried deep enough or long enough.

While scientists believe kerogen in sedimentary rocks to be the source of petroleum, they do not find petroleum in such rocks. Thus there is some question whether the final crude is formed in the source rocks or in the reservoirs where it is found.

As compaction takes place, sands, silts, muds, and clays, as well as deposits of calcium carbonate from animal shells and corals, greatly reduce in volume. The porosity of muds drops from an initial 70 to 90 percent (which is another way of saying that the initial mud was 70 to 90 percent fluid) to 10 percent or less in the resultant shales. Mud-born shale is impervious to fluid passage. The sandstones and limestones that result from compaction of sandy and carbonate sediments remain permeable, thanks to the larger particle size and initial open structure.

As the mud compacts over hundreds of thousands of years, it expels 15 billion to 20 billion barrels of fluids from each cubic mile. Compaction forces many times this huge amount through the bed from still deeper-lying sediments. At no later time is there anything like this enormous volume of migration and flushing action.

While the fluid is being forced out of the closely compacted mud into the less compacted and more porous sandstones and limestones, the organic constituents are subjected to slight but increasing heat and great pressure over an immensely long time. Very slowly the kerogen breaks down into simpler substances. It loses oxygen and nitrogen until only hydrocarbons remain.

Scientists once believed that intense heat in the interior of the earth produced petroleum from organic substances because they could duplicate such reactions in the laboratory. Yet they find substances in natural petroleum that could not be there if the material had ever undergone much heating. Now scientists believe that a little heat over a long time has had the same effect as much heat over a short time, and without destructive results. The time, of course, is measured in hundreds of millions of years.

As the hydrocarbons develop from the kerogen, molecule by molecule, they migrate with the water out of the mud and shale strata into the more open strata. If these strata are open to the sea — if they have no impervious cap on them — the hydrocarbons pass right through and are lost into the sea and atmosphere, to reenter the organic cycle. Far more petroleum has disappeared in this manner than has been captured in the earth.

If, however, an earlier layer of shale or salt or some other barrier interposes itself between the migrating fluids and freedom, progress stops. The hydrocarbons begin to collect, trapped against the barrier, and an oil pool begins to form. Meanwhile, the shale where the hydrocarbons originated reaches the limit of its compaction, becomes completely impervious, and ceases to pass fluid through or to expel it. The shale bed retains most of the organic material that the original mud held, but it contributes no more to the oil pool, nor is it itself a source of hydrocarbons.

Only in this way does a brontosaurus pass through a bed of algae and the bowels of the earth to reemerge 150 million years later as a gallon or so of crude oil.

83

CHAPTER 8 – GEOLOGY AND OIL

To oilmen, the location of an oil deposit is more geology than it is geography. While most people would say that the oil is on the North Slope of Alaska between the Brooks Range and the Arctic Ocean, the oilman places it in the Triassic and Jurassic under the Lower Cretaceous. Common usage says Venezuela; oilmen say Mesozoic and Tertiary. The general public says Saudi Arabia; oil geologists say Upper Jurassic.

These geological terms define a time in the history of the earth's existence. They are the labels geologists give to the consecutive layers of rock laid down over billions of years. These layers are the global equivalent of the rings of a tree. They reflect growth and change, good times and bad, and leave a coded record. Geologists break the code and lay out the record of the earth's past.

Interpreting the rings of trees, however, is easier than interpreting the layers of the earth. Like tree rings, rocks of a given geological age came into being at the same time all over the world. If undisturbed, the rocks of one era should lie in orderly fashion on top of those of previous eras. They often do, but the earth's crust has seen so many cataclysmic changes that many rocks now bear little resemblance to their original state. Faulted, upthrust, downthrust, folded, displaced, eroded, buried, and eroded again, what was then a prehistoric sea bed may be today's mountaintop, separated by miles from companion strata that were laid down simultaneously but now are so altered that only a trained geologist can decipher their relationship. Fortunately, each stratum retains its own indelible signature, its fossil record or "biofacies," which tell the geologist how old the rocks are and how they relate to the rocks surrounding them. Most important, the facies tell whether or not the rocks are likely to contain oil.

The creation of petroleum is a continuous and exceedingly slow process that proceeds in equilibrium with other slow geologic mechanisms, such as the deposition of sediments in seabeds and the movement of continents. In the billions of years that these processes have been operating, most of the oil and gas that was formed migra-

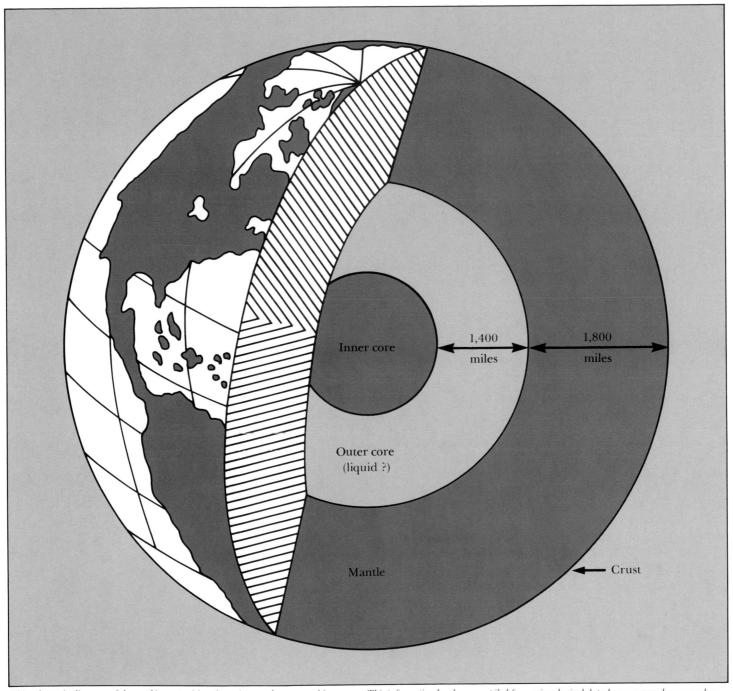

This schematic diagram of the earth's composition shows its mantle, core, and inner core. This information has been compiled from seismological data because man has never been able to penetrate more than a fraction of the earth's 40-mile-thick crust.

ted upward either to the sea floor or to the earth's surface. There it dispersed and reentered the cycle or was thermally decomposed or eaten up by bacteria. Only a very small percentage of all the hydrocarbons ever formed remained in the earth long enough to be found and exploited by man.

Paradoxically, geological events having nothing to do with oil help in its discovery. Oil becomes available by being trapped on its way out of its source beds. The geologist tells from the evidence where those traps most likely are. That evidence includes the present structure of the land, the age of the rocks, and the type of rocks. In the absence of surface indications such as tar pits, seepages, and gas vents, all of which can mislead, the oilman assesses petroleum potential by geological examination of an area.

The geologic events that trap oil result from tremendous forces that have been shaping the earth since its beginning. Scientists believe the earth to be about 4.6 billion years old. Consider it as a semiliquid ball on which float a number of large solid plates, called tectonic plates. These plates are in constant motion in relation to one another, like ice floes on a river. They are tens of miles thick and thousands of square miles in area. Made of igneous rock that derives from the deeper molten layers of the earth, they carry the continents, islands, seas, and oceans on their backs.

The continents and ocean floors, some 22 miles and three miles thick, respectively, are made of lighter rock than the plates. Like the plates, they are formed of elements from within the earth, their compositions altered by chemical and mechanical action.

Geologists and meteorologists, such as Alfred Wegener, long have theorized that since the earth began, the crust has been in ceaseless, infinitesimally slow motion. At first, the continents were united in a single land mass, which Wegener named Pangaea. In the earth's history, the continents have broken apart and reunited in new combinations; oceans have opened, been annihilated, and opened again. Europe, America, Asia, Africa, Australia, and Antarctica have all been part of the same land mass, not once but many times, and no doubt will be again.

A map of the world with its single continent, Pangaea, as scientists think it looked 200 million years ago.

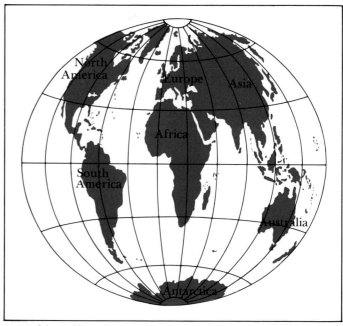

A map of the world as it looks today, with its seven major continents.

What geologic forces thrust up, meteorological forces tear down. The primal basalt, granite and lava — igneous rocks born of fire and heat — were ejected from the earth's interior to build the continents and islands. Then they were incessantly attacked by wind and rain, running water and ice, reduced to silt and sand, deposited in layers buried by succeeding layers, compressed to hard rock (lithified), and subjected to further heat and pressure. Mountains rose and fell, and continents collided. These ancient rocks, called "metamorphic" because they are metamorphosed, or changed, bear little or no resemblance to the igneous rocks from which they ultimately derived, or to the sedimentary rocks that were their immediate precursors. Mud became the sedimentary shale and the metamorphic slate. The carbonate shells of tiny animals became the sedimentary chalks, limestones, dolomites, and the metamorphic marbles. The metamorphic and igneous rocks are the basement complexes that form the cores of the continents and the roots of the mountains. Indeed, they are called "basement rocks." In the mass, they form the continental shields, or cratons.

Miners are very interested in igneous and metamorphic rocks because the processes that produce them also tend to form mineral deposits. However, to hit basement means despair to the oilman, because the conditions that produce basements assure that neither oil nor gas will be found, except through rare accidents.

The earth's first four billion years hold little interest for the oilman. They find no oil in rocks older than 600 million years — the pre-Cambrian rocks — because there was very little petrogenic life then and also because any oil there has long since been dispersed or destroyed. The oilman confines his interest to sedimentary rocks younger than 600 million years, those of the Paleozoic (early life), Mesozoic (middle life) and Cenozoic (later life) eras, within which geologists identify 15 distinct periods from the Cambrian to the Quaternary. Within each period, particularly the later ones, are epochs of still shorter lengths. The current epoch is the Holocene, which began a mere 10,000 years ago.

The beginning or end of an era, period, or epoch is determined by the appearance or disappearance of one or another type of life-form, as evidenced by fossils left behind in the stone. The earliest evidence of life is a fossilized bacterium 3.4 billion years old.

The oldest rocks identified are about 3.8 billion years old. This leaves about a billion years of the earth's existence unaccounted for. Despite this omission, there has been enough activity in the last 600 million years to keep petroleum geologists busy for generations to come.

Along with the development of the first vertebrates 400 million years ago, the first mammals 230 million years ago, and the first flowering plants only 100 million years ago, the ancient Atlantic Ocean closed up, the Appalachian Mountains rose, and Europe and Asia collided to form the Ural Mountains. Only 60 million years ago, Australia parted company with Antarctica and began moving toward Asia. And 50 million years ago, India collided with Asia, forming the Himalayas.

Vast beds of marine sediments on the coasts of ancient Asia and Europe became oil-containing basins on both flanks of the Urals. A sea bottom south of Asia became mountaintops between Asia and India. The Himalayas are still building. The delta of a great river that flowed westward to the Pacific across the then-connected continents of Africa and South America became the mountain-bound petroleum fields of Peru, Ecuador, Bolivia, and Colombia following the separation of the continents and the uplift of the Andes Mountains. Plate tectonics explains much that previously was baffling and has proved immensely useful to the petroleum geologist. Knowledge of the earth's tectonic history helps reveal how ancient marine sediments reached their present location and structure.

Petroleum originates with life in marine environments. The geologist and the paleontologist together determine from rock samples the life sequences and the disposition of sediments. Then they must determine whether the life residues and sediments eventually came to rest in such a configuration as to allow oil formation, oil migration, and, finally, oil entrapment in a place in which oil can be found and exploited.

The initial deposition of a sediment depends on the velocity and turbulence of the water that carries it and the size of the grain particles. Sand settles first. Fine clays carry farthest and settle last in quiet waters. Carbonates, the residue of tiny animals, remain in suspension longest. In great basins or river deltas, the various sediments lie down in even layers, sands closest to shore, silts and clays farthest out. As the shoreline builds up, the land subsides from the weight of overburden, and sand begins to deposit on previously laid down clays. Further additions cause the land to submerge, or perhaps the sea level changes, as it often has, and the sequence of sedimentation alters again. Built up over time, the result is thousands of feet of alternating layers of mud, sand, and chalk, which testify to the advancing and retreating shoreline, changes in water flow and climate, and movement of the continent from one part of the globe to another.

Over time, the sea into which a river flows evaporates and leaves behind immense plains of salt, sometimes called evaporites, that overlay earlier sediments. As years pass, additional sedimentation may accumulate over the salt beds, eventually burying them under more thousands of feet of sediments, a never-ending process.

One of the best places to get a graphic view of the process of sedimentation is the Grand Canyon of the United States. There, more than 700 million years of sediments lie evenly, one upon the other, undisturbed and open to view. Where the sea has advanced and retreated at intervals of a few million years is clearly evident, as is the way the Colorado River cut through the region. Some tectonic force uplifted the entire plateau from the sea basin in which the sedimentation occurred, but uplifted it slowly enough so that the river could cut its way, keeping pace with the rising land. The distance from the top of the sediments to the river's present level is about a mile, most impressive to the observer. But this is not a particularly thick bed of sedimentary rock. The thickest known, from the Cretaceous period, is about 64,000 feet deep, more than 12 miles of sediments laid down over 65 million years.

The Grand Canyon strata, moreover, remained flat, evidence of the quiet sea in which they grew and the

88

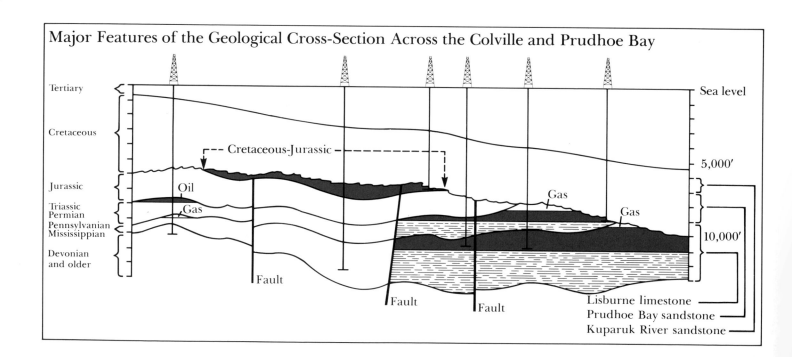

Major Features of the Geological Cross-Section Across the Colville and Prudhoe Bay

tranquil geologic history of the area. That very tranquillity means there is no oil. No geologic accidents occurred to trap any oil that might have formed there. Fortunately, most regions of the earth have not been so geologically quiet, and the strata have been thoroughly reshuffled to form traps of all sorts.

Commercial petroleum has been formed in rocks of virtually every age, but some ages have proved consistently more productive than others. The earliest oil-forming periods, the Ordovician and the Silurian, were probably periods of prolific generation of oil and gas, since the fossil record testifies to abundant life. But both periods later underwent many geologic changes that have caused most of their hydrocarbon deposits to disappear.

In later periods of the time scale, the chances of finding appreciable amounts of oil increase, though not necessarily proportionally. In the United States, most of the known oil occurs in the Permian and Pennsylvanian periods of the Paleozoic era, 65 million to 135 million years ago, and the Miocene epoch of the Cenozoic era, 12 million to 23 million years ago.

Obviously, there were tens of millions of years when conditions for oil generation in what is now the United States were not the most favorable. Yet there are noteworthy exceptions. The Triassic period of the Mesozoic, 181 million to 230 million years ago, has a poor reputation, but the Prudhoe Bay field in Alaska is a Triassic field.

Worldwide, nearly 90 percent of all the world's oil and gas lies in or has come from Mesozoic and Cenozoic strata and is, therefore, less than 230 million years old. The upper half of the Cenozoic accounts for 30 percent of all the oil ever discovered that is less than 23 million years old. Oil younger than a million years is rare. Any found in Quaternary deposits likely migrated there from older strata. For that matter, oil found in older strata may be older still, since it too may have migrated from even more ancient source beds to its present reservoir.

To find Paleozoic, Mesozoic, and Cenozoic oil in the same province is not uncommon. The widest age spread is found in the Rocky Mountain region, where oil ranges from the Paleozoic Ordovician to the Cenozoic Eocene. The major occurrences of Paleozoic oil are in North America and North Africa. Oil from the Mesozoic era, especially the Cretaceous, is nearly worldwide. So is oil from the Cenozoic. Interestingly, like wine, the older an oil is, the higher is its quality.

Depth of burial is also a factor. If source rocks are buried too deep and thus subjected to high temperatures, their hydrocarbons sometimes are cracked all the way to gas or even to elemental carbon. In shallow burials, the process never completes, and the result is oil shale instead of petroleum. Whatever the age of the rocks and sediments, the chances of finding oil are best at depths between 1,000 and 12,000 feet. Below that, the ratio of gas to oil increases rapidly. At depths greater than 20,000 feet, gas becomes the principal goal, since there is little chance of finding oil, regardless of geologic history.

Oil originates in the silt and mud that eventually become shale, the source rock, and then migrates, under geologic and hydrodynamic forces, into porous sandstone or carbonate reservoirs. Since oil and gas are lighter than water, they constantly try to rise, to float on the water. This driving force accounts for oil's constant migration until it is trapped or lost to the atmosphere.

Oil migrates in any direction and covers great distances because it follows the path of least resistance to the point where it either escapes or goes no further. It is trapped when it can go no further.. There are two kinds of traps: structural and stratigraphic. The structural type results from some cataclysmic geologic event that creates a barrier and prevents further migration. In this instance, the forces that act on oil are equal in all directions. The stratigraphic type is a natural development of the oil-bearing strata. In effect, it acts as a blind alley from which oil cannot back out. Structures are seldom all one type or the other, but usually combinations of the two.

By far the most common, and most easily found, structural traps are anticlines, in which at least 80 percent of the world's oil and gas has been discovered. An anticline is an arching of the strata caused by a salt dome thrusting up from below, or compression from earth movements that wrinkle the ground and produce an uplift to counterbalance the subsidence. Anticlines frequently have surface manifestations like hills, knobs, and ridges, but even when the surface is totally eroded,

Anticlines, Faults, and Salt Domes

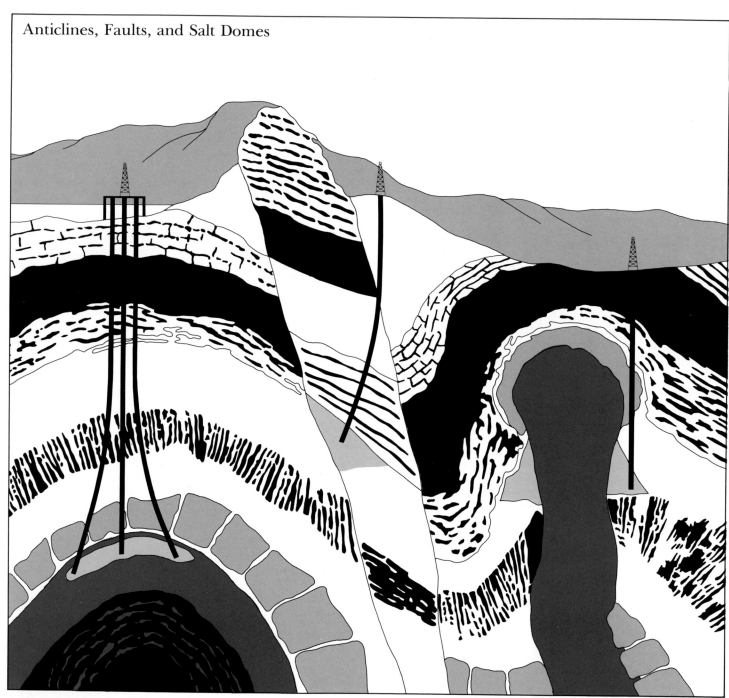

A geologic gross section showing how reservoirs of oil and gas are trapped in porous rock. Geologists look for anticlines (left), faults (center), and salt domes (right).

geologists using modern seismographic exploratory techniques detect subterranean anticlines fairly easily.

Ideally, an anticline forms a dome or roof of impermeable strata above a permeable oil-bearing stratum. Oil migrating from source rock, usually shale, moves upward through various permeable strata and eventually becomes trapped under the roof. Typically, such a trap has a gas space directly under the impermeable rock, a layer of oil and, beneath that, salt water.

Anticlines take many forms and sometimes are very large. Not all contain oil. Some arches several hundred miles long in the United States and Russia do not contain oil in their entire length. Geologic events have faulted or broken them, allowing the oil to escape. Accumulations are generally confined to a few particularly favorable areas along the ridge.

Domed anticlines generally result from an upwelling beneath them, such as a salt intrusion. Erosion cuts off the top of a structure, leaving deposits of oil only on the flanks; or the structure may be open at one side, yielding a formation that resembles a nose, which is what oilmen call it. A nose will have no oil unless some other structure blocks the open side.

Some anticlines are draped rather than folded. The original sediments were laid down over preexisting features in the basement rock, such as an ancient and subsided mountain range or a granite intrusion. Some fields in Kansas and the Texas Panhandle, called ridge fields, are laid down on ancient mountains, then compacted and further folded.

The opposite of an anticline is a syncline, a downfold in stratified rocks. Such a structure — an upright bowl — cannot trap oil.

Another important structure is the fault trap, often found in company with other types of traps. A fault is a break in the continuity of stratified rocks or even basement rocks. Forces on either side of a fault move in different directions or at different magnitudes. Eventually, the force becomes greater than the rock can resist, and the rock breaks. The opposite faces of the break slip against one another, and related layers of the strata are displaced from their original positions.

Some faults are very small, resulting from simple tension or compression, breaks caused by synclines and anticlines through the folding and wrinkling of the sediments. Or they can be hundreds of miles long, with horizontal displacements of many miles and vertical displacements of thousands of feet. Matching up the original configurations in very large faults is a test of the geologist's ingenuity.

Sometimes, forces are so great that one side of a fault will lift up, rear high above its mate, and curl over it like a wave on a beach. This "overthrust" places older strata on top of younger strata and thoroughly confuses the geologic record. There is much such overthrusting in the Rockies and the Alps.

To the petroleum geologist, faults are significant for two reasons. On the negative side, faults break open other types of traps and prevent oil from accumulating. Such faulting isolates the oil reservoirs in the long anticlines of Kansas and Texas and elsewhere. But by moving an impervious stratum across an open-ended permeable one, a fault can form a trap for oil and prevent it from migrating farther. An anticline nose can become an effective reservoir if a fault blocks it before the oil can get away.

The plane of a fault is usually slanted. One block is above that plane and the other below it. If the block is downthrust in relation to the block below the plane, the fault is said to be normal. If the upper block is higher, the fault is reversed. If the plane is at a shallow angle, less than 45 degrees, the result is a thrust fault in which one block thrusts over the other. A depression bordered by parallel faults is a graben. A long, deep graben covered with sediments is a rift valley, the most famous of which cuts halfway across Africa.

Faults are very complex structures, reflecting all the geological and meterological forces at work in a region. By fracturing rock, they can provide oil reservoirs where none would otherwise be, in basement rocks, for example. They break up reservoirs into noncommunicating sections and are thus important to oil-field development. In conjunction with other structural and stratigraphic factors, they can help produce oil accumulations. And they can also form traps on their own, as they do on the Texas Gulf Coast, where the growth of faults has bent the adjoining strata and produced, in effect, anticlines that trap oil.

Salt behaves like a plastic; it flows under pressure. Beds of salt covered by other rocks squish toward the surface in great blobs or domes sometimes miles wide. As they rise, these domes push aside and break the overlaying strata, which form a cap on top of the dome and lie at angles against it where the salt pushes through. Salt is also impervious. By rupturing the strata and then sealing them at the upthrust edges, salt can help trap any oil the strata may hold. Spindletop was a salt-dome field, as are many along the Texas Gulf Coast and elsewhere in the world.

Stratigraphic traps take many forms. One type is a region bounded on one or more sides by zones of low permeability. This change in permeability within otherwise orderly structures impedes the migration of oil and forms the trap. The change results from hydrodynamic and physical forces. Since there are no outward manifestations, such as a salt dome, anticline, or fault, which act as clues to the oil geologist, stratigraphic traps can be very hard to find. Much of the remaining potential oil to be discovered may be in huge stratigraphic traps that will be very difficult to locate.

Primary stratigraphic traps result from the sequence of deposition of sedimentary rocks, including source rocks. At the time of deposition, these beds are laid down in just the right sequence to provide a zone of permeability surrounded by impermeable strata. Where these traps are three-dimensional and bounded by air or water at the time of formation, they are called "paleogeomorphic," or "buried landscapes." These are the reefs, sandbars, and stream beds that were buried and eventually became reservoirs.

Not all oil originates in sediments, though all originates in marine environments. Some prolific petroleum structures once were reefs, gigantic structures built up over millions of years by tiny animals. They range in age from pre-Cambrian to Tertiary. Their soft parts became the raw material for the petroleum, while their carbonate shells became both source rock and reservoir.

The size of these reefs is staggering. They may extend for hundreds of miles along ancient coastlines, like

92

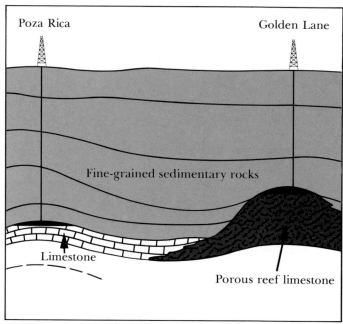

A geologic cross section of the so-called Golden Lane oil field in Mexico.

the Barrier Reef off Australia. One Permian-age reef in Texas and New Mexico is 400 miles long and 1,200 feet deep. More important, as each new generation builds on top of a reef, the dead layers sink beneath the weight. Eventually, the accumulated weight causes the underlying basement rock to subside, just as sediments do, and the ultimate depth of the porous limestone ends up many thousands of feet down, far deeper than the original shallow sea.

Some of the most productive oil fields in the world — Texas, the Golden Lane, North Africa, and the Middle East — are ancient reefs. In Libya, one measured only two and a half by three miles and held two billion barrels.

Another prolific trap is the sandbar or sandstone lens, a sandstone body enclosed in clays and shales. These often retain the original shape of the stream bed or beach that formed them and can be long, sinuous strings of oil-bearing strata. Kansas has a prominent example called the "shoestring sands." They often lie more or less parallel to ancient coastlines, ripples of clean sand left where the waves and emptying streams deposited them.

Sandstone traps usually have a lenslike contour, convex on the top and pinching out at the edges. Some very large reservoirs taper from considerable thickness down to a point where they literally pinch out between impermeable strata. The huge East Texas field is an example. Geologists hope to find similar deposits elsewhere, in buried submarine canyons at the edges of the continental shelves and even onshore.

Another very important type of formation is an unconformity. It is called a stratigraphic trap even though it arises from structural changes in the region. When old strata are finally eroded, resubmerged, and covered over with new strata (which probably have a different orientation from the old), at the interface is an unconformity. If these events happen in the right sequence and the new sediments place an impermeable cap on the old, there is a chance that migrating oil generated in the old rocks will be trapped at the interface. Indeed, traps can form above and below the unconformity as both the old and the younger strata follow their own destinies. The Prudhoe Bay field is an unconformity with traps above and below the interface. The older strata, of course, sometimes seal off the younger and prevent the escape of oil through the sediment above them.

The facies-change trap is another type of stratigraphic trap. Facies or lithofacies define the local rock type. In this type of trap, sediments move from the initial deposition point. Sediments carried in a river, for example, initially contain particles of many sizes. As the water velocity diminishes, the coarsest sands come to rest first, and the resulting deposit shows a gradual change from very coarse particles near shore to very fine and silty mud at the farthest edge. In such strata, there is no sharp distinction between permeable and impermeable. Such strata of the same age produce important traps if combined with structural features that seal off the permeable limestone or sandstone. A later upthrusting of the sediment that places the impermeable edge above the permeable, while overlaying the whole with shale, is such a feature. The sand or limestone then pinches out in its own stratum.

Today, enough oil has been found in enough places to give oilmen a very good idea of where it should be and an even better idea of where it will not be. Probably, there is none in the newly formed deep-ocean bed or in ancient metamorphic and igneous basement rocks or in salt beds. Certainly there is little below 20,000 feet, because it is too hot at that depth.

But there are local exceptions. The deepest producing oil well is more than 21,000 feet deep, in Louisiana. At the same time, even when the by now well-known barren areas are excluded, many regions considered productive turn out not to be. Salt domes provide an excellent example.

When Spindletop came in after five tries and showed the potential of salt-dome structures, oilmen rushed out to drill other salt domes. Some proved productive, but many more did not. Most of them somehow failed to trap oil. In many, there was none to be trapped. In others, the strata through which the dome pushed were hard, brittle, and fractured, thus allowing any possible oil accumulations to escape. At Spindletop, the strata were more flexible and softer, forming seals around the intruding salt and retaining the oil on the flanks and top.

Probably the most critical factor in the accumulation of oil is timeliness. If a structural or stratigraphic feature occurs too late to catch the oil during its migration from the source rock, then, however perfect a geological feature it may be, it is an empty trap. The only way to find out if it is full or empty is to drill it. There is no other way. That is why the oil business is risky — and expensive.

93

CHAPTER 9 –
SEARCHING FOR OIL

In the earliest days of the industry, wildcatters drilled where oil seeped to the surface or appeared in brine wells. Or they drilled randomly. Random drilling is not very scientific, but in the nineteenth century it produced a lot of oil. It also broke many more wildcatters than it enriched. Thus, oilmen soon looked for more reliable ways to determine, before drilling, whether there might be oil.

As oil was demonstrably a product of geological forces, it seems evident that the oilman would have turned without hesitation to the geologist. Not quite.

Early oilmen disdained geologists, preferring their own instincts, which invariably told them to drill where others had succeeded. Because the first oil was found in the valleys of northwestern Pennsylvania, oilmen looked in valleys when they went exploring. Only later did geologists show that many valleys were actually the eroded and breached tops of anticlines — not valleys at all but subterranean hills that contain oil.

Geological knowledge grew as rapidly as the hun-

dreds of wells drilled immediately after the Drake discovery. While oilmen soon determined what sort of terrain was likely to yield oil, and while they could read the subsurface structures by the chips their bails brought up, others of a more scientific bent tried to correlate the physical evidence. A geologist proposed the first anticlinal theory concerning the entrapment of oil as early as 1860. Even then, some geologists theorized that sandstone was porous enough to form oil reservoirs without fracturing. Since many states had formed geological surveys as early as the 1830s, knowledge of different types of rock, sediments, and underground formations was already well developed.

In time, both geologists and oilmen learned to recogize what type of valley was an anticline and where a hill or mound was an anticline and not a syncline. An excellent example is the Sheepherder's anticline often found in the Rocky Mountain area. Half a century after the Pennsylvania boom, oil was found in valleys where shepherds grazed their sheep. The setting frequently was

This oblique aerial photograph of the Kūh-e-Kurken Zard anticline in southwestern Iran shows that erosion of the hills has exposed about 6,000 feet of sediments.

one of flat countryside interrupted by an elongated anticline that featured an encircling ridge of upturned sandstone, like a bubble frozen in the act of bursting. Plant growth helped identify these structures. The sandstone rims often had juniper trees that contrasted sharply with the surrounding sage.

The surface geologist, the only kind there was in petroleum's early days, determined the shape and structure of exposed rocks and projected those data into the subsurface. This worked well as long as there were rocks at the surface to observe and measure. Later, geologists began to drill shallow holes and dig trenches to get more data, although none of the new information led specifically to oil deposits. No geologic, geophysical, geochemical, or occult method yet devised actually pinpoints an oil reservoir of a particular size in a particular place. But exploration techniques have improved to a point where the chances of finding oil are much better than they used to be.

Without scientific means (in other words, in rank wildcatting), only about one in 30 wells drilled in the United States produces oil or gas, and not necessarily in significant commercial amounts. With geological techniques, the ratio jumps to one success in ten tries. More scientific geophysical exploration boosts the rate to one in six, and a combined geological and geophysical approach brings it to one in five, pretty good odds. Of course, a geologist has to cover a lot more territory these days to find new areas with reasonably good prospects. Reaching those areas and maintaining crews in them is far more difficult and expensive than before. The easy oil is gone.

Surface geology is still a most important tool for the oilman, although petroleum geologists now use techniques and devices their predecessors lacked. And they are helped considerably by better transport — by land, sea, and air.

The geologist's standard implements are his legs and eyes and a little pick for chipping out rock samples. In addition, he has long used airplanes to examine larger areas and to see the general topography, which may reveal the underlying structure. Woman's Pocket, an important Montana anticline, is clear from an airplane. But to a person walking in the area on foot it is just

another hill or valley. So a geologist turns to other tools.

A high-altitude survey is generally the first step in oil exploration. Until the space age, that meant aerial photography. Today, high-altitude means Landsat satellites orbiting 570 miles up, repeatedly photographing practically every spot on the globe. These pictures, covering 115 miles on a side, show immense detail about the earth and its geological features. Computer processing of the images reveals far more about the ground than is apparent in the initial photo. Detection of Sheepherder's anticlines is easy now.

A new satellite that takes extremely high-resolution stereo pictures has been proposed. Stereosat distinguishes objects as small as a house or a large rock. This $100 million vehicle was designed with the help of the Geosat Committee, which is made up of 100 companies that specialize in oil and mineral exploration.

Aerial surveys still have their place. The method is to equip a plane with wide-angle stereoscopic cameras and to fly back and forth over an area taking overlapping strips of pictures. Later, these pictures are studied with stereoscopic viewers to help geologists prepare detailed topographic maps.

Aerial maps, from plane or satellite, indicate features of potential value far more clearly than any number of searchers on foot could find, and at much less cost. The aerial survey makes it easy to airlift geologists and crews to remote areas and also provides valuable information that will be useful when supply roads, camps, pump stations, and wells are to be installed. Many previously explored regions are now almost routinely remapped by air to reassess their potential.

New technology has greatly enhanced the value of aerial mapping. Side-looking airborne radar (SLAR) yields excellent pictures in all kinds of weather. It uses different light-wave lengths to differentiate surface and interior features in ways no land-based method can equal.

Geologists working in the old way — slogging through swamps, tramping across deserts, skimming over ice- and snow-covered tundra, hiking between mountain passes and even walking along city streets — confirm the results of aerial and satellite surveys. These

ARCO explorationists hike through the jungle in Magdalina Valley, Colombia.

geologists pay attention to minute detail, which reduces the great sweep of the airborne or space camera. The combination of aerial surveys with their minute recordings provides smaller-scale, exact geologic maps.

While the three million wells drilled in the United States have yielded immense amounts of geological information, the data provided by any particular hole are unique to that one hole. Another well drilled 100 feet away may come up with different data. To learn more, geophysical or geochemical methods help the geologist infer from the visible the nature of that which is invisible.

Since 1825, prospectors knew that underground bodies produce measurable effects at the surface. Swedish prospectors used compasses to detect magnetic changes or anomalies in the earth's magnetic field, hoping to discern the presence of ore bodies. The Brunton Compass still is used by surface geologists to measure the dip of beds.

Geologists have also learned that subsurface structures minutely affect gravity as well as the earth's magnetic field. By measuring and correlating these changes with known rocks and structures, they can map buried topography and find oil-bearing structures that offer no surface indication.

A way to measure variations in gravity was devised as early as 1877, and the first practical gravity meter was developed in 1899. The device works on a simple premise. Different materials cause gravity in an area to vary slightly from the norm because they have different densities. Igneous rocks are denser than sedimentary rocks. Thus, a person standing on a mass of granite weighs infinitesimally more than he weighs when he stands on a bed of shale or sandstone. Where igneous rocks are closer to the surface than elsewhere, the pull of gravity is measurably greater.

A geologist runs a traverse with a gravimeter. He observes and records the variations in gravity along the traverse. Then he draws a profile of the basement rocks. Where the basement rocks approach the surface, he surmises that they pushed up the overlying sedimentary rocks to form a dome or anticline where oil might be. A gravity-plus area, an area with higher than normal gravity, is an encouraging sign to the geologist.

97

Other materials, notably salt, produce gravity-minuses because they have less mass than that of sedimentary rocks. The gravimeter, or gravity meter, has been particularly useful in finding salt domes. The first gravity survey of a salt dome was made in Texas in 1917. Indeed, the first important geophysical discovery in the United States occurred in 1924, when gravity measurements uncovered the Nash Dome in Texas. So many salt domes were discovered along the Gulf Coast in the next few years that many thought that geophysics would completely displace surface geology.

The gravimeter acts like an extremely sensitive balance or scale. It is, essentially, a weight at the end of a spring. The spring gets longer in high-gravity areas, shorter in areas of gravity-minus. Small and portable, the meter weighs only about five pounds. It must be used, however, under extremely stable conditions, which limit it to one measurement at a time. On land, it does not give a continuous record of an area because it measures gravity at a number of different points, from which the

geophysicist then plots a gravity map. Modern gravimeters are so sensitive that they can measure differences of one part in 100 million of the earth's gravitational field.

Until recently, wave motion made it difficult to use a gravimeter on a ship. But now, mounted on a gyroscopically stabilized platform, the instrument allows continuous gravity measurements as the ship traverses an area of interest.

Magnetism helps the oil geologist understand the significance of the gravimeter's measurements. The earth's magnetic field varies locally under the influence of differences, much as gravity does. The anomalies may be ore bodies or rocks of various kinds that have wide-ranging magnetic properties. Igneous rocks are magnetic, some more than others, while sedimentary rocks are not. Thus, local variations in magnetism can be measured, and the presence of subsurface structures that might contain petroleum can be inferred.

Magnetism was first used to find iron as far back as 1825, and the technique was common by 1870, when two geologists developed the first really practical magnetometer. It was not until 1924, however, that the magnetometer found wide use for mapping geologic structures. First used for this purpose in the southwestern United States, the device proved superior to the gravimeter because it allowed the geophysicist to run traverses and obtain a continuous log or record. Movement did not affect the readouts and, in time, scientists took their magnetometers aloft, trailing them behind aircraft to continuously measure and record the magnetic variations of large areas. In 1939, Gulf Research developed a more practical airborne instrument, but its first widespread use was as a submarine hunter in World War II. It was first used to find oil in 1944.

For all the proved value of gravity and magnetics, they remain relatively coarse tools. Their findings need to be put through a finer screen to pay off significantly in the hunt for petroleum. That screen is seismography, listening to the earth, and it is the most effective and important of all geophysical techniques. For those who know how to listen and what to listen for, the earth, when prompted, has much to say. There are even indications that the latest seismographic methods may ultimately locate hydrocarbon deposits precisely. The oilman's dream come true!

98

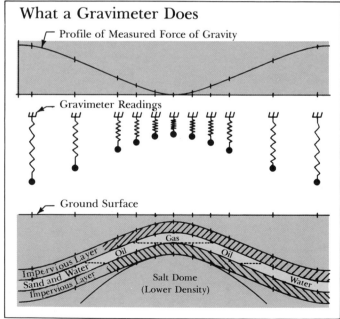

What a Gravimeter Does

Profile of Measured Force of Gravity

Gravimeter Readings

Ground Surface

Gas

Oil

Oil

Impervious Layer

Sand and Water

Impervious Layer

Salt Dome
(Lower Density)

Water

Figures 1-9 above represent observation stations where the gravimeter is installed. The attraction of the gravitational force of underlying strata (lower panel) is measured (middle) and then plotted on a curve (above).

Like other geophysical techniques, seismography is not particularly new. In 1848, Robert Mallet, an Irish geologist, proposed that artificial earthquakes produced with exploding gunpowder should be recorded with suitable instruments miles away, so as to survey and map "formations constituting the land. . . and the bottom of the great ocean." In 1849, Mallet made the first geophone, a device for observing sound reflections in the earth.

Practical applications, however, were slow to develop, and seismic exploration was not taken seriously until 1919. In 1923, in Mexico's Golden Lane, Atlantic Refining Company finally drilled a well based on seismic data. It was dry. However, another seismic-staked well proved successful in the same area in 1924.

The seismic method is simple in concept. Different rocks transmit, reflect, or refract sound waves at different speeds. A great noise or vibration at the surface sends sound waves into the earth in all directions. When the waves hit reflective interfaces, they reflect to the surface at a distance and angle from the sound source that indicates the depth of the interface. The sound wave also reveals additional information about intervening layers.

For many years, seismic crews used dynamite to generate underground waves, but today they use mechanical thumpers, particularly in urban areas. At sea, they use electrical discharges and clappers. Anything that imparts a sharp jolt to the earth does the job.

To make a survey, the seismic crew sets out a number of shot holes and a carefully patterned array of geophones to pick up the reflected or refracted sound waves and to transmit the data to recorders. As the crew sets off carefully timed shots, geophones pick up the data and transmit them to the recording medium.

In earlier days, seismic data were recorded in meaningful squiggles on photographic paper for visual interpretation. Today, the signals go straight onto magnetic tape and into a computer, which performs the analysis. Because the quantity of data has multiplied significantly, computer analysis provides vastly improved subsurface

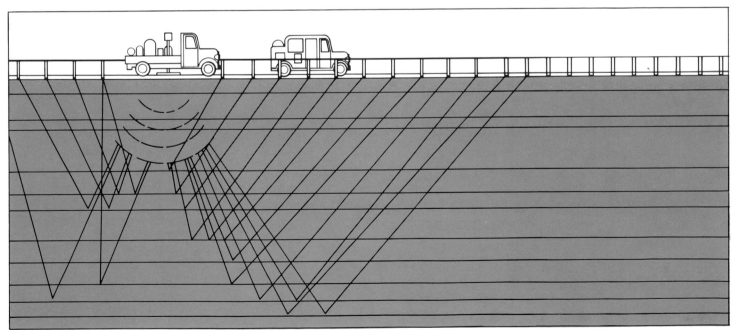

In one method of creating seismic waves, a vibrator truck (left) creates shock waves that penetrate the earth's crust and reflect from subsurface rock layers back to the surface, where they are recorded as seismograms in a second truck (right). The amount of time it takes the wave to bounce back from the strata gives a clue to the depth of each layer.

A seismic crew sets off small explosive charges in Montana. The resultant energy waves probe for promising formations far below the surface. The data gathered suggest to a company whether or not it should invest in drilling a well at that location.

information, so much so that a great deal of old seismic data is being reformulated for similar processing.

Evidence suggests that seismic methods can actually locate hydrocarbons deep in the earth. Sometimes, a data print-out, usually a dense mass of lines, contains faded areas or bright spots. These indicate significant differences between sound velocities from adjoining areas. Frequently, these bright spots turn out to be gas in young sediments, although other factors also cause the spots.

Where there is no reflecting stratum, seismography is useless. The sound waves keep on going into the earth. Lens-shaped formations can greatly confuse the echoes, as can many similar strata in the shot area. In loosely consolidated areas, such as the Texas Gulf Coast, the results can be fuzzy.

Shooting is costly. A seismographic crew covers 25 to 50 square miles in a month at a cost of up to $1 million in salaries, equipment, and computer analysis. So oilmen reserve seismography for areas that already show promise geologically.

There are other less-used, less well-regarded ways to look for oil. One other method measures the electrical resistance or conductivity of the earth to map changes and thus discloses the presence or absence of suitable formations. Electrodes are placed at intervals in the earth, a current is injected at a central point, and the differences in resistance from injection point to electrodes indicate the shape of the substructures. Once in vogue, the method is now little used because it is not very accurate.

Another method is geochemical analysis. The theory here is that microscopic amounts of hydrocarbons must migrate to the surface above a reservoir, where sufficiently sensitive techniques should detect them, or that certain bacteria will live there, feeding on the hydrocarbons. Opinions differ about where to look for these traces. One school says they should be directly over the high point of the pool. Another says that the high point will, in time, be effectively sealed and that any accumulations will be found around the edges in a halo effect.

Another view holds that hydrocarbons will escape along fault lines and come to the surface nowhere at all near the pool.

The theory receives respectful attention, but most oilmen continue to rely on geophysical, rather than geochemical, methods.

Still another method relies on the natural radioactivity of rocks. In some areas, the radioactivity is greater than it should be, specifically, in halos around many oil fields. Hydrocarbons trap or isolate the flow of radioactivity from the earth, and their accumulations are therefore characterized by radioactive lows surrounded by highs. Using Geiger counters, scintillators, and other measuring devices, the prospector makes maps of the area that, in theory, indicate the presence of hydrocarbons. Again, the theory receives respectful attention. No oilman overlooks an idea that promises to tell him where the oil is. But the money goes to the improvement of proved methods.

Of course, there have always been doodlebugs, diviners, and water witches who claim to find oil through the special properties of willow wands, electricity in their bodies, or ectoplasm. In the early days, when prospecting was mostly guesswork, anything that promised an advantage got a hearing, and doodlebugs thrived. Then, their record was probably as good as any other; today, it is poor.

Someday, we may find a way to determine exactly where oil is. Until then, there is only one sure way — drill.

Drilling an exploratory well is both a geological and a geophysical operation. The geologist examines material brought up from the well to confirm the geologic profile provided by other means, while the geophysicist continues to evaluate downhole conditions by physical characteristics. Actually, many skills are involved. The geologist checks the rocks; the paleontologist, the fossils; the geochemist, the chemistry; and the radiologist, the radioactivity.

Drill cuttings are sampled at regular intervals and keyed to the depth from which they came. They are

101

In the old days, explorationists such as these used divining rods, guessing tricks, superstition, and even dreams to decide whether or not to drill a well.

examined by the different scientists, who reconstruct the formations through which the drill has passed and establish from microfossils the ages of the strata. Of course, they are also examined for shows of hydrocarbons.

From time to time, drillers may use a special bit that drills out a long, solid core instead of chips. This core allows the geologists to examine the strata in proper sequence and thickness. Coring works best in formations that are neither too hard nor too soft. If they are too soft, drilling fluids wash away the material; if they are too hard, drilling time is excessive, and the cores are costly and difficult to extract.

The physical evidence that the drill brings up is important, but the basic data come from bore-hole logs, a geophysical way of telling what is down the hole. Electrical well logging goes back to 1911. Conrad Schlumberger, one of the most famous names in petroleum, then professor of physics at the École des Mines in France, figured that electrical measurements offered possibilities for oil exploration as great as the magnetic and gravimet-

ric methods then being discussed. Schlumberger's approach was based on the electrical resistivity of rocks.

His early work showed some success in finding salt domes, but his method was eclipsed by gravimetry, magnetics, and seismics. Then he showed that electrical methods could produce useful downhole data. In 1927, Schlumberger's people performed the first downhole electrical log on a 1,500-meter well in France. The log, which used a makeshift sonde, or electrode in a tube, was a chart of the electrical resistivity against depth. It indicated the nature of the structure wherever a reading was taken as the sonde was withdrawn from the bore.

Today, electrical logs are routine and highly automated. Current passes into the ground, through the resistive medium, and into the sonde. The result is a number of charts that show the varying resistance, the conductance, and the self-potential of the strata surrounding the well at every level. Geophysicists correlate these readings with rock type, porosity, permeability, water salinity, and the presence of oil and gas.

Geologists examining core samples at Roundup, Montana, see the strata in its proper sequence and thickness and thus can determine the nature of the formation.

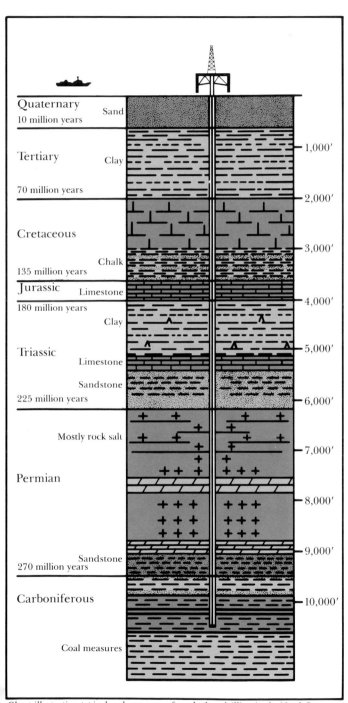

Quaternary	Sand	
10 million years		
Tertiary	Clay	1,000'
70 million years		2,000'
Cretaceous		3,000'
	Chalk	
135 million years		
Jurassic	Limestone	
180 million years		4,000'
	Clay	
Triassic		5,000'
	Limestone	
	Sandstone	
225 million years		6,000'
	Mostly rock salt	
Permian		7,000'
		8,000'
	Sandstone	9,000'
270 million years		
Carboniferous		10,000'
	Coal measures	

Chart illustrating typical rock sequences found when drilling in the North Sea.

In its simplest form, a sonde consists of three electrodes. An electrical current passes from an electrode near the surface to one of the three, and the electrical potential between it and the other two is measured. If the mud in the well has a lower conductivity than the surrounding rock, most of the current passes through the rock. As the sonde moves up the well bore, it provides a continuous log of the electrical properties in the surrounding rocks.

There are other types of logs. The sonic log measures the velocity of sound in the rock and helps delineate the characteristics of the rock material. It is very useful for detecting shales and for relating seismic data to bore-hole findings.

Another log is the radioactivity survey. Different rocks have different radioactive signatures that the log can identify. When a radioactive source is used to irradiate the walls of the bore, this kind of log provides information on porosity and permeability and distinguishes gas from oil, also.

These various logs tell much about the well being drilled, but more important is what they tell about the area being explored. A well either will have oil or will not. But if it brings in commercial quantities and thereby becomes a discovery well, the log tells what to look for in step-out and development wells that will be drilled to determine the overall size and potential of the discovery.

Step-by-step and over many years, using methods ranging from satellite photography through the ever finer screens of magnetic and gravimetric surveys to seismic shooting and finally drilling, oilmen will find new oil fields. It will take more years to develop these resources.

103

CHAPTER 10 –
RESERVOIRS

Although geophysical methods greatly enhance the chance of finding oil, they do not always tell as much as oilmen would like to know about the microcharacteristics of the rocks in which the oil resides, or of those surrounding the deposit, or of the oil itself.

These microcharacteristics — the porosity and permeability of the rocks; their chemistry and that of nonhydrocarbon fluids in the earth; the chemistry and viscosity of the oil; the local pressure; the relationship of oil, gas, and other fluids to one another in the strata; and even the nature of drilling fluids used in drilling the well — all these complex relationships together determine how much oil can ultimately be recovered from a reservoir. Though there may be plenty of oil there, these factors may determine whether oilmen can get any out at all.

A few of the methods previously discussed can indicate some of these vital reservoir characteristics. None are conclusive or even very accurate. It is like trying to determine the nature of the moon's surface by telescope from the earth. Until a hole is drilled, oilmen cannot be sure that the rock is porous and permeable limestone or porous but impermeable chalk.

Two factors more than any others determine the present and future productivity of a reservoir rock. These are the porosity and permeability of the formation.

Porosity is a measure of the open space in the rock, like the holes in a sponge or a foam. Expressed as a percentage that ranges from virtually zero to, rarely, more than 50, it is an absolute measure of the maximum amount of oil a reservoir of a given volume can contain. It does not measure the amount of oil that ultimately can be recovered from it.

Permeability is a measure of the ease with which a fluid can move through a rock or other barrier. Permeability and porosity, both functions of holes in rocks, are usually related, though not always. A sponge, for instance, is both very porous and very permeable because its body holes are both large and interconnected. Some types of foams, however, may be very porous, yet have zero permeability. In this case, every void is a single cell that is not connected to any other. So it is impermeable. Some rocks behave like sponges, and some, like foams. Oilmen prefer the sponge type, but most rocks behave in an infinite range of variations somewhere between the two.

Geologists express permeability in a unit called the darcy, named after Henri Darcy, a French engineer who promulgated the law of permeability while working on water filters in 1856.

One darcy permeability permits one cubic centimeter, or milliliter, of a fluid with a viscosity of one centipoise (specifically, water at 68 degrees Fahrenheit) to flow in one second through a volume one centimeter long by one centimeter in area at a pressure differential of one atmosphere. This is to say that the pressure on one side of a one-centimeter-thick filter or piece of rock is 14.7 pounds per square inch (psi) greater than that on the other side, at this flow rate.

At first glance, this does not seem like a very high

value. One cubic centimeter of water per second is a mere dribble. But a rock of one darcy permeability is actually quite permeable as reservoir rocks go, often made more so by the fact that some crudes are less viscous than water and exist in reservoir rock at temperatures considerably greater than 68 degrees Fahrenheit. Viscosity decreases rapidly with higher temperature, which makes the hot rock still more permeable. In practical terms, a ten-foot-thick reservoir rock with one darcy permeability will allow 150 barrels per day to flow into a well bore when the well pressure is then only ten psi less than that in the formation.

But permeabilities as high as one darcy are seldom found in practice, so permeability is usually measured in thousandths of a darcy, or millidarcies.

Absolute porosity and permeability of a rock are determined not in the reservoir but in the laboratory, where conditions are artificial. Rock, chemically and mechanically stripped of its fluids, is submerged in water under appropriate pressure, and fluid is forced into the

105

Permeable, porous

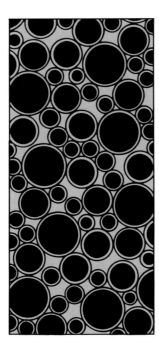

Nonpermeable, porous

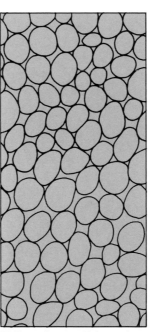

Well-sorted grains, high porosity

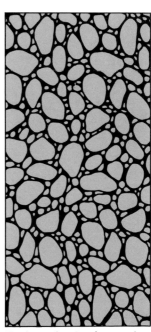

Poorly sorted grains, low porosity

pores until the rock accepts no more. When the rock absorbs no water, it has zero-effective porosity. If 50 percent of the water by volume disappears into the rock, then it has 50 percent porosity. This measure has nothing to do with the size of the rock's pores or its constituent particles, but reflects only its empty space. The finest sand or silt can have the same porosity as that of a pile of boulders. The difference is that a man can crawl among the boulders, while bacteria might find silt pores a bit tight. Porosity does, however, reflect the shape of the particles and how they are packed. A volume full of cubes packed face-to-face will have zero porosity, while the same cubes corner-to-corner would form a very porous mass.

While laboratory tests do not mirror the world found several thousand feet underground, they give a point of reference against which the variations between laboratory values and those found in the field can be measured.

Those variations can be immense. Many factors, few of them obvious, can make the effective porosity and permeability differ substantially from the laboratory absolutes. Only detailed examination of the actual reservoir rock will yield the information the oilman needs to determine the potential of a particular well and, of much greater long-term importance, to guide him in the management of the entire reservoir.

Oil-bearing rocks, originally deposited as sediments, were once composed of innumerable individual particles whose size and shape reflected the manner in which they were formed. Though squeezed, compacted, and cemented, as well as altered by chemical and thermal action over the eons, the primal particle shape remained to influence profoundly the behavior of the rock today.

Sand, for example, comes in a variety of shapes and sizes. A deposit of sand which became sandstone and a reservoir rock — about 59 percent of all world oil production comes from sandstone — might have been formed in several different ways, each of which affects the size, shape, and even the mixture of the sand grains, as well as the present-day properties of the sandstone.

In the initial formation process, the grinding of rock upon rock produced irregular and angular grains. Grains carried great distances in water and subjected to the honing effect of other grains became much smoother, the sharp edges worn off, but still irregular in shape and varied in size. Wind-blown sands, as might be found in desert dunes, tend to be very fine, smooth, and regular in shape, homogeneous in composition.

Carbonates, the limestones and dolomites, home of 40 percent of the world's petroleum, also settled as particles in colloidal suspensions or clouds of molecular-sized grains. They were chemical precipitations from the water that dissolved them. Carbonates also derived from the skeletons and shells of tiny animals, from coral reefs, and from beds of larger shells. All carbonates are organic in origin, whether reprecipitated from water or primarily sedimentary. Whatever their origin, they often underwent considerable change with time. Those derived from colloidal precipitates tend to be quite dense, with low porosity and permeability. Those which resulted from thick beds of sedimented shells remain soft, chalky, and very porous. They are also more like foams than sponges and thus not very permeable.

Chalks seldom prove good reservoir rocks, despite their porosity. Yet, again, there are exceptions. The Austin chalk trend in Texas has proved a prolific and recently appreciated oil source. What is exceptional here is that the chalk is fragmented so that the oil could migrate into it, not that the chalk itself is permeable. Many oil deposits have not formed where conditions were otherwise favorable because the likely reservoir rock was not sufficiently permeable. The forming hydrocarbons could not migrate into the rock and eventually were lost.

Oilmen talk in terms of original and secondary porosity and permeability. Original porosity and permeability are properties that have remained basically unchanged from the time a stratum was sedimented and compacted to the present time. Until recently, sandstones were generally said to have only original characteristics because the individual sand grains undergo little chemical or mechanical change over time. However, geologists now know that even sandstone will change with time and that there is great potential for secondary porosity in the rock.

Of course, the initial values of sandstone vary more

Grains of sandstone cemented together look like a city of conical skyscrapers as cratered as the moon's surface in this picture enlarged 6,000 times under the electronic microscope. Petroleum geologists gauge the production potential of reservoir rocks from information like this gleaned from the microscope.

because the pores in the stone became filled with something, such as other minerals or water. Carbonates, for example, easily precipitate in the pores of sandstone and greatly reduce or even destroy the porosity and permeability of a formation that in the beginning was highly porous and permeable. Such cementing is not confined to geologic history. Improper drilling or production methods and accidents that allow incompatible formation water or drilling fluids to enter reservoir rocks also can cause cementing. In extreme cases, water intrusion ruins wells and even destroys entire reservoirs.

Carbonates generally have low initial porosity and permeability. They develop secondary characteristics through fracture, jointing, solution in water (which literally melts holes in the rock), recrystallization, or combinations of all of these. Carbonates are far more active chemically than is sandstone. Dolomites result from the chemical substitution of magnesium for the original calcium in limestone, forming a magnesium carbonate from the original calcium carbonate.

Where water is present in a carbonate formation, there is a continuous process of solution and deposition or recrystallization. If solution is greater than deposition in any zone, porosity will be developed between the crystal grains. An important type of this porosity is found in dolomite zones that occur in conjunction with limestone deposits.

Porosity and permeability are much more difficult to measure in limestone than in sandstone. Sandstones tend to be homogeneous with little change in their characteristics from one part of the stratum to another. Limestones and dolomites, susceptible as they are to local chemical and mechanical action, can vary greatly within the same stratum.

What really matters to the oilman is permeability. If the pores are not interconnected and continuous, as in shale, clay, anhydrite, cemented sandstone, and chalk, the rock will be impervious and useless as a reservoir, even though it is possibly good as a seal or cap rock on a reservoir.

In a reservoir, permeability is usually greater horizontally than vertically. Grains of sand, clay, lime, or any other mineral are seldom spherical but are longer in one axis than in the other and also tend to be oval or flat rather than circular in cross section. When the grains were initially laid down, they tended to line up in the direction of water flow. They settled lengthwise on the flat sides, like planks in a stream. Thus, in the resultant rock, permeability with the grain is often several times greater than that against the grain. This is extremely important to proper management of the reservoir and productivity of the individual well.

While laboratory tests of rocks from an exploratory well may assure the oilman that the reservoir is porous and permeable and that oil is present, he still will not know whether he has a productive well or a producible reservoir. Neither porosity nor permeability is constant. As soon as he taps the strata, things begin to happen that change the properties they had while undisturbed. Knowing about downhole conditions helps explain these changes.

Pressure in a well, or in subsurface strata, generally equals the pressure exerted by a column of salt water between the stratum and the surface. Called "hydrostatic pressure," this pressure equals about 46 psi for every 100 feet of depth. At 10,000 feet, it would normally be about 4,600 psi; at 20,000 feet, 9,200 psi. It is not necessarily equal to the pressure exerted by the weight of the surrounding rock, but is much as a building that stands up without regard to its contents. Because rock is heavier than water, even salt water, rock pressure can be much greater than hydrostatic pressure. Hydrostatic pressure, however, does help to cushion the rock against its own weight, and changes in the fluid pressure bring related changes in hydrostatic pressure.

If the fluid pressure supports the rock at a pressure greater than hydrostatic for the depth, the rock squeezes out the fluid until it reaches equilibrium with the depth. The rock goes on squeezing until it supports itself without regard to hydrostatic pressure. If at some depth the rock pressure becomes so great that it crushes the rock to porelessness, it will squeeze out all fluids.

Although hydrostatic pressure in the rock is nor-

Blowouts such as this one, which occurred in Corpus Christi, Texas, around 1932, were costly in terms of life and property. Now they are controlled by use of blowout preventers.

108

mally proportional to the depth, the oilman occasionally encounters geopressured zones, where pressures are very much higher or lower than they should be. Pressures exceeding 10,000 psi have been encountered, sometimes with unexpected and disastrous results. These abnormal pressures, either high or low, are leftovers from a time when the particular formation was buried much deeper or much shallower than it was when discovered. Surrounded by impermeable strata that prevented fluids from escaping or entering to equalize pressures to normal hydrostatic levels, the zone became sealed.

Oil is not the only fluid in rock, and it was not the first. It shares the pore space with water.

Most reservoir rocks are water-wet. The initial sediments were laid down in water that surrounded every particle and grain. The pressures of sedimentation squeezed out most of the water, but residual water still occupied the pores when hydrocarbons began migrating into the rock. Many believe this water, called "connate water," is the residue of the primal sea. Others consider

this point of view improbable. They contend that tremendous volumes of water must have flowed through the rocks before the hydrocarbons arrived.

Most connate, or interstitial, water is salty, usually saltier than seawater. A few reservoirs — notably one in Venezuela and others in the Rocky Mountain area — are associated with freshwater. This indicates that they probably were formed under different circumstances from those associated with saltwater reservoirs.

A small number of reservoirs are oil-wet. Scientists think they were formed as aeolian deposits (wind-blown sand dunes) and were never submerged in the sea. Eventually, they were buried underground, where oil migrated into them. Oil-wet reservoirs present far fewer problems to the oilman than do water-wet ones. Unfortunately, they are scarce.

Normally, three fluids are present in a petroleum reservoir: water, oil, and gas. Water is usually pictured on the bottom, oil in the middle, and gas on top, each sharply delineated from the others. The oilman hardly ever en-

109

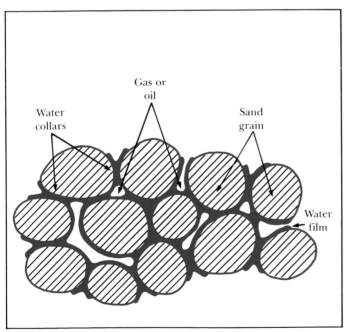

This enlarged section of a reservoir containing gas or oil illustrates the distribution of water and hydrocarbons. Small pores and too much water can make the rock act as though it were impervious.

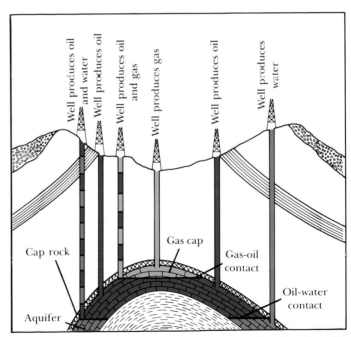

This cross section of an oil reservoir shows the arrangement of gas, oil and water in typical dome-shaped structures.

counters such an ideal situation, although in some types of fractured and cavitied limestones with very large interconnecting pores (and very high permeabilities and porosities), the actual is not far from the ideal. These are vast sponges that readily yield their contents, legendary producers of 50,000 and even 100,000 barrels per day.

In most reservoir rocks, however, the pores and passages are microscopic. Fluids behave differently there from the way they behave in large spaces. Capillary action and surface tension annul or greatly change the effects of pressure and density, or pressure-density differences. These are the forces that cause fluids to cling to and even climb vertical surfaces when gravity should make them fall.

Capillary action pulls the actual water level in a reservoir well above the oil level. Fingers of water intrude into the oil in a transition zone that can be quite thick in low-permeability formations. Besides, surface tension and the water-wetness of the rock grains keep a certain minimum film of water clinging to the grains throughout

110

the reservoir. This is the irreducible minimum water saturation. In large pores, the space occupied by the film of water may be insignificant, but in smaller pores, it can occupy all the space and make the rock impervious in effect. Connate water can also form a bridgehead for most of the water below the oil interface. If it is not carefully managed, it can lead to premature water production from the formation.

The moment the drill penetrates a formation, all this changes. The well bore introduces drilling fluids which, if carefully chosen and formulated, should cause no harm. More significant, the pressure in the formation changes, though drillers try to minimize that effect. As pressure in the hole declines, either because fluid is being withdrawn from the formation or because a high-pressure zone has been opened to a low-pressure one, the fluids in the rock and the very rocks themselves expand. Both porosity and permeability change. Paradoxically, the drop in pressure can actually cause a formation to seal itself off.

Gas is always present with oil. Under reservoir pressure, the gas frequently does not exist separately but is dissolved in the liquid, just as is carbon dioxide gas in a bottle of soda water. When the pressure drops, the gas comes out of solution, again just as in a newly opened soda-water bottle. In the bottle, the gas rises to the top and separates from the water to form a distinct gas phase. This also happens in oil reservoirs. However, there is also a gas-oil transition zone, as there is with oil and water, in which the two phases coexist. The oilman must control this zone to conserve pressure in the reservoir and to prevent a general separation of the gas phase, a frothing effect in which gas bubbles fill the tiny pores and passages and destroy permeability, producing in the reservoir a kind of vapor lock.

Drilling or producing operations themselves can damage the well bore and reduce flow. This abnormal permeability problem, called "skin effect," has any number of causes: contact with fresh water makes clay swell; minerals precipitate from incompatible waters or drilling fluids; the oil and other fluids form emulsions; spherical water droplets block the passages; drilling mud, bore cement, and finely ground material from the drilling itself enter and clog the pores. There are remedies for

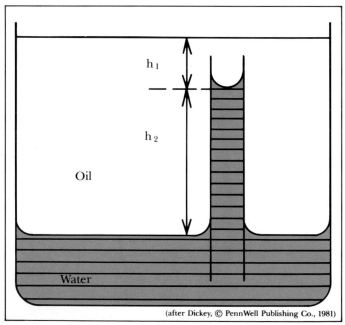

(after Dickey, © PennWell Publishing Co., 1981)

Capillary action affects the way oil and water behave in the reservoir. Under certain conditions, it pulls the water level above the oil level so that fingers of water intrude into the oil as demonstrated by this laboratory experiment.

most of these ailments, but it is better to avoid them if possible.

The nature of reservoir rock itself can cause other problems. While oilmen usually refer to a specific reservoir as a "sand," they really mean a sandstone that millennia ago was a sandbar, beach, or other deposit. With time, the individual and loose grains of sand consolidate. Oilmen call this a "consolidated formation."

Once in a while, however, oilmen run into a sand that is still a sand — not consolidated at all, or only very loosely so. By their nature, such formations are very permeable and porous and much less affected by the various factors that can plug the pores of a consolidated rock. If they contain hydrocarbons, they are likely to be prolific producers. Some of the Gulf Coast and particularly the Caspian Sea fields occur in loosely coordinated formations.

Unfortunately, these formations sometimes produce as much sand as they do oil. The Baku wells were famous for the sand volcanoes they built around their outlets. The quantity of oil and the pressure behind it were enough to drive the sand ahead of the oil or drag it along. Sand in such fields is an expensive nuisance for which modern technology has remedies. But many unconsolidated formations produce sand at such a rate that attempts to produce oil are not worth the trouble. In some instances, the sand erodes pipes and valves severely, even cutting holes in them.

Petroleum itself is not the only thing the oilman considers when he decides whether or not a particular reservoir will prove economic. Petroleum chemistry plays a part as well. Metals, salts, sulfur, and other ingredients in the oil interact with other fluids and with the formation rocks under changed conditions and pressure. Occasionally, these reactions produce emulsions and precipitates that plug the formation and make the oil more difficult to recover.

Viscosity, however, is far more significant than chemistry. Viscosity is perhaps the most important factor in determining the ease with which oil is pushed through the pore spaces of the reservoir rocks. If the oil is too thick, it will not flow. Viscosity also determines the degree to which oil will continue to stick to the grains in the rock and to resist moving out, becoming residual oil that is recoverable, if at all, only through costly secondary and tertiary methods.

Today, trillions of barrels of high-viscosity (or heavy) oil remain locked in reservoirs all over the world. The technology to recover this oil is available, but, in most instances, it is not yet economic to do so, and that is the key. There is no minimum rate of production for a viable oil well. If the producer can market the crude profitably, a well that produces a few barrels a day (a stripper well) can still be acceptable in the later stages of a field's life.

Indeed, with recent rises in the price of crude, producers are not abandoning stripper wells at the rate they were a few years ago. Paying careful attention to reservoir characteristics during the exploratory stage can help assure a profitable old age.

111

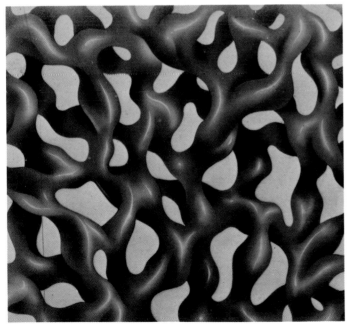

If the sand grains and salt water were stripped away from a section of oil-bearing rock, the remaining oil when magnified, would look like this.

CHAPTER 11 –
LANDING THE LEASE

Who actually owns most of the world's oil deposits? Surprisingly, perhaps, not the privately owned petroleum industry. The oil companies own very little underground oil in the United States and almost none in the rest of the world.

In most countries, subsurface minerals, including oil, belong to the state. Even though many countries previously permitted private land ownership, now they have nationalized ownership of their minerals. Mexico is a good example of a country that once allowed foreign companies to own privately vast oil properties — the land as well as the rights — but eventually nationalized its industry and reserved to the state the rights to all petroleum in the ground.

Most of the world now resembles Mexico. Overseas, the picture is altogether different from that in North America, where private ownership prevails, and even from what it was as recently as 20 years ago.

From the beginning, when oil companies could not negotiate with private individuals for rights to drill, they

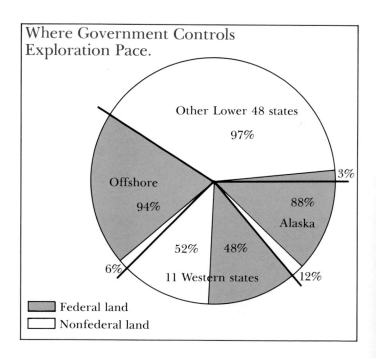

Where Government Controls Exploration Pace.

Other Lower 48 states 97%

3%

Offshore 94%

88%
Alaska

52%   48%

6%

11 Western states

12%

Federal land
Nonfederal land

negotiated concessions with governments. Very often, particularly in the Middle East, that government acted as a single person, who granted the concession in return for a royalty on any oil produced. These concessions were granted for a term of years, often as much as 50 years and, occasionally, in perpetuity.

As concessions negotiated in the early years of the century and after World War II have expired, they have been renewed on terms increasingly favorable to the countries and unfavorable to the oil companies — if they have been renewed at all. The trend has been toward total nationalization, where the oil companies operate as contractors, not as concessionaires. Where concessions remain, as in the North Sea, terms may be so onerous that the companies would find contract operation more profitable.

The United States and Canada remain two of the few countries where both the land and the minerals under it still are held largely in private hands, whether corporate or individual. The states and the federal government also own vast tracts of land, including all offshore areas below the low-tide mark. Many states hold mineral rights on some lands, even though the land at the surface is in private hands.

Some oil lands are owned outright by oil companies as corporations and oilmen as individuals, but such lands yield an insignificant amount of the daily oil and gas production compared to total production in the United States or the world. Almost all oil is owned by someone else: farmer, widow, orphan's home, land company, speculator, investor, developer, bank, school district, land-grant college, city, town, or state — not to mention one or another arm of the federal government. Any entity capable of owning land can own the oil and other minerals under it, though the owner of the surface rights may not have the subsurface rights. Some previous owner may have sold or given away the rights or otherwise separated them from the land.

The only way the oilman can drill for oil is to obtain a lease from whoever owns the land or the mineral rights. Until he gets the oil out of the ground, the oilman does not own it. Sometimes, the landowner does not own it either.

To assemble promising exploration tracts in the United States, the petroleum industry must deal with millions of individual landowners as well as government agencies, both as landowners and as regulators. In some instances, the oilman prefers to deal with the government because, however difficult negotiations with the government are, they are still easier than dealing with hundreds or even thousands of individual landowners.

Through the years, a great body of law, custom, and practice peculiar to oil properties has developed. In the early days, negotiations between landowner and prospective oilman were simple and straightforward. The oilman obtained the right to drill in return for a consideration and a royalty on the oil produced. Such leases were usually based on long-standing rules governing the leasing of rights to solid minerals, rules which in the United States were usually based on English common law.

Of course, nothing in law or land ever remains simple. Misunderstandings arise among friends, interests change, and often unforeseen circumstances arise which, in the absence of specific agreements or legislation, can be resolved only in the courts. In the early days of oil leasing and production, much was learned by error and trial, often in court. In fact, the early, simple contracts soon became more complex as the parties sought to protect themselves against a whole variety of contingencies.

Though oil leases were originally based on mining law, oilmen and landowners soon found that oil does not behave the same as gold or coal. It moves. The mobility of oil and gas has led to great changes in law affecting oil lands and leases. Early court decisions produced the "rule of capture." This rule was responsible for the destruction of many an early oil pool and has never been repealed, although state regulations that effectively fence it in are now widely in force and have been upheld by the courts.

Under the rule, a single well on a tiny tract could drain an entire field by pulling oil from adjacent areas. This meant that a landowner owned the oil under his land only if he could capture it. If his neighbor drilled a well first and removed all the oil, the idle owner had no recourse. About the only restriction on drilling was that a neighbor could not drill a slanted hole that crossed the property line underground.

Obviously, the only defense the landowner had when drilling rigs appeared on neighboring property was to elect to drill for himself as many wells as the land could hold and as close to the property line as possible. The result was a forest of derricks on 100-square-foot plots, familiar from old photographs and movies. Such undisciplined exploitation often left more oil in the ground than it recovered.

Only concerted efforts by the industry, the states, and the federal government eventually gained some control over these chaotic and wasteful practices. Today, most states allow only so many wells per unit of land. Each unit may be considerably larger than the tracts held by individual owners within it. The oil company must still acquire leases from the owners of each small tract in order to assemble a unit worth drilling. Rule of capture notwithstanding, such unitization makes it possible for each owner to benefit from a productive area, even though the leaseholder does not drill any wells on his land.

The need to deal with hundreds or thousands of owners makes the landman a very important asset, whether he works for an oil company or as an independent. It is a highly speculative and expensive business, promising either great rewards or a long, substantial drain on funds. Since it may not take much money to tie up leases in marginal or untried areas, or to make an astute investment in a key tract, many independent landmen seek leases at good prices, then sell them at better prices if and when prospects in the area improve. Of course, if prospects do not improve, and if the independent landman cannot interest a buyer in his leases, the cost of maintaining them can ruin him.

A landman tries to lock up all the likely land in an area as quickly as possible, before his rivals find out what he is doing, buy up some leases on their own, and drive up the price. When that happens, a landman trying to assemble a large tract may have to buy some leases at exorbitant prices from holdout landowners or rival landmen to complete his package. It is much easier for a company to deal with a government or single large landowner or leaseholder, though no less costly. Likely land is high wherever it is and whoever the owner.

An oil lease is a peculiar instrument. Its form and substance are now fairly standard, shaped through trial and circumstance over the years. It is a contract between the owner of the oil rights and the leaseholder. It gives the leaseholder the right to drill for oil on the land, and to use a reasonable amount of space for the drilling rig and ancillary structures, such as mud and slush pits, tool housing, perhaps crew housing, access roads, tankage, and pipelines for moving the oil out. The driller may not drill within a certain distance, generally 200 feet, of an occupied dwelling, though he may drill a slant well that takes oil from under the dwelling. It took a court case to settle that point, as it has with virtually every aspect of oil land law.

In the early days, oil leases were for a specified period of time, usually five or ten years. This method had worked for solid mineral leases, but proved unsuitable for oil. A driller could bring a well into production just as the lease expired, and the rights would revert to the landowner. While driller and landowner haggled over re-

Professional oil scouts such as the ones pictured here were first used at Cherry Grove, Pa., in 1882. They watched what was going on in town and at everyone's oil wells so that their company might gain an advantage over the others.

Wooden derricks elbowed residences for space in Los Angeles, as in this 1901 view of Court Street. Oilmen had yet to solve some of the problems of coexisting in an urban area, waste disposal being the most difficult problem.

115

newal or went to court, a driller on neighboring land could drain the pool. Solid minerals, by contrast, stayed in place during such arguments.

From about 1880 to 1900, the no-term lease was popular. These were open-ended and given virtually in perpetuity. The leaseholder could put off drilling forever, as long as he paid his rental, and the landowner could do nothing. It was a very good deal for the lessee, but the lessor earned no royalties if the lessee chose not to drill.

Eventually, the courts took the view that there was an implied covenant by the lessee to develop the lease in a reasonable time and that the lessor could refuse the rental payment and demand the start of drilling or cancel the lease. Today, the common form is the fixed-term lease with a "thereafter" clause. The fixed term might be ten years, during which the lessee must develop the property. If the lessee finds oil or gas, the lease continues "as long thereafter as oil or gas is produced from said premises." In general, such a lease protects the interests of all parties.

Thereafter clauses can differ markedly and can greatly affect the term of the lease. There are four major variations: the lease will remain in effect thereafter when

1. Oil is found in paying quantities
2. Oil is produced in paying quantities
3. Oil is merely found
4. Oil is merely produced.

The effect of the third variation is to give the lessee an unlimited lease if he finds any oil at all; there is no return to the landowner since no oil need be produced. And if the oil is not produced and sold, not only are there no royalties but also there is a possibility that the reservoir will be depleted before that particular well beomes a producer. Such leases obviously can cause extensive litigation in the future, especially if the rights of the original lessor are passed to other persons, such as heirs.

Any tract of land selected for drilling must be properly located and described. It is essential to know who owns precisely which few square feet, how that person came to have title to it, what is the history of the title, and precisely how the land is located with respect to other lands around it. Consider, for example, the offshore ter-

ritories or a poorly surveyed wilderness in deep mountain territory. Add to that the many different systems of land conveyance and measurement that have prevailed in different parts of the United States. The significance of the problem grows. No oil company wants to sink a great deal of money into a well only to have its lease challenged later because of an improper survey 150 years before, or because an Indian chief, for example, did not have the right to transfer the land at some earlier time.

The real incentive to the landowner is the production royalty. Rental payments are nominal. They reflect the potential of the land, and range upwards from a few cents per acre. In really prime areas, of course, where large and unexpected strikes have occurred before independent landmen or companies could get the land under lease, bidding can reach hundreds of dollars an acre. That is rare, however. The oilman does not want to buy the land, and its long-term value is as land, not oil property.

The landowner may also receive a bonus on signing the lease, though the bonus is more akin to a year's rent in advance. There is also a peculiar payment called a "consideration," mentioned in the lease as the binding transfer in return for the lease. It amounts to only one dollar and is seldom paid.

Since the usual lease is a negotiated contract between the landman or company and the landowner, individuals can and do change provisions to suit their circumstances. Over the years, however, a standard of sorts has evolved. A lease usually consists of the following main clauses:

1. A granting clause that specifies the minerals subject to the lease and the surface or operating rights the lessee may exercise in drilling or in producing oil and gas.

2. A legally sufficient definition of the lands and minerals leased.

3. The terms of the lease, usually a specified number of years with a specified rental for each year, and the thereafter clause.

4. The rental to be paid each year before oil and gas is discovered, sometimes called "delay rental" since its payment enables the lessee to delay drilling.

5. The royalty, the fraction of the value of the oil and gas produced to be received by the lessor.

In private dealings, the royalty is usually one-eighth, though in some areas it can reach one-sixth. Occasionally, not-too-bright landowners have been known to demand as much as a tenth or even a twentieth, and, occasionally, not-too-ethical landmen have been known to accommodate them.

In return for allowing the oilman to drill, the lessor receives one-eighth of the gross value of the oil and gas produced from his property. He may take that value in cash or in kind. Since most landowners do not want to engage in the oil business, the usual royalty payment is in cash. And since a royalty clause is a negotiable asset with a possible life of many years (some fields have produced for nearly a century), the lessor can sell it, assign it, leave it in his will, or divide it.

While company land departments and hordes of landmen deal mainly with individual landowners, the largest landowners are governments. Moreover, as heavily developed private lands have diminished their potential for new discoveries, government lands, onshore and offshore, have now become the prime prospecting areas for explorationists.

The U.S. government, like any other landowner, demands both a bonus and a royalty on its oil lands. But unlike the amount paid for private lands, the amounts paid out for unproved government lands can be enormous.

Also, the history of government lands may be even more confused than that of private land. Starting with the Continental Congress, the federal government acquired vast tracts through annexations, conquest, and purchase, all of which had different histories and traditions of ownership and assignment. And in bitterly fought legal battles, it also acquired huge tracts, notably some "tidelands," as close-in offshore lands are called, from the states. In addition, both the federal government and the states gave away huge blocks to the railroads to encourage the building of transcontinental lines and to homesteaders to encourage the settlement of the wilderness. Then there are the Indian lands, another story altogether.

117

The famous Million Dollar Elm at Pawhuska, Okla., was the place where Osage Nation oil leases were auctioned off. The bids went as high as $1,900,000 for a 160-acre tract, making the Osages the richest Indian tribe in the country.

Frank Phillips (bottom right), a cofounder of Phillips Petroleum, is made an honorary chief of the Osage tribe. Because the Osages controlled such huge reserves of petroleum, oilmen worked hard to curry favor with them.

The rules governing leasing of federal lands are based on the Federal Oil and Gas Leasing Act of 1920, plus its many subsequent changes and amendments. It is a body of law built on as much litigation as that governing private oil leasing. Such leases figure prominently in discussions these days because the untapped oil frontiers are usually under federal control and lie in difficult and costly areas, not to mention environmentally controversial ones. An example is Baltimore Canyon, off the East Coast, where oil companies bid tens of millions of dollars for the right to drill through the ocean floor in an untried area. If oil is found there in commercial amounts, then the oil companies will pay the government a substantial royalty according to the amount of oil produced.

In Alaska, where the Prudhoe Bay field was found on government land, the oil industry will pay more than a billion dollars in royalties and taxes to the state, in addition to the hundreds of millions already paid for the leases.

On oil and gas produced in the ten years between 1968 and 1978, the petroleum industry paid the federal government more than $24.7 billion in bonuses, lease rentals, and royalties. More than $18 billion of that went for bonuses and rentals before the leases yielded a drop of oil. In the same ten-year period, the industry paid state governments more than $13.4 billion in bonuses and rentals, royalties, and taxes.

Regardless of who owns the land, it is clear that gaining the right to drill on it is not cheap. And that is only the beginning of the expenses.

118

Outer Continental Shelf Planning Areas

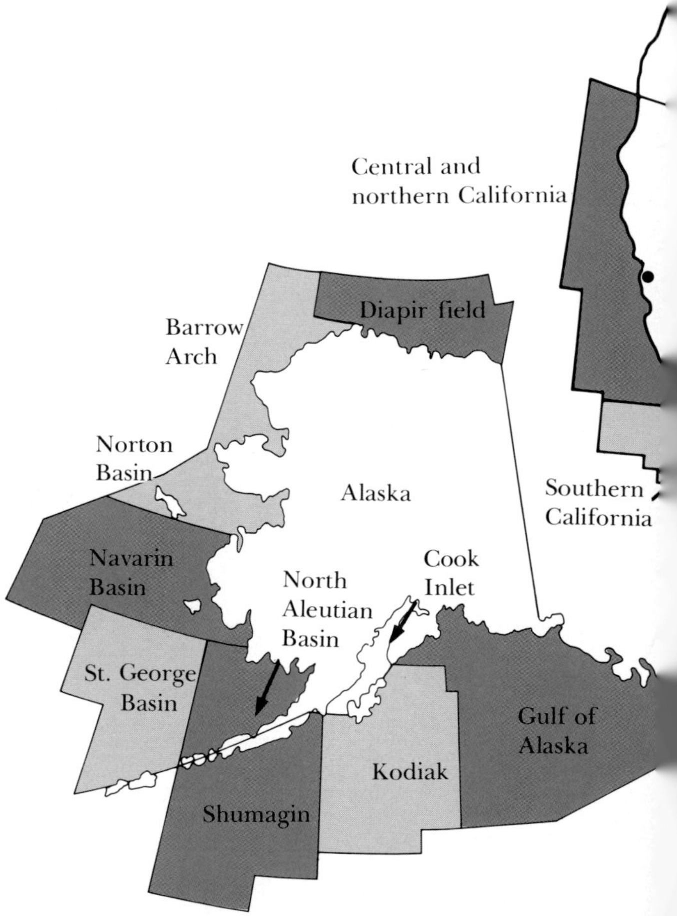

Boston

New York

North Atlantic

Mid-Atlantic

Norfolk

Charleston

Savannah
Jacksonville

South Atlantic

Mobile

New Orleans

Corpus Christi

Western

Central

Eastern

Tampa

Gulf of Mexico

isco

Barbara
Angeles

119

CHAPTER 12 –
DRILLING THE OLD WAY

In the early days of oil, an individual wildcatter with little money and a rickety rig could sink a hole in virgin territory and strike it rich. Today, when lease sales on inaccessible and unproved lands cost millions of dollars in up-front bonuses alone, a chance at the really big play is beyond the small operator. Yet the independent oilman is an important factor in the industry. The independents number in the thousands and still account for a third of our domestic oil production.

Many marginal and relatively shallow fields have a small output per well. Some were once good producers; others need newer technology to be profitable; many simply never had enough potential to warrant development by a major company with all its overhead costs. Such fields attract independents who can work at relatively low cost with simple equipment and are content with the income from a few wells making a few barrels per day. After all, there is always a buyer for oil, whatever the quantity. If older and cheaper equipment and technology can be used to recover it, so much the better.

An interesting thing about technology is that the new seldom totally replaces the old. In many applications, the original approach, suitably updated with modern materials and design, may still be the most effective way. So it is in drilling for oil, water, or gas. Where the pay zones are relatively shallow, say less than 4,000 feet, and the structures through which the drill must penetrate are known and well consolidated and not likely to collapse in on the bore hole, the old-fashioned cable rig or percussion drilling method may still be the best choice.

The spiritual heirs of Dad Joiner pilot some of these rigs, forever in search of Bonanza. But many more are on contract, working over old wells and developing small, shallow, and marginal fields where the relatively simple and low-cost cable rig still has the economic advantage — and will keep it.

Drake used a cable rig. Almost everyone who drilled water wells before him or oil wells afterward used a cable rig. Spindletop showed the value of the rotary rig in soft formations but did not displace the cable rig, which to this

In the early days of oil, drillers like these who came to Pleasantville, Pa., in 1868 could sink a hole, build a rig, and strike it rich. The lure of oil in Pleasantville was so great that the population jumped from 1,000 to 3,000 in that year.

day drillers use almost exclusively for water drilling, often for shallow oil wells, and particularly for reworking existing wells. It is cheaper.

The cable technique is very simple and very old. Men have been making holes this way since they first learned that there were worthwhile things in the ground deeper than men could dig for with their hands and tools. The Chinese even used percussion rigs to drill for hydrocarbons over 2,000 years ago.

When most people think of a drill, they think of rotary motion. They envision a bit that goes around and around, like a wood or metal drill. For drilling into the earth, they see a giant auger similar to those used to drill post holes or perhaps bore holes for building foundations. The cable drill is nothing like that, nor, for that matter, is the rotary drill. Auger bits will not work in rock.

The earth's soft surface into which people turn the garden spade or contractors sink building foundations is not typical of the deeper layers in which oil and gas are found. At depths greater than the weathered layers of

surface soil, 30 to 40 feet at most, the layers turn to rock. Oil and gas lie in or under rock of varying hardnesses, ranging from barely consolidated "cavey" sands to impervious and very hard granite, basalt, and shale. These cannot be penetrated with an auger. The oilman has to hammer his way, much as a mason uses a sledge and a star drill to make holes in masonry walls and brick.

The cable drill batters its way hundreds and even thousands of feet through solid rock. In its simplest form, the drill bit is a hardened length of pointed steel rod connected to a cable or pole. The cable comes up out of the hole and over a sheave on a trestle. A group of men pull the cable until the bit is a few feet from the bottom of the hole, then let go. The bit falls to the bottom of the hole and hammers the rock. Each cycle chips away a little more of the rock and makes the hole a little deeper.

The spring pole is a refinement of this primitive method. The cable or rope attaches to one end of a long, springy pole, which is anchored at the other end and rests on a fulcrum, like a fishing pole in a cleft stick or a diving

122

Two thousand years ago the Chinese used a rig similar to the cable rig. The nineteenth-century cable rig was steam-powered. The twentieth-century cable rig uses an internal-combustion motor. But the ancient Chinese used human or animal power, as depicted above.

springboard. Men at one side of the fulcrum put their feet into stirrups attached to the pole. They kick the pole down smartly, which slams the bit into the bottom of the hole. Then, just as smartly, smoothly, and in rhythm, they reverse the direction of the pole. The cable comes out of the hole and lifts the heavy bit until gravity drives it back into the bottom of the hole. This delivers a terrific impact to the rock. The process acts as sort of a vertical crack-the-whip.

The trick is to time the cycle so that the bit has maximum energy at the instant it touches the rock and is removed as soon as the energy is expended. Another is to avoid backlash, which is both wasteful and dangerous.

Though quite efficient in its use of rhythmic energy, the man-powered spring pole rapidly exhausts its power source. It takes relays of men and boys to keep one going. Because of the peculiar reciprocating motion of the spring pole, animals could not be effectively harnessed to it. Their function was to turn the windlass or bull wheel that raised and lowered the ever-lengthening cable with its tool string. The advent of steam power was necessary to allow even minor advances in this 2,000-year-old technology. Drake's drill was a steam-powered cable rig. Little has changed since. The Colonel would easily recognize all parts of a modern cable rig except perhaps the gasoline or diesel engine driving it.

Nor have many other things changed much since those early days. While only 10 percent or so of the wells drilled today use the cable method, aspects of drilling a well, whether cable or rotary, are the same. One, for example, is the naming of the well. This ritual is as formalized as a child's christening, and just as important, since it establishes how the well will be known forever in official records and informal discussions.

A well is named after its driller, the owner, or other identification of the land or plot on which it is drilled. The name is followed by the number of the well in the drilling sequence. The first well drilled by the Acme Oil Company on a plot leased from James H. Kennelworthy would be the Acme Kennelworthy No. 1. So it would be known officially, though informally it would be called Kennelworthy No. 1. In Alaska, the North Slope discovery well on state-owned land was ARCO Prudhoe Bay State No. 1.

Another well by the same driller on the same lease will be No. 2, and so on. A different driller on the same lease starts over with No. 1.

Once a drilling site is determined from geological indications, the lease is acquired, and the drilling regulations are confirmed, a way must be provided to get to the site and build the structures needed during drilling .

Getting to the location may be a problem, depending on the part of the world in which it is located. A "location" is a block in which local regulations permit the drilling of a single well. A "lease" or "tract" is divided into many locations, like a checkerboard. Each square is a location. Acme Kennelworthy No. 2 may be drilled a location or two away from Acme Kennelworthy No. 1, which could be a distance of several hundred yards or even a mile or more. The actual wells might well be drilled from the same spot, a common practice in such hostile environments as offshore or the Arctic; but through slant drilling, the holes will bottom in different locations.

Men and materials must be able to move to the well site, whatever its location. If the well is on the state capitol lawn in Oklahoma City or in downtown Los Angeles, access is easy, though the driller must be especially solicitous of the neighbors. But if it is offshore, in a swamp, on top of a mountain, in the Arctic, in a tropical jungle, or in an arid desert, getting there may be more than half the battle. Access that was suitable for seismic crews and even test-drilling rigs carried in rugged but light off-the-road vehicles will not do for the large amounts of heavy equipment that must go in. A major road-building project sometimes is the first part of the job, even the biggest part of the job, if the field does not meet expectations.

In places where oilmen cannot build roads — offshore and in the Arctic ice, to name two such places — the logistics are a story in themselves. The very first step in drilling a well is simple pick-and-shovel work. There is no ceremony at the groundbreaking, though the money poured into the hole, regardless of its eventual success, frequently is many times that spent on a building whose birth is attended by government officials and celebrities.

Today, of course, the pick and shovel are likely to be a backhoe or bulldozer. The object is the same: to dig a

Oklahoma City was unique because it was one of the few fields located in a large urban area. Before zoning laws were enacted and enforced, wells were drilled wherever the driller thought he would find oil, as in this photo showing the Oklahoma Capitol surrounded by oil rigs.

hole, called a "cellar." The cellar is 8 to 10 feet square and 6 to 20 feet deep. In unstable soil, it is lined with concrete, which also provides a firm anchor for the top of the well casing. The cellar allows space under the rig for assembling valves, pipe, and fittings; for jointing and unjointing pipe and tools; and for constructing the foundations of the derrick and the rest of the rig.

In the past, the driller put in heavy timber or concrete foundations called "sills," on which the legs of the derrick and the derrick floor stood. In those days, the derrick was a wood or steel structure up to 100 feet tall. It provided clearance and facilities for handling long strings of pipe casing and tools. Today, the derrick is usually a portable steel frame erected at the site and dismantled when no longer needed. But it serves the same purpose.

The crown block stands at the top of the derrick. It is a huge pulley with at least three sheaves. Over these sheaves run the casing line for raising and lowering casing into the well; the sand line for handling the bailer which removes sand, cuttings, and fluids from the hole; and the drilling line, or wire line, at the end of which are the drilling tools. The bottom of a stationary derrick is usually enclosed with canvas or other sheeting, which forms a room on the derrick floor.

Standing near the derrick are racks for casing pipe, the enginehouse, a covered passage between the derrick and the enginehouse, a forge, and a toolhouse.

Inside the derrick stands a vertical beam called the "Samson post." Balanced on it is the walking beam, one end of which is positioned directly over the center of the derrick floor, where the well is drilled. The other end of the walking beam is connected to a vertical rod called the "pitman," which, in turn, is connected to a steel crank. These terms, interestingly, stem not from oil-patch usage but from the earliest steam engines.

The crank is attached to a huge wheel called the "band wheel," which the engine drives either through a transmission or by a belt. The engine turns the band wheel, which causes the crank to reciprocate, which pushes the pitman up and down and causes the walking beam to seesaw on the Samson post. The tool cable attached to the other end of the walking beam goes up and down in the hole. The tools at its end hammer at the rock below, just as they did for the ancient Chinese and for Colonel Drake.

A big drum called the "bull wheel" also stands on the derrick floor. The drilling line is wound on it. This cable, reeved over the sheaves in the crown block and clamped by the temper screw to the end of the walking beam, suspends the tools in the hole and serves to pull them or run them. The temper screw, seven or eight feet long, allows the driller to keep pace with the deepening hole by manually adjusting the position of the clamp that secures the wire line.

Another cable-wound drum, called the "calf wheel," stands nearby. This cable is the casing line, which is used to run or pull casing. A third, smaller drum, called the "sand reel," handles the sand line, which raises and lowers the bailer. All three drums, or wheels, are driven off the band wheel through belts, ropes, chains, or sprockets.

Of course, the rig just described is almost obsolete today, though the terminology remains. Most modern

125

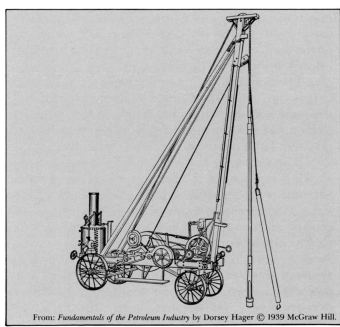

After years of building stationary rigs, drillers decided to build rigs that could be dismantled and moved from one location to another. This early portable rig has the same features as a modern one: it is mounted on wheels, collapsible, and quite small.

cable method. The Depression of the 1930s idled rigs, and then World War II drove overworked existing rigs to collapse, while shutting off the supply of spare parts. At the end of the war, the industry's drilling facilities were worn out. They were replaced with the most up-to-date technology, which largely meant rotary rigs to cope with the more difficult drilling tasks the industry faced. Technology had already erased the old distinctions between hard-formation and soft-formation drilling equipment.

One of the earliest and most important innovations had already been made, at Spindletop, where Lucas used a fluid as an integral part of the drilling process for the first time. The fluid, a suspension of clay in water that was aptly named "mud," provided a compensating pressure in the well bore that helped keep the walls from sloughing off or caving in.

At the time, Lucas's primary concern was to keep the bore open. He was unaware of the many other wonderful benefits that mud bestows upon the driller. For example, Spindletop was a gusher — a blowout — something that modern oilmen consider very bad form, if not criminally wasteful and dangerous. Had Lucas and his crew known how, they could have used their drilling mud to overcome the pressure in the well and contain it. But then, they might not have invented that remarkable assemblage of valves known as the "christmas tree."

The Spindletop rig, primitive as it was compared to today's rigs, had all the elements of a modern rotary rig in the same way the Wright brothers' airplane had all the elements of modern aircraft. The differences are of degree, not kind, but the degree is considerable.

The Lucas rig had a bit, a drilling string of pipe, rotating equipment, hoisting equipment, circulation equipment for the drilling fluid, engines, and transmissions. Modern rigs have added installations for pressure control, such as blowout preventers (BOPs), as well as sophisticated instrumentation and automatic control. Power plants producing thousands of horsepower replaced the steam engines of a few hundred horsepower. Fluid-torque-converter transmissions have taken over from belts, pulleys, and chains. But the principle remains the same: to turn a weighted bit at the bottom of

126

This modern rotary rig in California is drilling ahead. Its drill pipe and casing are ready for use. The next drill pipe joint leans against the rig (center), waiting to be added to the string as the hole becomes deeper.

previous casing and thus must be smaller in diameter. A troublesome well that requires setting much casing rapidly narrows until it becomes too narrow, at less than 5.25-inch diameter, to drill any farther. Many a well has thus been abandoned before reaching design depth. Wells as small as three inches have been completed, but at such a small diameter, the driller cannot get enough weight on the tools to continue drilling efficiently. Other problems abound as well.

The serious drilling comes after spudding in. It begins with assembly of the tool string. At the bottom of the string is the bit, a very large, chisel-shaped block of steel or iron as much as six or seven feet long and up to two feet wide. The exact shape of the bit differs according to the strata encountered and the task to be done. Bits have a variety of names, such as "Mother Hubbard," "California," and are twisted, straight, eccentric, bull-nosed, and star-shaped. The last is a giant version of the mason's star drill. These bits drill straight or crooked holes or holes larger than their own diameter.

Backing up the bit is the drill stem or auger bar, a length of iron twenty feet long and smaller in diameter than the bit. The jars follow the stem. Jars are heavy, elongated links, like chain links. They are named for their action. Because they are loose, they allow the driller to impart a jarring action to the bit, both to apply more force on the downward stroke and to jar the string loose on the upward stroke if it becomes stuck.

Another drill stem follows the jars. Called the "sinker," it attaches to the cable through a swiveled rope socket. The swivel allows the tools to turn without twisting the cable. The tool dresser and driller screw together all these items with heavy jacks and wrenches, often power assisted. The joints must be tight. The entire assembly weighs two to five tons and is 20 to 40 feet long. The driller does not want it to come apart several thousand feet underground.

Once the string is made up, the driller lowers it into the hole, not quite to the bottom, and clamps the cable to the temper screw at the top of its travel. He then slacks off the cable from the bull wheel so that it is now suspended from the walking beam instead of the crown block. Drilling commences.

Driven by the engine through the various linkages, the walking beam seesaws, lifting the tools from the bottom and banging them down again about every two seconds. The driller adjusts the tool depth with the temper screw as the hole goes deeper. The adjustment is critical. If the cable is not quite long enough, the tools will not strike with maximum impact and may even miss from time to time, or "peg leg." Too slack a cable also wastes energy and, worse, makes a crooked hole. Today there are instruments that tell the driller the tension on the cable and help him make the proper adjustments. There are even automatic systems that slack off cable and maintain proper tension as the hole goes deeper. But any experienced driller prefers the personal touch. He judges the state of his tools by feeling the cable and sensing its vibration.

As a cable lengthens to a few hundred feet, it begins to act like a huge elongated spring and by itself becomes a very active part of the drilling string. As the tools are pulled off the bottom of the hole, the spring jerks them to

127

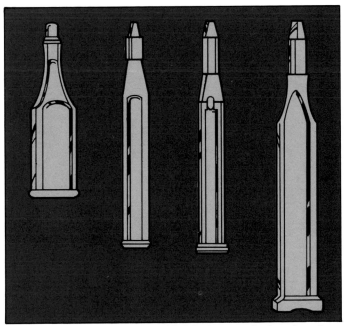

Since 1900, cable-tool bits have changed little in shape, but they have changed dramatically in material and the exactness of manufacture.

the top of the cycle. When the direction reverses, the spring then slams the tools into the bottom with many times the energy of their weight alone. The longer the cable, the greater this effect. The driller tries to control the operation so that the walking beam and the cable are in perfect rhythm, allowing gravity and resilience to do most of the work.

Depending on the hardness of the strata, the temper screw soon reaches the limit of its travel, and the hole becomes cluttered with cuttings. These cuttings cushion the blow of the tools and reduce drilling speed. The driller then unclamps the cable from the temper screw, pulls the tools from the hole with the bull wheel, and ties them out of the way. The bailer, a cylinder up to 40 feet long with an open bottom that can be closed with a ball-type valve, is lowered into the hole on the sand line. It picks up cuttings and water and cleans out the hole. Cuttings are dumped into a slush pit alongside the rig and carefully examined for signs of oil or gas, and also for classifying and verifying the strata.

While the bit is out of the hole, the driller examines it for wear. If it is worn, he changes it. He may also change to a different type of bit, either because it is better suited to the strata or because he wants to straighten the hole or widen it. Worn bits are "dressed," that is, heated in the forge, hammered into shape, and tempered. "Sharpened" is not the right word, because they are not really very sharp, but are hard and properly shaped to withstand the abrasion of the rock and to pulverize it.

Unlike rotary drilling, which uses a heavy drilling fluid to support the bore, flush the cuttings, and lubricate the bit, the cable system uses relatively little water at the bottom of the hole to suspend cuttings and lubricate the bit. If nature does not add water from natural aquifers, the driller must do so with a hose or the bailer.

This relative dryness is one of the advantages of cable drilling, particularly for exploratory wells. There is no foreign material in the well to block off, or perhaps inadvertently seal, strata through which the bore passes. Cuttings are not contaminated, which allows geologists to make accurate logs from actual samples more easily. And if there is a hydrocarbon show, it is not masked. Drilling mud has masked many a pay zone that a rotary drill

passed through without the drillers being the wiser. With a cable rig, it is hard to miss even a minor show. On the other hand, should the string penetrate a high-pressure zone, there is nothing to hold the pressure back but the weight of the tool string and cable.

A driller prefers to send his tools straight through rock without casing, letting the bare rock form the wall of the bore. It is cheaper, but is not always feasible. Usually the drill penetrates aquifers or soft strata that cave in the hole. The drill may even encounter hydrocarbon-bearing strata that the driller wants to bypass for the moment. In all these situations, he must set casing.

Wells have been cased from earliest days. The year 1903 marked a most important advance in technique when casing was first cemented in place. Then the driller simply dumped cement into the bottom of the cased hole from a bailer. When the cement set, it effectively sealed off the casing, and the cement plug at the bottom could then be drilled out. The technique walled off the bore from the strata and permanently isolated problem areas.

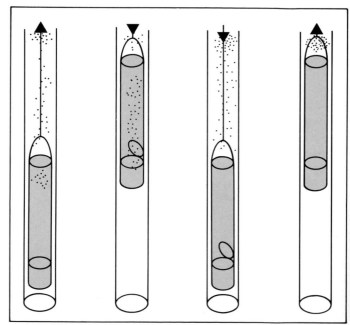

Bailer operations, illustrated here, clean cuttings and water out of the hole. The bailer, a cylinder up to 40 feet long, is lowered into the hole on the sand line. Its opening is controlled from above by a ball-type valve.

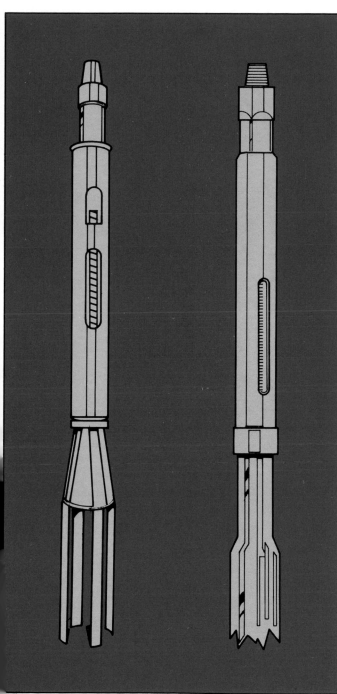

An improved technique uses plugs ahead of and behind a slug of cement which is forced under pressure into the hole. When it reaches bottom, the cement squeezes around the outside of the casing into the annulus between the casing and the actual bore hole, more effectively sealing the casing in place.

Drilling seldom proceeds without incident. Cables break; tools unscrew from the string; walls cave in. If the hole is not yet very deep when such an accident occurs, it may be cheaper to abandon it; move, or "skid," the rig; and start over. But if there is already a large investment, the driller will try to salvage the hole.

If a cave-in has trapped the tools, the driller tries to jar them loose and then drill out the cave-in and case it. If the tools break off, he fishes for them, using various types of special implements designed to grasp or screw into the broken ends so that they can be retrieved. If there is not too much debris in the hole, he tries to drill through it. He may even try to drill around it, though that is more easily done with a rotary rig than a cable rig.

Cable drilling was the dominant method up to the 1930s. Long before then, the technology of cable drilling had gone as far as it could, while that of rotary drilling was fast developing. The deeper wells needed in various strata not suitable for cable tools required the newer technology and gradually pushed cable rigs into the background. The two methods continued to compete side by side for many years. But rotary rigs drilled deeper. The deepest cable-drilled well was completed in New York in 1953 at 11,145 feet. It took two and a half years to drill, which illustrates the competitive disadvantage of the method. In good strata, a cable rig can drill about 60 feet a day. Under ideal conditions, a rotary can drill up to 2,000 feet a day. At depths greater than about 4,000 feet, cable drilling becomes increasingly difficult, while a rotary drill is just beginning to hit its stride.

Yet cable drilling has not disappeared, nor is it likely to. Though the forests of derricks that hid the East Texas pines are long gone and will never return, their descendants, packed into fleets of battered trucks, still roam the High Plains and the hills of Tennessee.

"Going fishing" does not signify fun in the oil industry. It means that the cable has snapped and the driller is looking for the lost drill bit with special tools like these, equipped to grasp or screw into the broken pieces so that they can be retrieved.

CHAPTER 13 –
DRILLING THE MODERN WAY

When Captain Anthony Lucas brought in Spindletop so spectacularly in 1901, he did it with a lot of luck and a rotary drilling rig.

Spindletop was the first gusher ever seen in the Western Hemisphere, and it was born in a type of field and strata that had been little explored up to then. The first four holes Lucas drilled not only produced no oil but also were difficult holes to drill. The soft, sandy ground continually collapsed on the bores, and the cable tools of the day found it hard going. Lucas and the Hamill brothers, his crew of innovative drillers, decided to use a rotary rig whose bit could chew its way through the soft formations.

Rotaries had been around for some time and had been used to drill for water in Louisiana as far back as 1821. A powered rotary drove a diamond-studded bit in 1860. But although they had been widely used to drill water wells and a few successful oil wells, rotaries were still a rarity in the oil patch, where the cable rig was king.

Even after Captain Lucas's success, the cable rig remained the overwhelming choice until long after Spindletop. In most cases, the rotary just did not do the job as well or as cheaply. Two men and a boy could run a cable rig, and the boy was expendable, leaving only the driller and the tool dresser. A rotary rig, on the other hand, normally needed five or more people per shift: a driller, a tool pusher, a derrickman, two roughnecks, and perhaps a motorman.

Moreover, cable tools were speedier and more effective. They drilled faster in the hard rock and at the shallow depths most commonly exploited up to the turn of the century. After 1901, the oil world began to find uses for both rigs, and rotaries became much more common. Cable rigs continued to be used in the hard formations found in most of the country, and rotaries began to be used in the soft strata of the Gulf Coast and some other areas. Some rigs, called "California combination rigs," used both types. They were used where formations changed greatly within the same well. Neither type drilled much deeper than 4,000 feet.

Depression and war finally vanquished the older

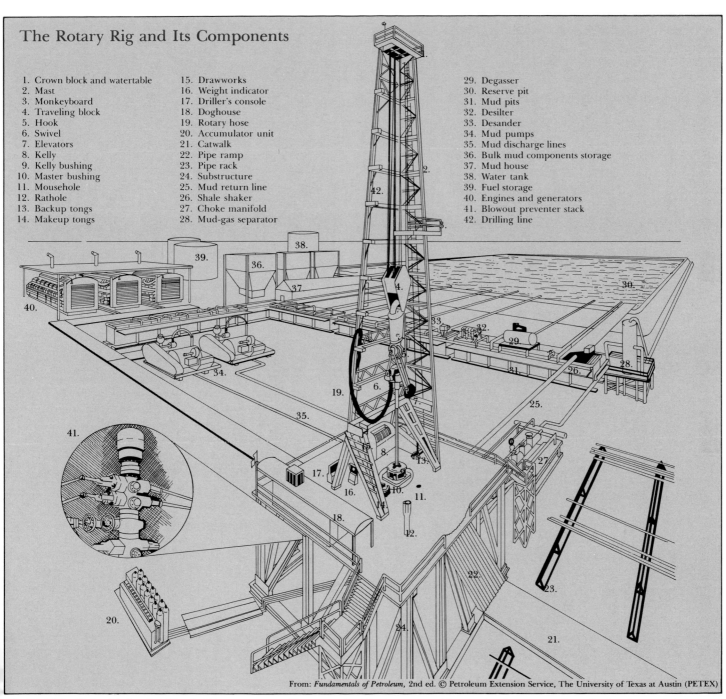

The Rotary Rig and Its Components

1. Crown block and watertable
2. Mast
3. Monkeyboard
4. Traveling block
5. Hook
6. Swivel
7. Elevators
8. Kelly
9. Kelly bushing
10. Master bushing
11. Mousehole
12. Rathole
13. Backup tongs
14. Makeup tongs
15. Drawworks
16. Weight indicator
17. Driller's console
18. Doghouse
19. Rotary hose
20. Accumulator unit
21. Catwalk
22. Pipe ramp
23. Pipe rack
24. Substructure
25. Mud return line
26. Shale shaker
27. Choke manifold
28. Mud-gas separator
29. Degasser
30. Reserve pit
31. Mud pits
32. Desilter
33. Desander
34. Mud pumps
35. Mud discharge lines
36. Bulk mud components storage
37. Mud house
38. Water tank
39. Fuel storage
40. Engines and generators
41. Blowout preventer stack
42. Drilling line

131

From: *Fundamentals of Petroleum*, 2nd ed. © Petroleum Extension Service, The University of Texas at Austin (PETEX)

After World War II, the oil industry's drilling facilities were worn out. They were replaced for the most part with rotary rigs, the most up-to-date technology, because the industry faced more difficult drilling tasks. The rotary rig is factory designed to perform one function—to make a hole—and it does so very efficiently at the rate of up to 2,000 feet a day.

cable method. The Depression of the 1930s idled rigs, and then World War II drove overworked existing rigs to collapse, while shutting off the supply of spare parts. At the end of the war, the industry's drilling facilities were worn out. They were replaced with the most up-to-date technology, which largely meant rotary rigs to cope with the more difficult drilling tasks the industry faced. Technology had already erased the old distinctions between hard-formation and soft-formation drilling equipment.

One of the earliest and most important innovations had already been made, at Spindletop, where Lucas used a fluid as an integral part of the drilling process for the first time. The fluid, a suspension of clay in water that was aptly named "mud," provided a compensating pressure in the well bore that helped keep the walls from sloughing off or caving in.

At the time, Lucas's primary concern was to keep the bore open. He was unaware of the many other wonderful benefits that mud bestows upon the driller. For example, Spindletop was a gusher — a blowout — something that modern oilmen consider very bad form, if not criminally wasteful and dangerous. Had Lucas and his crew known how, they could have used their drilling mud to overcome the pressure in the well and contain it. But then, they might not have invented that remarkable assemblage of valves known as the "christmas tree."

The Spindletop rig, primitive as it was compared to today's rigs, had all the elements of a modern rotary rig in the same way the Wright brothers' airplane had all the elements of modern aircraft. The differences are of degree, not kind, but the degree is considerable.

The Lucas rig had a bit, a drilling string of pipe, rotating equipment, hoisting equipment, circulation equipment for the drilling fluid, engines, and transmissions. Modern rigs have added installations for pressure control, such as blowout preventers (BOPs), as well as sophisticated instrumentation and automatic control. Power plants producing thousands of horsepower replaced the steam engines of a few hundred horsepower. Fluid-torque-converter transmissions have taken over from belts, pulleys, and chains. But the principle remains the same: to turn a weighted bit at the bottom of

This modern rotary rig in California is drilling ahead. Its drill pipe and casing are ready for use. The next drill pipe joint leans against the rig (center), waiting to be added to the string as the hole becomes deeper.

132

a hole so that it makes more hole, to provide a means of removing and replacing the bit, to maintain a suitable drilling environment in the hole, and to help the crew control the drilling and complete the well.

A rotary rig is much larger and more complex than a cable rig. The drill bits are fitted to the end of a length of stiff pipe rather than to a flexible cable that can be wound on a drum. Drill pipe comes in 30-foot joints and goes into the hole one or more joints at a time. But when the crew takes it out, as it must to change the bit, set casing, cement, or run logs, the crew pulls it in as long a "stand" as possible, three or even four joints at a time. The longer the stand the crew can handle, the more efficient the operation. A three-joint stand is a "tribble," and four is a "fourble." Since a fourble is 120 feet long, the derrick or mast must be taller than that. Modern derricks are more than 200 feet tall.

Rotary derricks must withstand much greater loads than that of a cable rig. The derrick supports the entire drill string during drilling, a weight of several hundred tons, and must also support racks of drill-pipe stands when the string is out of the hole. These racks may hold 30,000 feet of pipe weighing 200 pounds or more per foot. And with all that weight, it must also withstand wind loads of over 100 miles an hour.

Derricks still are built piece by piece where oilmen expect to drill several wells from a single location, such as offshore or in the Arctic. But the familiar shape has largely disappeared from more accessible regions. It has been replaced by the collapsible portable drilling mast. Since this structure must do the same job under the same conditions as a stationary derrick, it is equally massive. It has about the same relation to a portable cable rig as a Boeing 747 does to a Piper Cub.

Rotary or cable, the mechanics of site selection are the same. But site preparation and setup differ. In rotary drilling, there is no cellar. Instead, bulldozers scrape out a large pond, or pit, near the well site, which is then lined with plastic. This mud pit is a holding place for the drilling fluid and is sealed from the ground under it to prevent seepage into any water table. Sometimes the mud pit is a large tank set on the ground. Occasionally, there is another pit, the sump, to hold water, or perhaps the crew

The draw works consists of a heavy drum around which the drilling line is wound, a series of shafts, clutches, and chain and gear drives that control drilling speed.

133

drills a water well to service the site if water cannot be piped in. Close by the well bore is another hole, called the "rathole," used to store the kelly (see below). Sometimes a "mousehole" is used to hold lengths of drill pipe awaiting hookup. Sometimes, the rathole and the mousehole are one and the same.

A length of conductor pipe up to four feet in diameter is set, either by drilling out a hole for it or by pile driving. This pipe serves as a guide between the rig deck and the point where serious drilling begins. It also acts as an overflow basin for the drilling fluid and prevents near-surface sand and gravel from caving into the hole.

The rig itself sits on the substructure, an elevated platform eight to ten feet high, which the crew assembles at the site from erector-set-like sections.

On the substructure are mounted the derrick, or mast, and the draw works, all part of the same system for handling drill pipe and casing. The crown block tops the derrick. It is much more massive than the one used in cable rigs and with many more sheaves. An equally massive traveling block hangs from the crown block. A cable — the fast line — runs from a drum in the draw works, reeves through the crown and traveling blocks ten times or more, and, finally, anchors as the deadline to the substructure. Fixed at the anchoring point, the weight sensor tells the driller how much pressure he has on the bit. Bit weight is the single most important measurement in rotary drilling, yet the industry did not devise a satisfactory way to measure it until 1925.

In the middle of the drilling platform sits a steel disk larger than three feet in diameter that turns on bearings in oil. In the center of this rotary table is a square hole into which fits the kelly bushing, itself a steel block with a square or hexagonal hole through it. The kelly (no one knows the origin of the name) fits in that hole. It is a hollow 40-foot-long square or hexagonal bar, connected by threaded couplings through short subs, or stubs, to the swivel above and the drill string below. The kelly moves freely up and down in the bushing. But when the driller pulls the kelly out of the table, the bushing comes with it because of a stopper block fastened below.

The way a rotary works is simple: the engines, through transmissions, turn the rotary, which turns the kelly, which turns the drill string, which turns the bit at the bottom of the hole. The swivel allows the kelly to turn while the hooks and traveling block that support the entire drill-string assembly remain stationary.

A heavy hose which loops to the side of the rig connects to the swivel where it connects to the standpipe. The standpipe connects through hose and piping to the mud, or slush, pumps. These pumps force the drilling mud through the system at high pressure, volume, and velocity.

In the bore itself, drilling fluid lubricates and cools the bit and the drill string. The mud makes it easier for the machinery to turn the string, helps center it in the bore, and protects it from abrasion against the rock walls. The clay in the mud cakes out on the wall of the bore to form a protective surface. This prevents sloughing and seals off small pores through which unwanted water might enter or into which fluid might be lost. Of course, this same sealing effect runs the risk of masking the presence of hydrocarbons.

The hose that supplies mud to the drilling operation is looped over the hook and connected to the side of the rig. Mud cools and cleans the bit, prevents cave-ins, and transports cuttings to the surface. In addition, its great weight prevents blowouts.

To check mud composition in the field, engineers must determine ion concentrations of chloride, sulfates, hydrogen, calcium, and magnesium. They use a portable laboratory like this one to make their on-the-spot tests.

The chief function of the mud is to balance the pressure in the well against that in the strata where the drilling is taking place. Usually, the pressure at any point is measured as the hydrostatic pressure. Since the mud completely fills the bore and is usually heavier than salt-water, its pressure will be greater than the hydrostatic. Occasionally, pressure in the strata will be higher or lower, or a gas pocket will cause a kick, which must be controlled to prevent a blowout. The driller can change the composition of the mud by adding heavier solids such as barites, by reducing the solids content, or by changing to mixtures of oil and water. Sometimes, he even adds asphalts to seal off shales and clays that swell in the presence of water and possibly block the bore and cause the drill string to stick.

Originally, drillers made mud by running cows through the mud pit. Today, mud chemistry and formulation is an art and science all its own. Mud viscosity and velocity in the well must be sufficient to suspend and carry to the top of the well all the cuttings made by the drill bit. The mud must also hold cuttings in suspension when drilling and circulation stop. At the well top, the mud exits and pours across the shale shaker, a screen that removes the cuttings, some 100 tons from the first 500 feet of hole alone. These cuttings are regularly sampled and logged to tell the driller in what sort of formation the bit is, or was an hour earlier. The screened mud then goes to the mud tank, where its level is carefully monitored. A rising level means rising pressure in the well and possible trouble. A falling level means lost circulation and possible trouble. Either way, change spells trouble.

Lost circulation means that, somewhere below, a porous or fissured formation, a thief formation, is sucking up the mud as fast as the pumps can supply it. The crew tries to control the situation by loading the mud with lost-circulation material (LCM), such as plastic strips, sawdust, and plastic foam, hoping that it will enter the formation and seal it off. But they must also examine the formation for oil, gas, or water. They know that it must contain one or all of these because there are no empty spaces down in the earth. If the LCM does not work, then the crew must case and cement the bore at that point. This costs time, money, and, not least, diameter.

Rotary power plants are diesel, gas, or diesel-electric engines, generally mounted near the draw works and rotary table. Diesel-electric power plants, located some distance from the structure, allow a certain flexibility and also spare the drilling crew some noise.

Rotary rigs are rated by their horsepower, which also defines the maximum depth a particular rig can drill. More than 1,500 horsepower from three engines is not unusual; more than 3,000 horsepower is routine for big rigs.

Where does all that power go? Mostly into drilling and fluid circulation. Twisting a string of heavy pipe two miles long and pressing against the rock with a force of 50 tons or more requires constant power, and plenty of it. Pumping thick, heavy fluid to the bottom of the hole at high velocity and balancing downhole pressures consume 2,000 horsepower alone. By contrast, the draw works requires relatively little power for handling the string and casing.

Strictly speaking, rotaries do not spud, although oilmen still call it "spudding" when a well is started. Because the bore of a rotary-drilled well gets narrower each time it sets casing, the driller starts with the largest feasible bit size, 24 inches for a deep well. The crew attaches the bit to the bottom of the kelly, linking it by an intermediate piece called a "bit sub." The threads on the bit sub — a very inexpensive piece of equipment compared to the kelly — take the wear, preserving the costly kelly, the drill collars, and the pipe.

After the mud flow begins, the bit is lowered into the conductor pipe. As the bit chews into the ground, the mud flows from the pumps through the piping and hose, into the swivel and down through the kelly to the bit. There it exits at high velocity, sweeping out the chips and carrying them to the top of the conductor pipe and back through the mud system.

When the kelly is drilled down — when the hole is about 30 feet deep — the driller pulls the bit from the hole. Using power tools, roustabouts unscrew the bit. The draw works lifts the kelly out of the way. The crew moves

the drill collar from the mousehole over the well, secures the bit to its end, and lowers the collar with the bit attached into the hole. These collars are not the drill pipe that make most of the hole. Rather, they are extra heavy lengths of pipe that weight the bit and stabilize the string, much like the drill stems used for the same purpose in cable drilling.

Slips, tapered clamps that hold the pipe at the rotary when it is not connected to the traveling block, support the collar as the kelly is stabbed into its threaded end. When the joint is made up, the kelly goes back into the rotary table, and drilling resumes. After another 30 feet, the crew again lifts the kelly and repeats the process, adding another drill collar.

After a day's drilling of 500 feet or so, the crew removes the kelly and stores it in the rathole. Then they pull up the string and hold it in place with slips as they disconnect the top three or four joints. Tribble by tribble or fourble by fourble, the derrickman high on his monkey board swings each stand over to the rack, called a "fingerboard."

Once the drill string comes out of the hole, the draw works sets the surface casing. This walls off the bore from shallow strata and aquifers. The first joint of this casing has a heavy shoe at the bottom that guides the casing string past irregularities and supports the weight of the pipe by floating it on the fluid in the hole. When set, the pipe is cemented in place.

Then the driller puts a new, smaller bit on the first stand of drill collar. The bit is the business end of the whole costly, complex assemblage. The early rotary bit was not an auger screw, nor did it look anything like a modern rotary bit. Called a "fishtail," it modified then current cable-tool bits. It was a rotating flared chisel that scraped at the rock as it turned.

The industry still drills very soft formations with a simple fishtail bit that is little different from a cable bit. But the bit for rock is a sophisticated machine in its own right. The invention of the modern rock bit by Howard R. Hughes, Sr., in 1908 started the revolution in drilling technology which today permits holes to be drilled deeper than 30,000 feet. Without tough, long-lasting bits capable of going rapidly through all types of formations with infrequent changes, this could not be done.

Hughes's original design did look something like a fishtail in profile but otherwise was most unlike it. He used two toothed cones opposed to one another. They were free to roll like wheels when forced against the rock as the entire bit rotated. The teeth of the cones continuously brought new metal surfaces into contact with the rock and relentlessly ground, chipped, and powdered the rock instead of scraping it away.

Hughes secretly tested his first bit at Goose Creek, Texas, and lost it. A little later he tried again, successfully. Within two years, the Hughes Roller Cone Rock Bit was the standard of the industry, and one of America's greatest fortunes was securely founded.

Roller bits now have three cones rather than two. For soft formations, they have long teeth. For very hard formations, they may have no teeth at all, just buttons of tungsten carbide. Typically, bits have jets between the rollers from which the drilling mud exits, clearing chips

In 1909, Hughes introduced this rotary bit, the first equipped with two-cone cutters. Its main innovations were the simplicity of its design and construction, its large cutters, and its ability to maintain hole diameter.

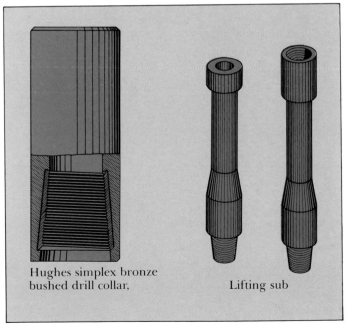

Hughes simplex bronze bushed drill collar,

Lifting sub

The Hughes simplex drill collar (left), used in the 1920s, was 2 to 3 feet long. By the 1930s, collars were lengthened (right) up to 50 feet to speed up drilling and keep the holes straight.

instantly from the bottom of the well. In soft formations, the jets help drill. If the formations are too soft, they may wash out the bore.

For extremely hard formations and for special purposes, the driller uses bits studded with industrial diamonds, the hardest substance known. These bits generally are not rollers, but are flattened cones the same diameter as the well bore, with grooves in their surfaces for the drilling fluid. Some such bits are hollow. These are coring bits, which allow the driller to cut out a solid core that he can bring intact to the surface.

Sometimes, special core barrels are substituted for the first two drill collars, and these bring back cores as long as 60 feet. Coring tells far more about subsurface geology than any other method. But it takes time and interferes with the real business of finding oil. So most coring is done on exploratory wells to establish the geology of a region and provide basic data to be correlated with other types of logs.

Once the casing is set, the drill string goes back in the hole, stand by stand. The derrickman takes each stand from the fingerboard and attaches it to the elevators, a set of clamps on the traveling block. The stand then swings over the hole, here the crew stabs it to the stand already held in the hole by the slips. They make up the joint with the aid of tongs, chain, and the rotary and then lower away for the next stand. When the bit reaches bottom, the kelly comes out of the rathole and is stabbed into the box of the last joint. The kelly bushing sits in the rotary table and drilling resumes. When this is done, the crew cuts off the conductor pipe at ground level and installs a blowout preventer on the surface casing.

There is always the risk that the bit will enter a high-pressure zone or gas pocket. The underground pressure from these areas can overwhelm the drilling fluid and even the weight of the drill string. During the pulling of the string, the action of the pipe occasionally reduces pressure in the well and temporarily allows subterranean forces to take over, threatening a blowout. The blowout preventer (BOP) is the operator's defense.

138

A wide variety of bits was needed to drill this deep well at the Mills Ranch field, Texas, in 1975. It found gas at a record 24,482 feet, far below the depth where oil is likely to exist.

This Hydril Blowout Preventer stack consists of one annular and two ram BOPs. Working together, they seal off the annulus and contain the kick. Since coming into use in the 1930s, BOPs saved countless resources, lives and money.

The BOP is a complex arrangement of valves, sleeves, and rams that closes the wellhead exactly when required. The drill string passes through the BOP; the mud flows from it. Any gases or fluids that develop in the well also exit through it. All tests are done with the BOP in place.

The BOP is a very heavy assembly, rated for many thousands of pounds of pressure. In normal operation, it is a valve that helps maintain proper pressure in the well. In emergencies, such as a kick, when gas pressure in the hole overcomes the pressure exerted by the weight of the drilling mud, the BOP closes, contains the pressure, and then releases it under control. For extreme emergencies, the BOP has hydraulic rams that squeeze the casing and the drill pipe to close off the bore.

The modern BOP is a highly reliable device. When wells blow out, people have failed, not the BOP. Someone installed something incorrectly or left something open that should have been closed.

As drilling proceeds, enormous weights build up in the bore. The driller adds between 20 and 30 drill collars for a deep hole. This 600- to 900-foot drill stand puts 50 tons or more on the bit. Then the driller continues to lengthen the string by adding drill pipe in 30-foot sections. These are made of high-tensile steel and are relatively thin walled and thus much lighter than the drill collars. Yet this pipe must withstand tremendous forces. A 15,000-foot string weighs 200 tons, and the topmost joint must support it all, with some flotation help from the drilling fluid. Indeed, deep drilling had to await the development of steels that could bear such loads.

Each time the rig drills down a kelly, the crew must add another joint of drill pipe. Fast-drilling formations keep the crew busy as it makes several changes an hour. In very hard strata, it may not make a kelly a day. Eventually, of course, the bit has to be changed. That means a complete round trip.

A round trip involves pulling the entire string from the well, replacing the bit, and putting the entire string back into the well. This procedure separates rotary drillers from cable drillers. A string and bit cannot just be reeled in. The entire string must come out, stand by stand, and go back, stand by stand. On a 15,000-foot well,

139

This crew began exploration drilling in Corral Creek, Wyo., in 1979, in the area of the Rocky Mountain Overthrust. Unfortunately, this hole turned out to be dry, so the crew had to plug it up, repair the land, and move on to the next site.

it can take nearly six hours to pull the string and another three to four hours to run it back. Pulling takes longer because of the great weight that must be lifted.

Single joints of drill pipe are quite stiff. They do not bend, twist, or break. But a 15,000-foot string of pipe is about as stiff as a strand of wet spaghetti. The turning movement of the rotary twists the pipe several times, like a long rubber band, before it starts turning the bit. If the string were a knitting needle, it would be 236 feet long. And only the driller's constant attention keeps the bit going in the right direction.

With such forces at work, things sometimes go wrong. Most frequently, the string twists off somewhere down the hole, leaving a lot of pipe, drill collars, and the bits. This is known collectively as the "fish." Some of the hole may cave in, trapping the string, or the bit may break up, a threaded joint may unscrew, or threads may strip out. These misfortunes have happened many times. The crew expects them. The techniques for dealing with them are at least as well developed as those for drilling itself.

Fishing expeditions are time-consuming and costly. No driller wants to write "Fishing" in the log day after day when the rig and crew cost thousands of dollars an hour. So the crew pulls as much of the string as it can, then goes back with special tools that grind down jagged surfaces and provide a good hold for other special tools to grasp either the inside or the outside of the pipe in the hole. With luck, the crew then retrieves the rest of the string. If it is stuck, special jars are attached to add over 100 tons of extra jarring force to the stuck string. Sometimes, of course, that breaks things up even more.

A junk basket or magnet recovers small parts. If all else fails, the driller uses explosives to make small parts smaller and then flushes them out with mud and drills through them. Given the cost of delay, the operator will not spend much time fishing. As a last resort, he leaves the fish in the hole, cements it, and deviates the well from a point above the break.

Oilmen normally try to drill wells vertically. However, the natural dip of formations coupled with the suppleness of a drill string often causes a bit to wander far off course. Spiral wells, wells that bottom far from their intended end points, and even wells with dog legs are sometimes the result when the driller tries to correct a deviation. Such crooked wells use more pipe and casing, miss intended formations, and are more likely to have stuck or broken pipe.

Costs dictate that some natural deviation be allowed. But appropriate bit pressure and drilling speed and mechanical assemblies that stabilize the drill string and prevent it from changing direction rapidly can control such deviation.

Sometimes the driller deliberately wants to change direction and to aim the bit at a specific point underground. Offshore, or in especially difficult terrain, it is very costly and sometimes impossible to prepare separate sites for every well. So oilmen prepare one site and directionally drill many wells from it, each bottoming at a point far distant from the others. Or the desired oil may be under a building or in a place that cannot be reached except by slanted drilling.

Occasionally, directional drilling is used for damage control. If a well blows out and does so much damage that control cannot be regained through the original bore, a new well is drilled some distance away connecting the new bore with the old one underground. Adequate blowout preventers drain off pressure from the damaged well. Then the driller pumps in cement and mud to contain the pressure.

Directional drilling is often the last resort when a runaway well has "cratered" — that is, blown out with great force accompanied by fire and explosion, so that it has destroyed all surface fixtures and made a large crater where the top of the well used to be. Cratered wells are extremely dangerous and costly because they cannot be controlled until the new well meets the old one far underground. That can take many weeks of difficult drilling. By analogy, in oil country when a good idea goes unexpectedly bad, it is said to have "cratered."

A directional well costs more to drill than does a vertical one. One reason is that the drill rate is slower, extending drilling time by as much as 30 percent because the driller must make frequent surveys to be sure he is on course.

The driller starts a directional well vertically and then cases it. Then he uses a whipstock, a guide shoe that

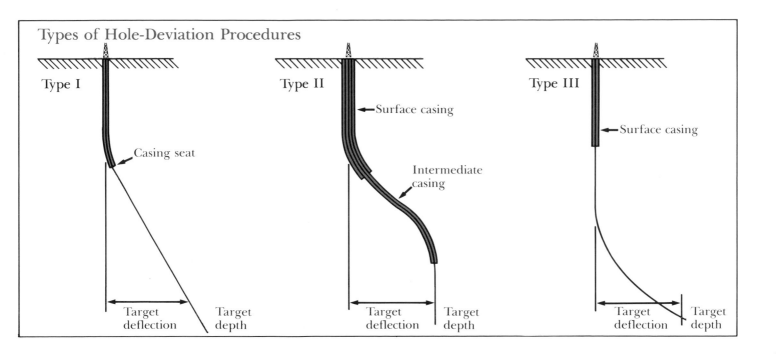

Types of Hole-Deviation Procedures

Type I

Casing seat

Target deflection   Target depth

Type II

Surface casing

Intermediate casing

Target deflection   Target depth

Type III

Surface casing

Target deflection   Target depth

This roughneck is attaching pipe to the drill string on the floor of a rig at the Mississippi Canyon Platform in the Gulf of Mexico. Joints of pipe range up to 30 feet in length and weigh several hundred pounds.

stays in place and bends the drill pipe until the bit is going at the correct angle. Various special bits and flexible components are added to the drill string during the turning phase, but once the correct angle is reached, the driller returns to normal bits, stiff collars, and pipe. Angles greater than 60 degrees can be built and held. At a depth of 10,000 feet, the bottom may be displaced as much as two miles away from the well site.

Some newer technology has proved to be of help in directional drilling. One example is the turbo drill. This is a downhole motor with a diameter the same as that of the drill string, five inches or so, and with the bit at its end. It contains a turbine driven at high speed by the drilling fluid being pumped through it. This device goes into the well at the end of the conventional drill string, but the string does not rotate or does so relatively slowly. All or most of the rotation is at the bit itself.

There are other promising innovations. Downhole electric motors have found some limited use, and even flame drilling, which uses a very hot flame to melt

142

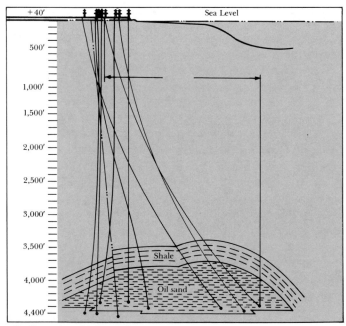

This illustration of deflected drilling at Huntington Beach, Calif., in the 1930s shows the criss-cross of wellbores that prompted the courts to step in and settle boundary disputes. Afterwards, well spacing and surveys were required.

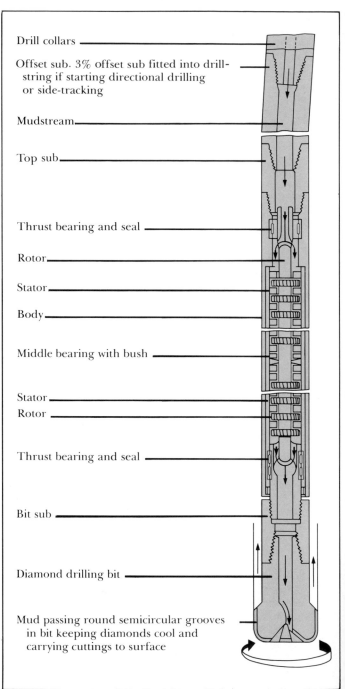

In turbo drilling, the drill bit is turned by a turbine. The turbine rotates much faster than a conventional rotary drill string, which makes it more effective for very hard formations.

through hard rock, has been tried. The turbo drill and electrically powered bits have proved useful for directional drilling, because it is easier to pull a stationary string than to push a rotating one. For that reason, many operators use them offshore, particularly in the North Sea and the Gulf of Mexico. They are also better for drilling very hard rocks and will save money on relatively shallow wells in hard formations.

There are many drawbacks, however, that may prevent these methods from ever supplanting the rock drill at the end of four or five miles of high-tensile steel pipe. One main disadvantage is that their bits wear rapidly, requiring more round trips. And they do not drill in soft formations any better than conventional techniques.

Continual advances in drilling technology allow drilling ever deeper in search of energy, though not always hydrocarbons. Techniques and equipment developed by the petroleum industry are also being used in the search for geothermal energy. Industry-developed technology and equipment also recovered parts of a sunken submarine from the deepest ocean and tried to drill through the crust of the earth at its thinnest point to further knowledge of the earth's origins.

But it is still hydrocarbons that oilmen seek and for which they primarily develop these tools. And the only place left to go is deeper — faster and cheaper. It is not the equipment design that limits drillers but material strengths. Just as the really deep wells of the last 20 years had to await space-age metals, the deeper holes in even less hospitable places wait upon still better and perhaps cheaper materials.

143

CHAPTER 14 – OFFSHORE

In the empty reaches of the South Atlantic, the Indian Ocean, or the frigid Arctic, commercial seamen encounter some unusual seagoing apparitions these days — some that resemble toppled Eiffel Towers plowing resolutely behind large oceangoing tugs, others that look like fat grain elevators planted incongruously amid the ocean swells.

These monsters are actually giant drilling platforms, on their way from assembly points in Texas, Norway, Scotland, or Japan to one of the many offshore areas where much of the world's search for new hydrocarbons has intensified in recent years. They provide solid evidence of man's ingenuity and resolution in reaching out toward new energy frontiers.

Oil geologists believe that offshore reserves are the last significant reservoir of hydrocarbons. They estimate that North America's offshore resources alone range between 12.5 billion and 38 billion barrels of oil and between 61.5 billion and 139 trillion cubic feet of natural gas. Americans already depend on offshore production for 22 percent of gas and 13 percent of crude oil. These areas grow more important to the energy future each day.

The offshore environment, however, is hostile to the oilman, posing new challenges to his technology and economics as well as to his courage. Wells drilled offshore cost up to 20 times more than equivalent wells onshore. Some areas where the industry regards oil prospects as highly promising are still beyond man's technological reach. And, as too often seen in accidents in the North Sea, the North Atlantic, the Gulf of Mexico, and elsewhere, the forces of nature can still overwhelm the ingenuity of man.

Yet even greater offshore exploration is essential to bridge the gap between today's heavy dependence on hydrocarbons and tomorrow's alternative energy base.

The offshore movement has cast the oilman in entirely new roles, not the least of which is that of a blue-water sailor who hauls these awkward rigs on voyages that rival in length those of an earlier age of exploration. The oilman also has had to learn how to live and work in precarious isolation, to probe formations lying beneath seawater, and to produce oil and gas from offshore wells in a manner that satisfies new requirements for safety and environmental protection. Fortunately, the industry has been practicing for quite a while; the first patent for an offshore oil rig was filed in 1869.

Once the bit reaches the ocean bottom, drilling for oil at sea is not significantly different from drilling on land. The geological structures containing hydrocarbons look just the same and are equally difficult to find. The driller employs a similar range of tricks and tools, from drilling muds to casings, as he would on shore. The major difference is the sea itself.

As early as 1896, the first oil from offshore formations was produced by slant drilling from onshore drilling sites. By 1910, a successful gas well had been completed a mile into Lake Erie. Offshore drilling from makeshift wooden platforms built a few yards from dry land was conducted near Santa Barbara, California, in

Tugs, workboats, and barges cluster around the central platform of the Ninian North Sea oil field. This 776-foot-high platform was towed 516 miles to its destination off the Isle of Skye.

Early prospectors in Summerland, Calif., noticed oil seeps when the tide receded. At the turn of the century, they built wharves perpendicular to the beach to support the first wells drilled in water.

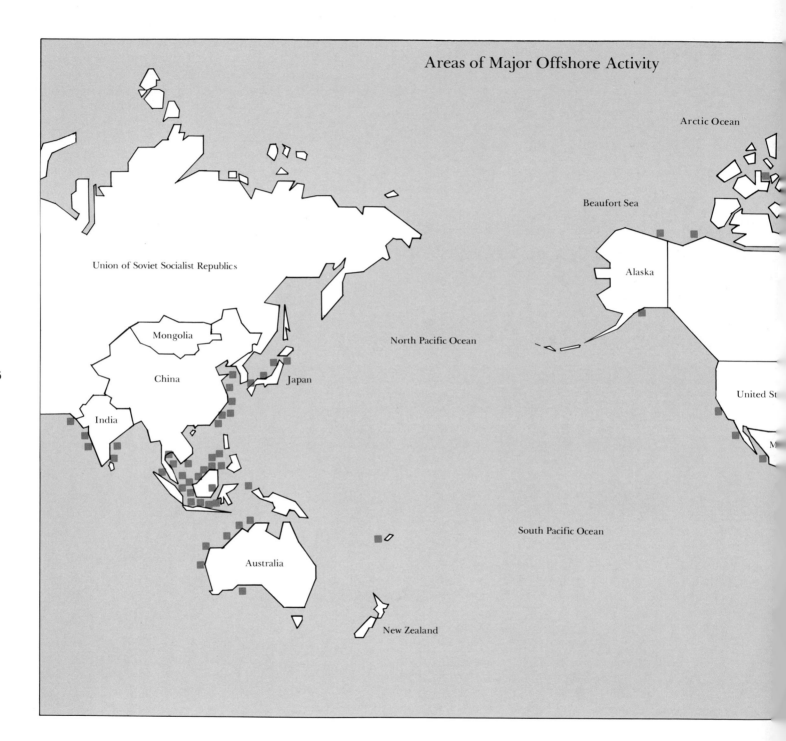

Areas of Major Offshore Activity

Arctic Ocean

Beaufort Sea

Alaska

Union of Soviet Socialist Republics

North Pacific Ocean

United St

Mongolia

Japan

China

India

South Pacific Ocean

Australia

New Zealand

146

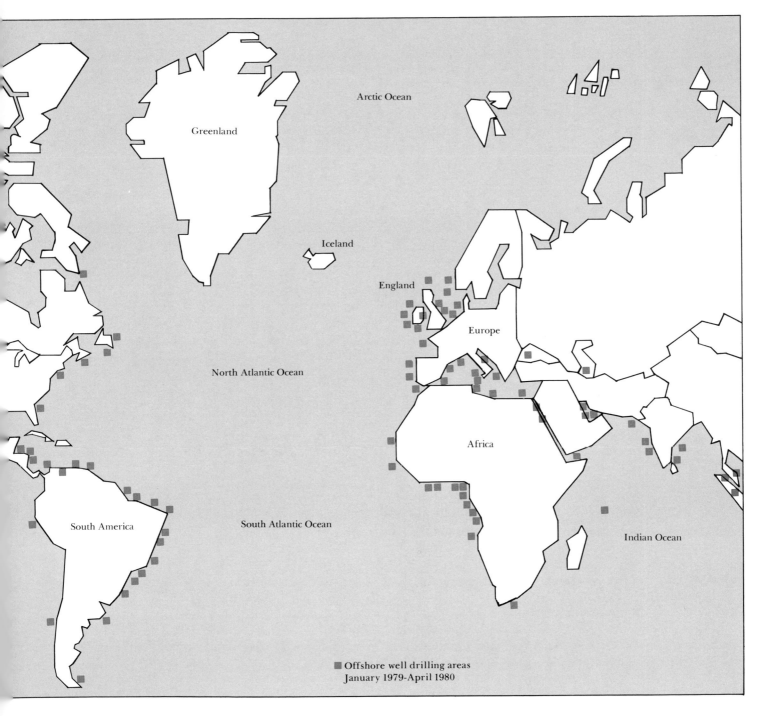

Arctic Ocean

Greenland

Iceland

England

Europe

North Atlantic Ocean

Africa

South America

South Atlantic Ocean

Indian Ocean

■ Offshore well drilling areas
January 1979-April 1980

the early years of this century. In each case, however, the drill rig was immobile and could be used only once.

Why not put the whole arrangement on a barge, ground it on site, and then, after completing the well, refloat the barge and move to another site? Thus, the idea of the submersible rig was born. It was developed by the Texas Company, in Louisiana, in 1934. The first well in open water was put down from a pier a mile offshore from Louisiana in 1937. Others soon followed. But for the most part, the plenitude of onshore production in those days made offshore development a curiosity rather than a practical prospect.

After World War II, the industry began looking to sea more seriously. Geophysical mapping of offshore areas became easier and more sophisticated. Geophysicists ran magnetic and seismic surveys from boats, just as they did on land, exploding small charges or air blasts against the surface of the water, then picking up reflections on hydrophones towed behind the survey vessel. Specially designed shipboard gravimeters and magnetometers helped them build accurate profiles of the sea bottom.

These surveys indicated the vast potential under the Gulf of Mexico, potential the industry was quick to test. The offshore era truly began in 1947, when the first well beyond sight of land was spudded 12 miles off Louisiana, but in water only 18 feet deep. It was (and is — the well still produces) a true offshore operation, built on a large platform supported on steel pilings 2 feet thick and 140 feet long, driven 104 feet into the sea floor.

On the top of the platform stood a 129-foot derrick capable of withstanding winds of more than 100 miles per hour. Typical of platforms in that era, it contained the derrick and draw works, while a 260-foot tender barge stored the mud pits, drilling water, fuel, and other supplies and provided living quarters for 30 men. Total cost of the platform: $230,000.

Offshore operations have changed enormously since then. Today, the *average* cost of a floating rig is $20 million, while the most expensive fixed platforms, such as those operating in the North Sea, sometimes run over $1 billion. One platform now operating in the Gulf of Mexico in water deeper than a thousand feet is nearly a quarter of a mile tall and weighs more than 40,000 tons.

For all their size and cost, such structures are not permanent. Once the reservoirs that they tap run dry, the platform becomes superfluous, not to mention a potential menace to navigation. The owner writes off the construction and operating costs over the limited life of the field, assuming it has one. If the reservoir proves disappointing, or nonexistent, the platform is written off as a dead loss, except, perhaps, to fishermen. Platforms, like wrecks, encourage prolific marine life and, occasionally, eventually become sea "farms."

Virtually all offshore activity to date has been in relatively shallow water, less than 500 feet deep, with a few ventures down to 1,000 feet or so. Oilmen are working on technology that permits much deeper drilling, since scientists estimate that half the likely hydrocarbon-bearing sediments offshore lie in water depths greater than 6,000 feet. Reaching into this frontier has top industry priority.

The difficulties of working in such conditions are formidable. Depending on the region of the world, offshore rigs have to withstand hurricane winds, ice floes, strong currents, wave action, and inhospitable bottom conditions. Submarine pressures are crushing at even a few hundred feet and reach almost 2,600 pounds per square inch at a depth of 6,000 feet. At the bottom, deep muds can swallow many feet of piling before solid footing is found. Rocky bottoms can make placement equally difficult.

Despite the variations in surface and subsurface conditions, any drilling unit requires some sort of stable platform from which to operate. Over the years, three distinct types have evolved: mobile platforms, drill ships, and stationary platforms. Each type has a number of variations.

Submersible barges were the earliest type of mobile marine unit, originally designed for the swamps and bayous of the Gulf Coast. These units were towed to the drilling site, ballasted, and settled on the bottom and, when drilling was complete, they were refloated and moved to the next site. As operations moved into deeper water, these rigs became more sophisticated, so that today's multihulled submersibles, which operate in waters

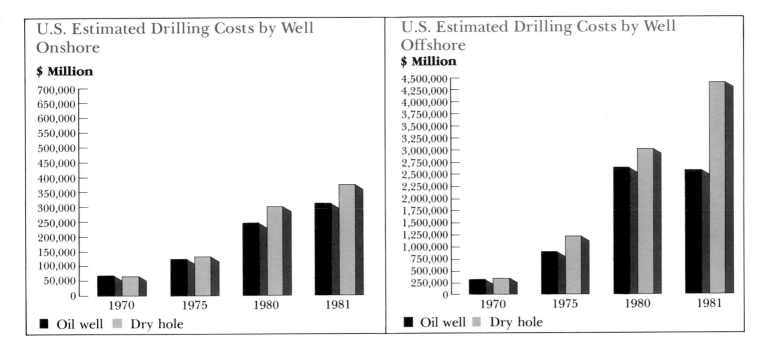

U.S. Estimated Drilling Costs by Well Onshore

$ Million

700,000
650,000
600,000
550,000
500,000
450,000
400,000
350,000
300,000
250,000
200,000
150,000
100,000
50,000
0

1970 1975 1980 1981

■ Oil well ■ Dry hole

U.S. Estimated Drilling Costs by Well Offshore

$ Million

4,500,000
4,250,000
4,000,000
3,750,000
3,500,000
3,250,000
3,000,000
2,750,000
2,500,000
2,250,000
2,000,000
1,750,000
1,500,000
1,250,000
1,000,000
750,000
500,000
250,000
0

1970 1975 1980 1981

■ Oil well ■ Dry hole

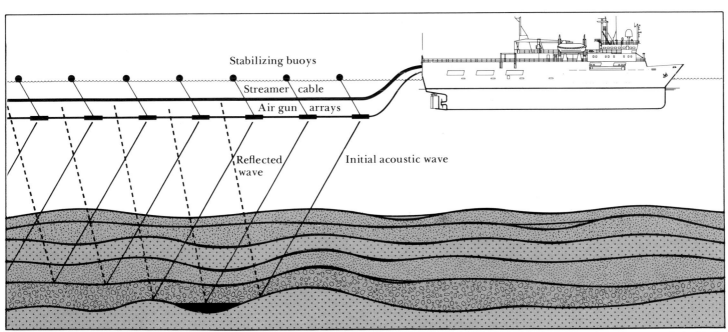

The ARCO Resolution *is one of the latest seismic vessels. Explorationists release acoustic charges from air guns trailing behind the vessel. When the energy waves echo back to the surface, a streamer cable picks up the signals and sends them to the* Resolution's *instrument room, where they are interpreted. Information gathered this way provides data for drawing contour maps of the geological formations beneath the sea bottom.*

up to 200 feet deep, bear scant resemblance to their primitive ancestors. These units have lower hulls which, when filled with water, sit on the bottom. There, through struts, it supports a second hull that remains on the surface as a working platform.

As the submersible rig has grown more versatile, it nevertheless has given way in most situations to semisubmersibles and jack-up rigs. As its name suggests, the semisubmersible gains stability by immersing itself partly below the surface of the water. This eliminates most wave motion and leaves a drill platform above the water. Several anchors are usually employed to hold the unit in place over the drill site. More sophisticated "semis" employ auxiliary thrusters to hold them stationary.

The first semisubmersible, *Blue Water No. 1*, started life as a full submersible and became a semisubmersible more or less by accident in 1962. Designed in the two-hull configuration, *Blue Water No. 1* was intended to operate in water up to 75 feet deep. After drilling was complete, the bottom hull was pumped dry and refloated, and the whole unit was towed to another site, the upper hull far above the water.

To save time, the rig's operators began moving her before she was fully floated. Someone noticed that even in a semisubmerged state, the rig retained almost as much stability as it did when it rested firmly on the bottom. After a few modifications, the concept of the semisubmersible rig evolved, and it was able to drill in waters 200 to 300 feet deep. Despite the fact that her new configuration permitted operations in most weather and wave conditions, Hurricane Hilda capsized and sank *Blue Water No. 1* in 1964.

Blue Water No. 1 has had many successors. Semisubmersibles do one-third of today's offshore drilling. The *Ocean Ranger*, the largest of these, was almost 400 feet long, 262 feet wide, and 337 feet tall. In 1979, it was towed on a 14,500-mile, four-month voyage from the Gulf of Alaska around the tip of South America and up to the East Coast of the United States for operations in the Baltimore Canyon area. The vessel was so stable that the crew could play billiards in the recreation room while outside the seas ran 30 to 40 feet. However, in 1982, while drilling exploratory wells off Newfoundland, it capsized

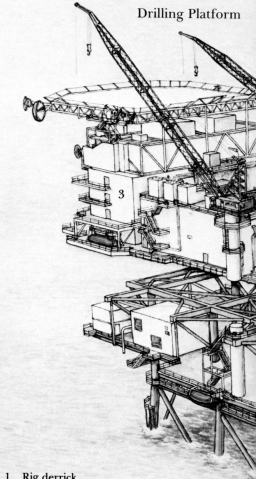

An Offshore Drilling And Production Platform

Drilling Platform

1   Rig derrick
2   Wellhead area
3   Accommodations section
4   Bridge from drilling platform to production pla
5   Processing equipment module
6   Utilities and water injection module
7   Switchgear and transformation module
8   Power generating module

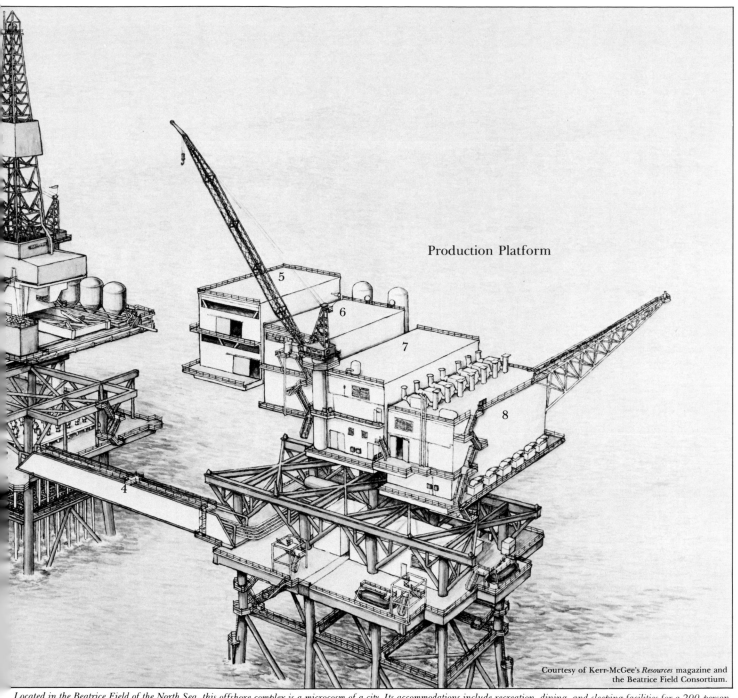

Production Platform

5

6

7

8

4

Courtesy of Kerr-McGee's *Resources* magazine and
the Beatrice Field Consortium.

Located in the Beatrice Field of the North Sea, this offshore complex is a microcosm of a city. Its accommodations include recreation, dining, and sleeping facilities for a 200-person crew. Its processing equipment includes oil, water, and gas separation facilities plus a produced water treating plant. Other facilities include a power generation package, seawater treating and injection facilities, plus pumps to ship the oil to shore. The platform contains a drilling rig that can be used to drill deviated wells from any of the 32 well slots.

in a winter storm, and eighty-four-men perished.

More than half of today's offshore operations employ jack-up platforms, large floating structures with long legs that can be extended to the ocean bottom. Once in place, the upper platform is jacked up above the water to the desired height. When drilling is complete, the process is reversed, the legs retracted, and the unit moved to the next site. Jack-ups operate in waters up to 300 feet deep.

Drill ships are the most versatile of the offshore units. They are self-propelled vessels of more or less standard ship hull design, except that they have holes (or wells) midship, called "moonpools," over which derricks stand. These ships are used primarily in areas too deep for other types of exploration drilling. They are expensive to build (more than $40 million each) and costly to operate (as much as $65,000 a day), but their mobility makes them a valuable tool in the oilman's kit.

These vessels have uses beyond oil exploration. The drill ship *Glomar Challenger*, for example, was used by the National Science Foundation to recover core samples from the ocean bottom at a depth of 23,000 feet.

How do these ships manage to hold position over drill sites at such depths? Surely no anchor chain more than four miles long would be practical or effective. Marine engineers solved the problem by building in dynamic positioning devices, similar to inertial guidance systems, that sense movement in any direction and use thrusters or propellers to compensate and push the ship back where it belongs. This method is so accurate that a ship can be held within ten feet of straight vertical from a well site several thousand feet below. If the ship is caught in a storm and forced to leave station during a drilling operation, marker buoys or wellhead transponders help it relocate the position and resume work.

Lateral stability is only half the story. Ships at sea also move vertically in response to wave and tide motion. Unless compensated for, such motion would have disastrous effects on drilling operations, alternately putting huge stress on the drill string as the weight of the ship came down on it, then pulling it away from the hole bottom when the ship was borne upward by the waves.

Several techniques are used to handle this problem.

153

The Glomar Atlantic is one of several self-propelled drill ships owned by the Global Marine Drilling Company and leased out to the oil companies for offshore drilling and exploration. It operates in water up to 2,000 feet deep, drilling through an open area in the ship called a moonpool located under the drilling rig (above). The ship is built to sustain long-term operations under demanding offshore conditions.

One is the motion-compensated crown block, a huge, complex, hydraulic-mechanical system that surrounds the normal crown block and allows it to ride up and down relative to the ship's movement. In this way, the block remains stationary relative to the drill string, which keeps steady pressure on the bit. Obviously, the block is far more costly than its landside equivalent — as is almost everything else associated with offshore drilling.

Another problem that confronts the drill ship operator is finding the bore hole once the ship leaves it. The task of locating and reentering the bore, which may be several thousand feet below and may measure only two feet across, is simplified by other devices.

Transponders and underwater television help the driller locate the general area of the bore, and an ingenious jet-assist device on the bottom of the drill string allows him to move it around freely as he probes for the opening. Then the bore hole is topped by an inverted cone about 16 feet across and fixed to the first piece of casing. The effect of the cone is like that of a funnel on the neck of a bottle. It gives the operator more latitude as he inserts his string.

Mobile platforms like drill ships and semisubmersibles are still the most effective way to conduct exploration and field development work offshore. Despite their great cost, the economics of mobile units still favor them over stationary platforms that cannot be recycled, or even more costly and technically difficult slant drilling from distant onshore sites.

Once a field has been brought in, however, the stationary platform becomes the most efficient way to produce. Stationary platforms fixed to the ocean bottom provide the most secure, stable conditions for working and living offshore. They enable large crews to live in comfort and operators to drill a number of directional wells that fan out over a large area within a given formation. Such platforms also allow for construction of oil- and gas-processing facilities on board, as well as limited storage facilities.

In a few years, some of today's offshore techniques

154

In November 1981, this 16-foot two-man submarine was brought to Santa Barbara, Calif., to map the sea bottom and pinpoint the location of a large natural-gas seep. Other methods, such as side-scanning sonar, were unsuccessful because the rising gas bubbles distorted the signals.

Emergency escape capsules like the one pictured here promote the safety of workers on offshore platforms.

effectively once he is submerged. Midget submarines manned by skilled technicians and scientists now extend the industry's reach even further.

Environmental and safety concerns continue to dominate debate about offshore petroleum development. Although pollution from offshore platforms is considerably less than that from pleasure boats, the infrequent accidents that do take place are spectacular and environmentally calamitous. So are the accidents that involve loss of life. Government and industry must approach these problems with a spirit of constructive cooperation.

As the last stages of the petroleum age approach, oilmen will press into more remote and hostile environments in pursuit of the energy to fuel modern industrial society. Offshore areas, including the most remote sections of the Arctic, represent the last major unexplored portion of the globe where substantial deposits of oil and gas may be found. Finding oil in these areas will be a supreme test of the vision and determination of the petroleum industry. Those who know the industry have no doubt that this will be accomplished.

will no doubt seem primitive. The industry already has the capacity to complete wells underwater, so that the only evidence of petroleum activity is a pipeline that emerges from the sea at some onshore point. As early as 1960, one operator set a complete production unit on an underwater well without direct human help. The unit produced for 17 years without a hitch, giving out only when its reservoir did. Since they are inobtrusive, these techniques will be used much more in offshore areas near heavily populated regions where community objections are strong on environmental grounds, despite their high costs.

No remote system can wholly substitute for men on the spot. Divers, for example, continue to be essential to drilling operations, particularly when trouble arises. Research and new technology have extended the reach of the modern underwater worker by allowing him to dive deeper and stay under longer. New electronic tools, including hand-held sonar, have helped him work more

CHAPTER 15 –
COMPLETING THE WELL

Long ago, completing a well merely meant striking a producing formation and letting the oil flow. In the shallow cable-drilled wells of Drake's day, there was little need for much more if there was pressure enough in the formation. The driller dug earth pits and allowed the oil to flow into them; later, he pumped the fluid into wooden barrels or tanks for sale.

In sandy, loamy, or crumbly soil, the driller usually set casing at the first show of hydrocarbons, then continued drilling into the producer formation. The well produced through the casing to the surface. But even by Drake's time, well-completion methods had become fairly sophisticated. As early as 1806, a salt driller had set casing to shield off surface strata, used packing in the well to control aquifers, and installed tubing to take production from the well without fear of contamination and to protect the bore from damage.

After a few blowouts, the early drillers started closing in their wellheads with oil savers and control heads which allowed the operator to divert the oil directly to tanks and pipelines. These early control assemblies were forerunners of today's massive blowout preventers (BOPs) and christmas trees. Intended for well and oil control, the early assemblies had about the same relationship to modern BOPs as kitchen faucets have to a fire hydrant.

Early production was also analogous to getting water from a well. Oil from open-hole completion, in which the driller set casing to the producing formation without cementing, either flowed naturally or was readily pumped. Completion entailed little more than stopping the drilling and providing a way to handle the oil at the top of the well.

If the oil did not flow, or flowed too little, the driller then opened up the formation by shooting the well with explosives. Explosives fracture formations to make them more porous and productive. They shoot holes in casing to allow selective testing and production. Explosives even shut off blowouts and put out fires. A huge charge of explosives detonated over a flaming well can extinguish the biggest fire, like blowing out a birthday candle.

At one time, drilling shallow wells and occasionally shooting them to stimulate production was enough. The oil flowed. But even in early days, drillers realized that proper handling at all stages would greatly improve both the productivity and the life of a well and save wells that might otherwise be lost.

A shallow bore in hard rock that traverses no aquifers and bottoms in a well-consolidated producing formation needs no casing or any other special care to protect it from water intrusion and make it a producer. If the formation is tight, a little nitroglycerine will loosen it up. But such ideal wells occur seldom, and then only in shallow strata.

The oilman's persistent bane has always been water, fresh or salt. Uncontrolled, water in an oil formation can ruin an otherwise productive well and even destroy an entire field. If the driller encounters water, he seals it off and prevents it from running down the bore to the hydrocarbon strata or even to barren clays and shales that

This fire occurred at a Shell oil well in Southern California on November 16, 1921. Prior to widespread use of blowout preventers, fires like these were a common danger at drilling sites.

This cementing crew at an Oklahoma well site in the late 1930s uses the sack cutter and jet mixer, two pieces of equipment developed by Erle P. Halliburton. In the 1950s, bulk transportation of cement speeded up the process considerably.

might swell and block the bore or stick the string.

By 1866, drillers were using casing and seed-bag packers to control water. The seeds in the bags expanded in water to form effective seals at the shallow depths and low pressures then common. When the driller encountered water, he filled the bottom of the bore with seed bags, ran casing through them, and continued drilling. When he reached hydrocarbons, he placed seed bags and casing before drilling into the producing formation. The packing sealed off the producing formation from water intrusion and ensured that the hydrocarbons went up the casing and did not bypass through the annulus.

Modern drillers generally use cement and casing to seal off formations. But today's deep wells often traverse several potential producing zones that the operator does not want to seal off permanently. In these instances, he uses special packers, like sophisticated rubber and metal donuts and plugs, above and below the formation. Such packers isolate the other producing zones until the driller wants to get at them at another time.

Well drilling was once a haphazard operation. Oilmen drilled until they hit oil, reached the limits of the rig, or ran out of money, whichever came first. There was time enough to worry about completing the well when the moment came. Now they plan and budget wells carefully. They have a target depth in mind, plus some idea of the intermediate formation and the bottom conditions. They plan the drilling and the completion and allow for possible variations.

When they reach the target depth, nine times out of ten they find nothing, or nothing useful, especially on wildcats. There may be ten million dollars and a year's work down the hole, but they put no more into it. The operator retrieves such equipment and tubing as he can — most of the expensive casing stays in the ground — and seals off the bore with cement plugs. It is all a dead loss or a costly lesson in applied geology.

If commercial quantities of hydrocarbons are found, the operator completes the well in a way that allows it to produce at its optimum rate for a maximum period of time, without damage to itself or the reservoir and with the least lifetime maintenance. Every such completion, even of adjacent wells in the same field, is unique.

158

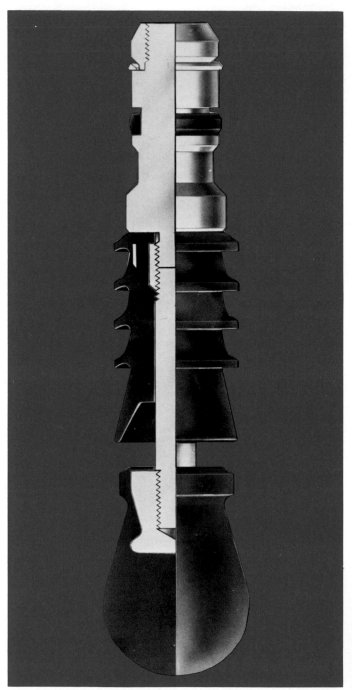

The five-wiper Omega latch-down cementing plug is one of several types used in cementing oil and gas wells. It follows the cement slurry and wipes the tubing clean. Other types precede the slurry and clear the pipe of the drilling mud.

The first decision is who will make the completion, the driller and his crew or a team of specialists brought in on contract. Generally, a completion calls for equipment lighter than drilling equipment. The operator may save money if he moves his large drilling rig to another location and starts a new well, while a contractor finishes up the one just drilled. Also, a complex multiple completion where the well will produce from several previously bypassed pay zones takes a lot more expertise than a relatively simple single completion.

The completion crew and engineers also need to know the downhole conditions — pressures, mud composition, type of formation, and potential flow rates. Tight, hard sandstone formations, for example, demand different approaches from those used in unconsolidated sands, limestones, and dolomites, each in the various types of geologic traps. The type of formation surrounding the trap also influences the manner of completion.

When a typical well has reached its design depth and struck hydrocarbons, it contains many strings of casing of progressively smaller diameters. Each of these strings has been cemented to seal off intervening strata and also may have packers where potential producing horizons have been temporarily bypassed. The well is full of drilling fluid, and its bottom-hole pressure is at least equal to the hydrostatic pressure for the depth and is possibly much greater. When the well bottoms, it is shut in to measure the pressure. That can be 10,000 or more psi at the wellhead.

Before the operator spends time and money completing a well, he must gain some idea of its potential. For example, he wants to know how fast the reservoir formation produces fluids. To obtain such information, he runs a drill-stem test.

For this test, the operator isolates a part of the hole containing an oil or gas deposit, then connects this part of the hole to the surface. The connecting tubing is the drill pipe itself. To the end of the drill stem, the driller attaches a drill-stem tester and a formation packer, which seals off the area of interest from the rest of the well bore. The tester is an assemblage of valves connected to a nipple that goes through the packer into the formation. The driller controls the valves from the surface.

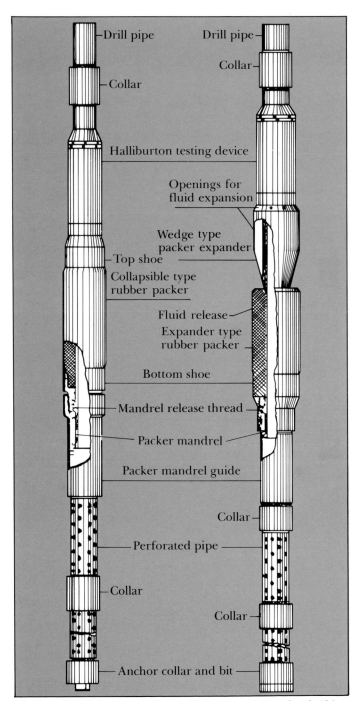

159

The Howco Full Hole Tester, introduced by Halliburton in 1926 and used widely throughout the 1930s, enabled the testing of a formation without setting a casing or bailing. Since then, the original design has been streamlined and improved.

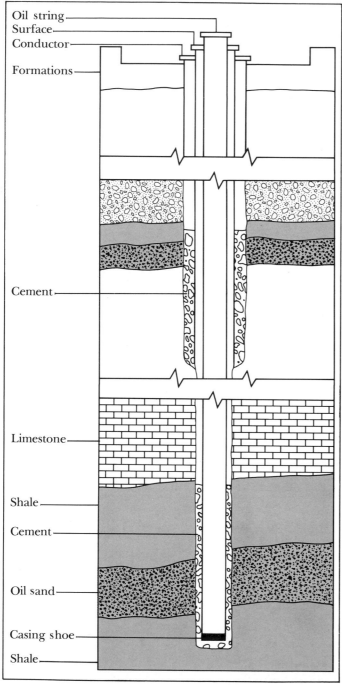

Oil string
Surface
Conductor
Formations

Cement

Limestone

Shale

Cement

Oil sand

Casing shoe

Shale

The casing protects the well from underground water and prevents loose shale and sand from falling into the hole. It also helps to control the flow of fluid from the well. The final casing for most wells is the oil string, illustrated here.

The whole assembly is lowered on the drill string, which has been emptied of drilling fluid. The packer seals off the formation, and the nipple enters it. The valves open, and the formation fluid rises to the surface, where the flow is measured through another valve or choke of known size, and the formation's potential is calculated.

When the well is ready for completion, the operator runs a final string of casing from the surface to the producing formation. This string is smaller in diameter than all previous strings and is the same size from top to bottom. If the producing formation is firm, as is a strong limestone or dolomite, the string may be run only to its top and cemented. The drill then bores into the formation itself. This is an open-hole, or barefoot, completion.

The barefoot completion, which differs little from that used in the earliest wells, has the advantage of being cheap and simple. It also allows greater flexibility in future well operations and makes the full well bore available to flow. But it has disadvantages, too. There is no way to regulate fluid flow into the well bore. It is hard to control gas or water production. It may require frequent cleanouts. And it needs more attention once completed. There are also many situations where open-hole completions cannot be used, such as unconsolidated formations and multiple-zone completions.

The operator has to support unconsolidated producing or surrounding formations. To do this, he drives the casing right through the formation, sets it below the producing horizon, and cements it in place. Then he perforates the casing to allow formation fluids to enter the well bore.

Perforating is accomplished with a special gun or a shaped-charge device. The gun, a circular container that fits inside the casing, has electrically fired cartridges that discharge small, high-velocity projectiles. These penetrate the casing, the cement around it and the formation itself. In this way, channels are opened up through which the fluids flow. The shaped-charge perforator creates high-temperature, explosive jets that punch and burn holes in the steel casing.

In loose formations, many wells produce more sand than oil, sand that can quickly chew up casing, tubing valves, and fittings before clogging the wells completely.

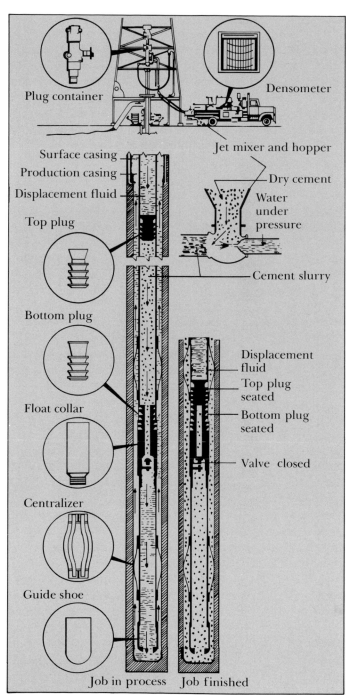

Plug container

Densometer

Surface casing

Production casing

Displacement fluid

Jet mixer and hopper

Dry cement

Water under pressure

Top plug

Cement slurry

Bottom plug

Displacement fluid

Top plug seated

Bottom plug seated

Float collar

Valve closed

Centralizer

Guide shoe

Job in process   Job finished

This diagram shows the step by step operation of a casing cementing job from the mixing of the dry cement to the point where the top plug reaches the bottom plug and the pumps are shut down and the slurry is allowed to harden.

The simplest sand control uses an open-hole completion. At the bottom of the casing, just above the formation, the driller places a packer and a hanger from which he suspends a slotted liner. This is actually a sieve that holds back the sand.

In very brittle formations, where the sand washes out despite the liner and threatens to collapse the overlying strata on the producing formation, the operator washes fine gravel into the cavity at the bottom of the hole, both to give support and to screen the sand. Various chemicals, usually resins, are also used to consolidate the grains of sand.

More complex, and also more common as wells are drilled ever deeper, is the multiple completion. In this case, a single well bore produces from several different horizons. This is not the same as a group of directionally drilled wells, where several wells slant off in different directions from the same drilling location.

Wells are completed with casing or tubing. A casing completion uses the final casing string as the production

161

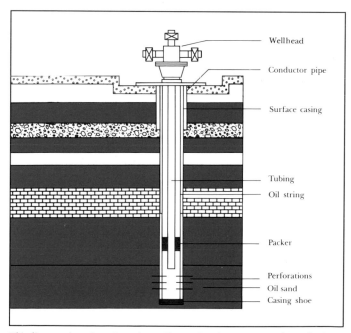

Wellhead

Conductor pipe

Surface casing

Tubing

Oil string

Packer

Perforations

Oil sand

Casing shoe

This diagram shows how the casing, tubing and packer are arranged in a flowing well. Tubing is the final string of pipe. In conjunction with the packer, it keeps well fluids away from the casing.

conduit to the surface. This one-way route — like a single straw between the reservoir and the surface — does not allow recirculation of fluids. A casing completion uses several successive parallel strings of casing within the bore, each one set into separate producing formations. Each casing string is carefully perforated so that the projectiles do not puncture neighboring strings.

In tubing completions, small-diameter production tubing is run inside the casing, from the surface to the bottom of the hole. The tubing hangs free but usually is set in a packer at the bottom. In either case, an annulus between the tubing and the casing allows recirculation of fluids such as corrosion inhibitors and solvents. This arrangement even allows the driller to run instruments into the well while it is flowing. Tubing is more common on multiple completions. Single, dual, triple, and more packers — the modifier refers to the number of holes in the packer through which tubing can pass — are set at the various producing zones. In effect, many different wells can produce quite different grades of hydrocarbons from widely separated horizons.

After the driller has placed all the downhole hardware and accomplished any necessary perforation, the well is essentially complete, although it may not as yet be ready to produce. Sometimes the pressure of drilling fluids still in the bore blocks the flow. Or the formation reacts with the drilling fluids and seals itself off at the face. Or a too-tight formation restricts proper flow regardless of pressure. In these situations, the driller cleans out the bore. He attaches a swab to a cable, pulls the bore fluids up, and dumps them into the slush pit. Often, swabbing and the powerful suction action that accompanies it are all the well needs to begin flowing. The driller may also use a bailer on a cable to bail out sand, mud, and residual fluids from the bottom of the hole. If these do not start the flow of oil, more heroic measures are called for.

In the old days, when these measures were needed, oilmen called in the nitro man. Now they call in the acidizer and the "frac" trucks.

Explosives and the petroleum industry have had a long relationship. The first reference to "shooting" a well occurred in 1860, when nitroglycerine was the fracturing agent of choice. Colonel E. A. Roberts patented a detailed shooting system in 1865. In 1866, he demonstrated the effectiveness of explosive fracturing when he shot a dry hole that promptly began producing 20 barrels a day. A year later, he shot the same well again and increased its output to 80 barrels a day. Torpedoes were soon exploding in wells all over Pennsylvania. Dramatic as the increases in flow on Roberts's wells were, they were minor compared to later developments in explosive and other fracturing methods.

Perhaps the nitro man was the first of the specialized oil-well-service people. His red wagon laden with flasks of liquid nitroglycerine was once a familiar sight as it trundled gingerly over oil-country roads. His was a lonely job. When he lowered his long metal torpedoes (each of which contained several hundred pounds of nitroglycerine) into the well bore, on-the-spot observers were scarce.

Nitro's reputation as dangerous was well earned. Many a nitro man disappeared with a flash and a bang. Many more lived to become oil-patch legends for their courage and resourcefulness. They did a necessary job, one that no one else would attempt, at a level of personal risk that outweighed any possible reward. Their service to the industry was immense.

Nitro men are rare now. They have been replaced by acidizers and frac trucks. Chemists have known for a long time that acid dissolves limestone and dolomite. Oilmen calculated that if they introduced acid into fissures in oil-bearing limestone rock the subsequent dissolution of the rock would increase greatly the formation's permeability and improve its yield. As early as the 1890s, patents were issued on methods using acids to stimulate well production. The first deliberate acid treatment occurred in 1932 and increased the well's production 20 times. Today, this technique is routine.

Hard, tight sandstones must be handled differently. Though porous, sandstone is sometimes impermeable because the grains are cemented together and the pores are closed off. The driller treats such formations with chemicals that dissolve the cementing compounds and increase permeability. If that fails, fracturing, or "fracking," is the last resort. Often this is done in conjunction with acid and chemical treatment.

At the turn of the century, torpedoes were delivered in wagons like this. By this time, torpedoes were filled with nitroglycerine, not gunpowder.

In 1864, Colonel E. A. Roberts applied for a patent based on his idea that the veins of oil-bearing rock could be opened up by explosion. His idea met rebuffs initially, but after its first successful test, torpedoes were used everywhere.

A cylinder about four feet long, the Roberts' torpedo was filled with gunpowder and lowered by wire into a well. An iron weight lowered on another wire detonated the powder. The explosion opened up fissures and drove out paraffin.

Modern fracturing uses pressured water, gels of various types, and other fluids that are compatible with the formation and the hydrocarbons. There is hardly a more dramatic sight at a well than a dozen or so huge frac trucks arrayed around the rig, each mounted with an immense superpressure pump, all roaring together as they strain to inject tons of fluid thousands of feet into the earth at nearly unimaginable pressures. The pressure they generate is great enough to lift and break up the producing formation for a wide area around the hole bottom. The increase in permeability and production of the formation is often enormous, provided the fractures stay open after the trucks leave.

When the fracturing pressure is removed, overburden pressure reasserts itself and sometimes reseals the formation as tightly as before. The contractor, therefore, pumps a propping agent down the hole with the fluid. This agent, usually sand or even glass beads, goes into the formation with the fluid and stays there. It props fractures open against formation pressure when the fracturing pressure is gone. These agents are often the key to a successful frac job.

164

After completion, the operator must still get the oil to the surface and then to the market. To do this, he first removes the blowout preventer and replaces it with the christmas tree. This is an assemblage of valves and piping arranged to manage each well's pressure, fluids, gas, maintenance needs, and method of production. It is tailored to the well by the petroleum engineers.

The christmas tree has a number of different-sized chokes, or "beans," through which the well produces a known quantity per day. A multiple completion well has additional valving and piping to handle the output from each producing horizon. The christmas tree controls the gas usually associated with oil production and includes stuffing boxes to admit simple maintenance tools and instruments to the producing bore, plus a pumping unit if the well needs one.

New wells in new fields usually produce by natural flow. Reservoir pressure forces the oil to the surface. Since nature provides this driving force at no extra charge, the wise operator tries to preserve natural flow as long as possible. The choke in the christmas tree limits the

This hydraulic fracturing job near Vernal, Utah, in 1978 shows the pump trucks connected to the manifold that carries a mixture of fluids to the well.

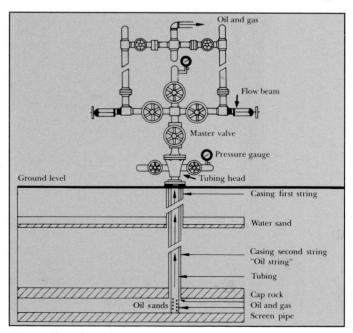

The assemblage of control valves, pressure gauges, and chokes at the top of a well is called the christmas tree. It controls the flow of oil and gas after the well has been drilled and completed.

The christmas tree, like this one at the Kuparuk River field on Alaska's North Slope, helps control the flow of oil to the pipeline, in this case, TAPS.

flow from these wells. They are seldom allowed to flow unrestricted.

When natural pressure inevitably drops, the operator helps the well produce. A gas lift, an extension of natural lift, often is used when large production rates must be handled. When the well stops flowing on its own, the operator pumps pressurized gas down the annulus around the tubing to its foot. The gas enters the well and flows back to the top, carrying the oil with it. Separators, aides in removing associated gas, take out the oil and recycle the gas. Gas-compression plants are costly, but, once installed, they are relatively inexpensive per unit of production.

Much more familiar to the casual observer is the mechanical lift of the bobbing horsehead. Throughout oil country all over the world, they are the most common sight. These apparitions are the power end of relatively simple subsurface pumps. Driven by gas, diesel, gasoline, or electric motors, the rocking beam operates a plunger pump at the bottom of the well. Screwed or inserted into the foot of the tubing, the pump is connected to the unit by thousands of feet of steel sucker rods.

Rod-operated plunger pumps have disadvantages. The string of rods is so stretched that the pump stroke is much less than the head travel. In deep wells, the pumping unit is very large to assure a relatively small stroke at the bottom of the well. Some rods inevitably break. While these units are relatively costly to run, their simplicity makes them the design choice in most places.

Today, most wells use a single pumping unit with its own motor. One early design featured a central engine that transmitted its motion to the individual wells over cables strung on poles, sometimes for hundreds of yards. At each well, the cable went over a pulley and down to the pump. As the central engine oscillated, the cables and their poles moved in precision.

Other techniques have been developed to drive pumps. One method, using circulating oil, drives them hydraulically. This system uses up in space and complexity what it gains in reliability. Another approach employs electrically driven downhole centrifugal pumps. In some instances, this method is the only way to make very crooked holes produce.

166

This pumper in the middle of a field of wildflowers near Bakersfield, Calif., is used when the natural pressure of the well has subsided.

To produce very heavy and viscous oil, the operator sometimes heats it by passing steam or hot water down the annulus. This, in turn, heats the production tubing. On occasion, more expensive measures are necessary, such as injection of steam into the producing formation itself. These methods are used more and more commonly today, as oilmen tap deposits once considered uneconomic because the oil was too heavy.

Wells seldom maintain production without problems. Sand clogs the bore. Wax collects in the tubing. Sucker rods wear and break as do other downhole parts. Corrosion renders parts, and even tubing and casing, useless. Productivity falls because of changes in permeability and pressure. Problems range from minor and irritating ones to major and very costly ones that require virtual recompletion of the well. Whatever the problem, its cure is the province of the well-service contractor. Applying the cure is called "workover."

In the simplest workover the operator goes back into the well with a cable rig to pull tubing, pumps, and other downhole equipment from the well. Then he inspects, repairs, and replaces the equipment and bails sand, wax, and fluid from the bore. Finally, he reruns the tubing, pump, and sucker rods; folds the rig; moves off; and resumes production.

Occasionally, the operator has to cope with water intrusion, which means setting new casing, perforating it, and squeezing cement into the offending aquifer to seal it securely. Or when the producing horizon becomes exhausted, he may wish to recomplete the well at a different horizon. Then he cements off the exhausted one and recompletes at a different, higher level. Or he may deepen the existing well. Because the distance traveled is often trivial compared to total depth, the operator uses cable tools rather than the rotary that drilled the original bore. With the lighter rig, he can get in and out much faster.

Gas deposits occasionally include exceedingly dangerous hydrogen sulfide gas. Drilling crews and local citizens have been killed when caught in a sudden cloud of the rotten-smelling gas as it erupted from a well.

Wells produce water, gas, and sand, as well as oil. So the field operator provides facilities to remove and mea-sure these substances before the oil leaves the field. Then he has to dispose of these in a desirable way.

Production from each well goes through a flow line to a common gathering station. There, separators remove the gas. The oil collects in tanks, where it is measured and the water decanted. Both gas and oil then go to larger treatment facilities, where they are further dehydrated. Condensates are removed from the gas, and the lethal hydrogen sulfide is reduced to a harmless state. The next step moves the oil into a tank battery, where it is stored and gauged for delivery to a field-gathering system for eventual transfer to a trunk line. Some gas may remain at the field for use as fuel and pressurizing agent, though for most of the gas the final destination is the refinery, a tanker, storage, or the customer.

Offshore and marine completions differ little from those onshore. Much of the work is done remotely and by divers. The difficulties and costs are much greater. Water and brine disposal offshore causes little problem as long as the waste is first cleaned thoroughly of oil. Frequently, the only safe disposal of waste water onshore is to drill another well and reinject it into an acceptable aquifer or use it for waterflood operations. In many instances, the cost of treating the water exceeds that of treating the oil.

CHAPTER 16 –
RESERVOIR ENGINEERING

Oil gushers seldom gush very long. Like punctured balloons, newly tapped deposits quickly release their overpressure. In a few days, or even in a few hours, the tremendous flow from a new well becomes a mere dribble or stops altogether.

In the early days of oil, gushers signaled success and brought great rejoicing. But perceptive oilmen soon realized that the most prolific gushers often gave way to penurious producers that yielded only a few barrels a day and then only with costly pumping. Oilmen soon learned an even harder lesson, that unrestricted production during a field's flush period sometimes damages producing formations to the extent that they no longer produce. Many an early field had to be abandoned before its time for lack of proper management following well completion.

Petroleum reservoirs are both unique and complex. Their fluids — oil, gas, and water — exist in an intimate relationship with one another. The character of the rock itself, the temperature and viscosity of the fluids, the pressure, and even the shape of the reservoir all profoundly influence fluid behavior, principally the manner in which fluids move through the reservoir rock toward a well bore.

The reservoir engineer works with the driller to design the best completion method for a well. He considers all the factors that exist in the untapped formation, as well as how they will change over the life of the field. He thinks in terms not of a single well but of all the wells, how they affect one another and how, taken together, they affect the life and performance of the particular reservoir. He sometimes decides that for economic reasons the reservoir is not worth developing.

Many of the scientific principles that underlie reservoir engineering, mainly geology and fluid physics, were developed in the 1930s. But just as oilmen disdained geologists when they thought anyone could drill a hole in the ground and find oil, so, too (until the years following World War II), the talents of reservoir engineers were not widely appreciated, and, until recently, they

In 1910, this drilling crew posed proudly next to a Long Beach, Calif. gusher. However, oilmen soon realized that gushers were not only wasteful, but were damaging to the long term potential of a reservoir.

were in somewhat greater supply than demand.

Early work in the field concentrated on determining the basic properties of reservoir rocks and fluids, the dynamics of fluid movement in a reservoir, and the manner in which this new knowledge could be used to determine the likely recovery from a formation. Reservoir engineers also provided valuable data that assured proper design of production systems, secondary and even tertiary recovery techniques, and the overall management of a reservoir.

Today, the reservoir engineer still works in the dark, as does everyone else in production. He uses inference, rather than direct knowledge, because his physical samples are limited to a very small portion of a field. He makes his decisions on information equivalent to a nationwide survey based on 10 to 15 interviews. He uses data derived from various logs. In his work, the computer has proved invaluable. The computer model of a reservoir has become his standard tool for managing that reservoir. Four primary driving forces in a petroleum reservoir are solution gas, gas cap, water drive, and gravity. Sometimes they exist together. Which one predominates may change over the life of the reservoir.

Solution gas is gas dissolved in the crude under reservoir pressure. There may be anywhere from 100 to 1,000 or more cubic feet of it per barrel of oil. When the reservoir is tapped and oil starts flowing, the gas comes out of solution and expands. Expansion pressure helps push the oil through the rock to the well. At the same time, gas bubbles begin to fill pores in the rock and, eventually, link up to make a path the gas can follow without pushing oil ahead of it. Oil production falls rapidly as the gas pressure dissipates.

A gas cap is a space over the oil filled with pressurized gas. Usually, a gas cap develops as the oil level decreases and gas comes out of solution. The expanding pressure of this gas also pushes the oil out of the well and is a more dependable driving force than solution gas alone.

Water drive originates below the oil. If there is a large aquifer beneath the oil deposit under the same pressure as the oil, the water expands slightly (as the oil flows out) and exerts a pressure on the crude, driving it

169

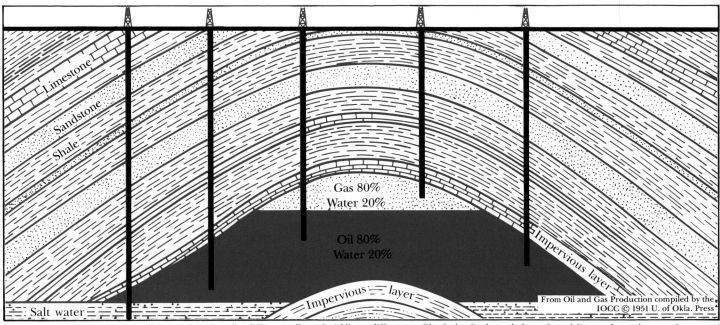

Limestone

Sandstone

Shale

Gas 80%
Water 20%

Oil 80%
Water 20%

Impervious layer

Impervious layer

Salt water

This geological cross section of an oil deposit shows how five different wells, each yielding a different profile of what lies beneath the surface, delineate the total extent of a reservoir.

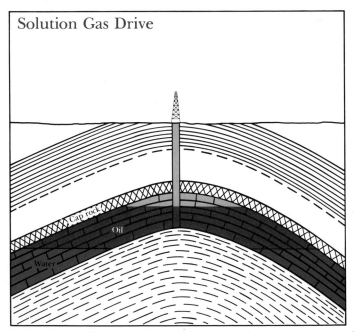

Solution Gas Drive

Cap rock

Oil

Water

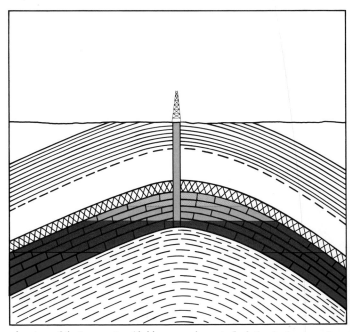

In cases where aquifer pressure is too low to drive the oil, oilmen regulate production by taking advantage of the pressure provided by gas coming out of solution in the reservoir. The diagram at left shows an early stage in this production. At right, the gas cap has enlarged and the casing has been extended to the oil level.

Gas Cap Drive

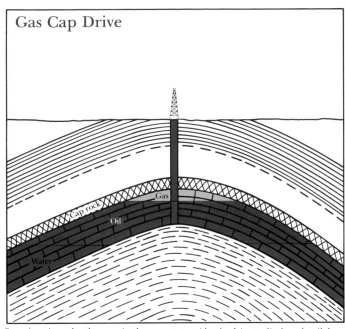

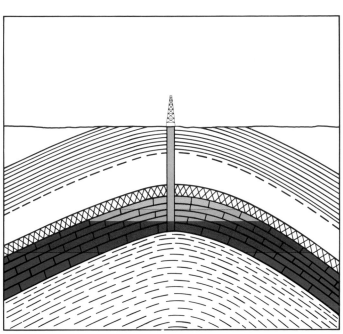

Sometimes in a closed reservoir, the gas cap provides the drive to displace the oil downward toward the producing well. Again, the production rate must be paced carefully. A well beginning production is diagrammed at left. After some time, the gas cap expands and the oil level is pulled down (right).

Water Drive

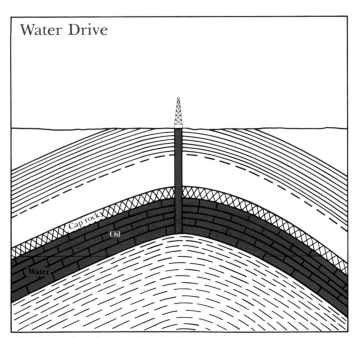

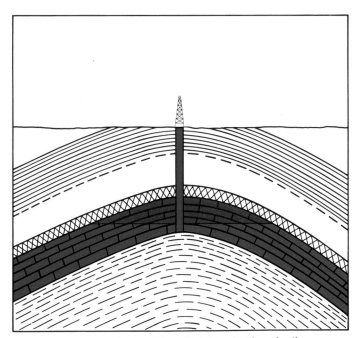

This diagram shows how water in a closed reservoir drives the oil up to the producing well. Eventually, the water reaches the foot of the well (right) and replaces the oil.

toward the well. Water drive is effective where the ratio of water to oil is very large or where a surface outcropping continually recharges the aquifer, bringing in new water to maintain the pressure.

Gravity drive depends on the differing densities of oil, gas, and water to cause the oil to flow naturally to the well. This force begins functioning as soon as the gas comes out of solution but is most noticeable after the reservoir has lost gas pressure. It is most effective in very thick formations and in those with a pronounced dip or slant as well as good permeability.

Other factors help. As pressure drops, the formation rock expands and adds its small driving force. The oil itself expands, as does the connate water. These changes hinder passage of the oil; both the expanding rock and connate water reduce effective permeability.

These various natural driving forces result in quite different recoveries. Solution gas drive normally brings up 15 to 30 percent of the original oil in place. Gas cap drive typically recovers between 30 and 40 percent. Water drive is more effective. It recovers 35 to 60 percent, with 50 percent the norm. Gravity drive, when it works, recovers 55 to 70 percent of the oil in place. Overall, however, despite very high recoveries from some reservoirs, only 30 percent or so of the oil in the ground is recovered by natural or primary methods. About 70 percent of all the oil discovered remains in the ground when the reservoir loses its natural drive, when oil no longer flows naturally to the well bore.

Whatever the driving force in a reservoir, the point of lowest pressure is the foot of the well bore. Fluids in the rock move toward that point, whether pushed by their natural pressure or drawn there by some artificial-lift method, such as a pump or gas lift. The fluid, of course, does not lie in a vast empty cavern, as in a storage tank. If it did, it would flow unimpeded toward the well and would drain all parts of the reservoir volume equally. In such a cavern, an operator would pump the oil out just as fast as he could pull it out — or as fast as pressure pushed it out — and drain it to the last drop, except for what clung to the walls. The National Oil Reserve, a solution-mined cavern in a Louisiana salt dome, is just such a reservoir.

However, if the cavern also contains water under the oil and pressured gas above it, the producer has to use some care. Because he wants all of the oil but none of the water, the well bore extends to the bottom of the oil layer but not into the water. He wants to control the flow of oil at a rate that gives maximum oil production without stirring up the oil-water interface. Also, the gas above the oil supplies the driving pressure. He does not want the oil to drain so fast that it creates a vortex around the foot of the well bore and the well begins producing gas with the oil.

Some reservoirs behave as though they were caverns filled only with oil, gas, and water. Among these are the reefs of Mexico's famed Golden Lane and many in the Middle East. These very porous, very permeable formations sustain very high production rates without reservoir damage. They produce at a volume virtually undiminished from discovery to extinction. Extinction comes only when the water layer reaches the well bore. Little recoverable oil remains in such a reservoir at the end of its life.

Most producing formations are more complex, however, as is the physics of fluid movement within them. The fluids find their way through tiny pores in the rock, and if the rock is water-wet or oil-wet, the ease of passage of one fluid or the other is affected. Viscous oil may resist flow and cling to the rock. It allows water and gas to bypass to a point far beyond the normal interface. Fingers of water or gas sometimes intrude into the oil phase so that the well produces water and gas in addition to or even instead of oil, even though there is plenty of oil in the reservoir.

The danger is not that oil production is too slow but that it stops long before it should. For example, in rock of low permeability, the well quickly cleans out the crude around its foot and then produces nothing until more crude oozes through the rock to fill the space. An operator tries to produce such a well at a rate no faster than it can sustain, no matter what the initial promise.

If the operator tries to run a well wide open, several things can happen. The well may refuse to be pushed, producing at its own rate. Low pressure created by excessively fast withdrawal in an unstable formation sometimes causes the strata to collapse. This can ruin the well permanently or, at the least, can necessitate costly rework. If the crude is heavy, less viscous water may move in to fill the space. If the oil-bearing formation is not water-compatible, the water can utterly ruin the producing strata and destroy not only the well but the entire reservoir.

Another outcome is that the well may begin producing only gas. Loss of the gas cap reduces the natural driving pressure and forces the operator to resort to artificial lift sooner than he had expected. Also, if the pressure is released too quickly, the gas dissolved in the crude may come out of solution, fizzing and frothing. Gas loss causes the reservoir to go flat.

Of the some 450 billion barrels of oil so far discovered in the United States, only 150 billion barrels have been considered recoverable. Of that, only 28 billion barrels remain. There are still 300 billion barrels which up until recently were too expensive to produce.

This oil stays in the ground because it no longer forms a cohesive mass in the formation. After primary

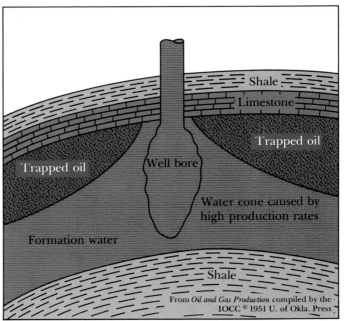

This diagram shows a result of producing a well at an excessive rate. The water finds channels toward the well opening, forcing the oil away from the bore.

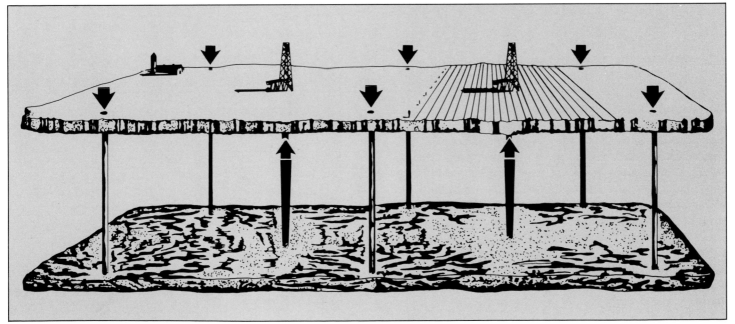

One way to enhance a reservoir by water flooding is to inject water at the corners of a square, at the center of which stands the producing well. The water, forced down into the producing sand, displaces the oil, forcing it upward toward the well opening.

recovery, the remaining oil exists as tiny droplets clinging to the grains of rock or isolated in the pores, surrounded by connate water or gas. What happens is something like pouring heavy oil from a bottle. Most of it goes out easily, but even washing with plain hot water does not remove the last traces.

Oilmen long ago realized that they could not recover all the oil at a reasonable cost. They also realized that if natural drive failed perhaps technology could provide what nature lacked. They reasoned that if natural water or gas pushed oil toward a well, then man-supplied water or gas would do the same. The waterflood secondary-recovery technique developed from this reasoning.

As early as 1865, oilmen in Pennsylvania discovered that water flooded into a depleted reservoir made the wells produce anew. Some evidence shows that the flood was accidental, but its effects were noticed. Deliberate experiments soon followed. However, because for years unwelcome water intrusion meant that a reservoir was nearly dead or stillborn, oilmen were understandably skeptical about schemes that deliberately introduced water into producing formations. Though the practice began on a limited scale in the 1860s, it was not widely used until after World War II.

Today, waterflood is so routine that it is often begun long before a field has exhausted its natural drive. Used this way, it is more an enhancement to primary recovery than a secondary-recovery method. When to start is an economic and engineering decision based on analysis of the reservoir's predicted performance and the costs.

An operator uses any fluid or gas to pressurize or push the oil. Unneeded gas is often repressurized and reinjected to maintain primary pressure. However, gas injection enhances primary production by only 5 to 10 percent, and gas itself is increasingly valuable, often too valuable to put back in the ground. In times when liquefied and pressurized petroleum fractions were surplus, they were often reinjected. But now they also are too valuable in the market. Steam or other heat sources are used for heavy oils. But 90 percent of all secondary recovery uses water. It is cheap, readily available, easy to handle, and effective.

The process is simple in principle. High-pressure

174

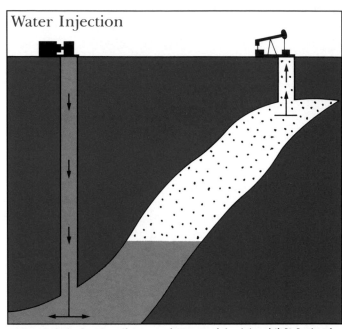

Water Injection

This method of enhancing oil recovery shows water being injected (left) forcing the oil from underneath up towards the well bore.

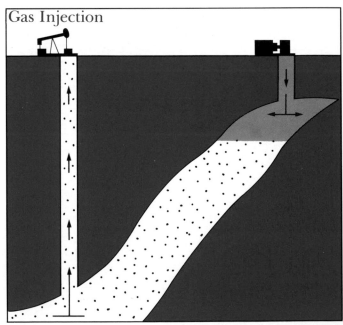

Gas Injection

Another method of enhanced oil recovery shows gas being injected (right) from the top downward on the column of oil, thus maintaining reservoir pressure.

pumps force water, which has been treated with suitable chemicals to make it compatible with the rocks, through injection wells. The injection wells, which are either newly drilled or converted production wells and, thus, just as costly as production wells, are strategically spaced over the field, alternating with production wells. The pumped water spreads out in the formation in a front or bank that pushes oil ahead of it. The fronts from several wells combine to wedge the oil out ahead. The entire reservoir then breaks up into minireservoirs, each with a producing well at its center.

Eventually, the water front reaches the production well, which begins producing water with the oil. In time, the front completely sweeps its area, and the well produces only water.

But there is still plenty of oil in the reservoir. A good primary production combined with a good waterflood results in a cumulative production of 50 to 60 percent of the oil in place. Until recently, it was not worth the expense to attempt to recover the remaining 40 to 50 per-

cent by so-called tertiary methods. Now it is becoming more worthwhile all the time.

The only other secondary method of consequence is steam injection. In some California reservoirs producing very heavy and viscous crudes, steam injection is the primary method and steamflood the secondary one.

In steam injection, also called "huff and puff," boilers at the surface burn oil to make steam — one barrel of oil for every three produced — and inject it into the well. After injection, the well is closed in and allowed to soak for a few days. The formation around the foot of the well heats up, and the thick oil thins out and flows readily. The well then produces at an acceptable commercial rate, often for months, before the cycle has to be repeated.

Eventually the field must be steam-flooded. This method is similar to waterflood, except that the injection wells introduce steam, which condenses to hot water in the formation. Incoming steam drives the hot water as a front that heats and displaces the oil. Steam flooding has recovered 50 to 70 percent of the oil in place.

175

Chemical Flooding

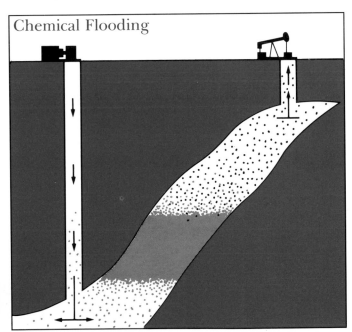

In chemical flooding, flow is enhanced by the injection of a detergent solution which contacts the trapped oil and interacts to release it. Polymer-thickened water injected behind the detergent then drives the oil toward the producing well.

Steam Injection

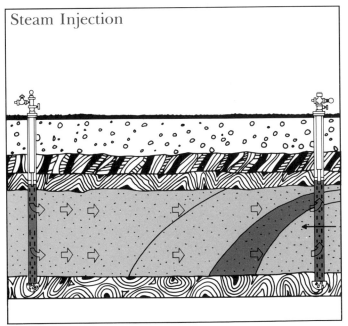

The injection of steam to enhance production has two effects: as the steam turns to hot water, it reduces the viscosity of heavy oil trapped in the sands. At the same time, it pushes the oil toward the well.

Fortunately, waterflood, gas injection, and even steam injection are relatively cheap ways to boost flagging production. Many a reservoir produces on secondary recovery for a much longer time than its short period of primary production. Since World War II, many abandoned fields have been brought back to useful production through such methods, their stripper wells averaging less than three barrels per day each. There are some independent companies whose primary business consists of buying exhausted properties and making them producers again through secondary recovery methods.

Beyond secondary recovery, there are ways of recovering the rest of the oil. If even a fraction of the 300 billion barrels now considered economically unrecoverable were recovered, the United States would gain considerable breathing space as well as independence from outside sources. Unfortunately, though the technology to strip reservoirs clean exists, it is too expensive to use. It costs more to produce a barrel through tertiary recovery or enhanced oil recovery (EOR) than the oil returns on the market. Still, as prices increase, some methods look marginally promising.

The three principal methods of enhanced oil recovery are as follows:

1. Chemical recovery, which uses surface-active agents (surfactants) and detergents to wash the oil from reservoir rock.

2. Miscible, which uses various solvents, liquid, and gas to flush the oil from the rock.

3. Thermal techniques, which are used to recover heavier crudes.

One of the oldest tertiary techniques using a surfactant resembles a laundering operation. Here, the surfactant, something very much like soap or detergent in its chemistry and mode of action, helps water wash oil from rock surfaces and out of pores. The first patents for surfactant treatment were issued more than 50 years ago, but the real work did not begin until the late 1950s. The technique is still in the research stage because the chemicals are costly. In any but the most favorable geologic locations, great quantities of detergent must be used to increase production significantly.

The technique is similar to a waterflood. Water in-

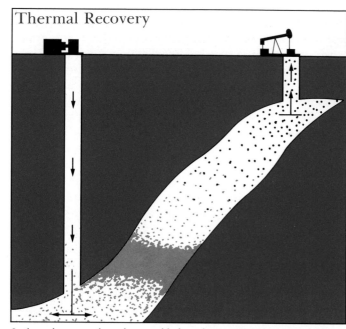

Thermal Recovery

In thermal recovery, the underground hydrocarbons are ignited, creating a fire front fed by air. This pushes the oil toward the producing well.

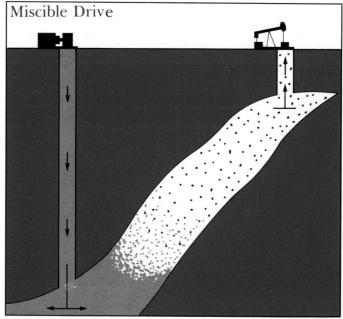

Miscible Drive

In miscible drive, LP-gas such as propane or carbon dioxide (CO_2) is injected as a slug. This mixes with the oil to free it. A driving fluid injected behind the CO_2 slug forces the build-up of displaced oil to the well.

jected into the reservoir contains a detergent or other chemical chosen for its compatibility with reservoir rocks and fluids. It forms a detergent front backed by a slug of water or brine thickened with a polymer. The slug helps keep the detergent front intact as it moves through the formation and flushes the smallest traces of oil from the rock.

Much can go wrong when this method is used. A cubic foot of rock may contain only a quart or so of oil, while the surface area of the rock's pores and passages may equal 20 to 30 football fields. The surfactant adsorbs on all that surface. Also, clay or other highly reactive minerals may leave very little of the agent free to wash out the oil. This situation is similar to hard water preventing soap from sudsing.

Problems such as these led to the development of a second method. To combat adsorption, a micellar type of surfactant is applied. Here, the molecules cluster in groups called "micelles," each of which holds a droplet of oil at its center. Aided by other chemicals in the solution, micelles are less likely to be adsorbed and lost to the process.

Miscible processes use a solvent to dissolve the oil in a fluid of lower viscosity, so that the fluid combination moves easily through the formation. Light hydrocarbons are obviously suitable, but their high cost rules them out. Interest now focuses on such gases as nitrogen and carbon dioxide. Nitrogen, of course, is almost as free as the air, and there are huge natural deposits of carbon dioxide in New Mexico and Colorado.

While neither nitrogen nor carbon dioxide is miscible with reservoir crude, each gas, when mixed with crude, extracts a range of hydrocarbon components that form a less viscous solvent front. This front travels through the formation from the carbon dioxide injection point to the production well, sweeping crude with it. Unfortunately, because the much lower viscosity of the solvent bank causes it to finger through the heavier crude, the method is not very efficient. The operator tries to counter by alternating water slugs with gas in the water-alternating-gas (WAG) approach.

Despite its drawbacks, the carbon dioxide solvent method appears to be such a promising recovery technique that ARCO has constructed a 400-mile pipeline to carry carbon dioxide from Colorado to West Texas at transport pressures of 2,000 pounds per square inch. There, oilmen are using about 20 trillion cubic feet of carbon dioxide to recover the three billion to five billion barrels of oil still in the ground, a giant oil field.

A third EOR method is thermal. Steam flooding in combination with chemical and solvent techniques are tertiary techniques as well as secondary ones. But a more radical and seemingly more simple approach is *in situ* combustion, creating heat in the reservoir by injecting air or oxygen and burning some of the crude in the ground. This technique is called fireflood.

Its concept is simple. The operator injects air, or air and water, and ignites the crude at the injection point. Pressure of the incoming air forces the flame front out from the injection well. Its heat travels ahead. The combination of heat, steam, and flue gases forms a slug that thins the oil and forces it to the production well.

Though simple in theory, this method is difficult in practice. Firefloods suffer from many operating problems, including sand production, corrosion, emulsions that reduce production, and control problems. Most operators dealing with heavy crudes prefer steam stimulation.

Naturally, the viability of all these processes turns as much on considerations of cost as on those of geology and technology, especially of geology. The reservoir rock, temperature, pressure, and composition of fluids must be considered in selecting the appropriate technique. Under some conditions, no technique is feasible.

Polymers, surfactants, detergents, and treating chemicals are expensive, costing as much as $20 per barrel of oil recovered. Since these chemicals are frequently derived from oil, the cost of treatment keeps pace with the price of the recovered crude. Despite the huge increase in the price of crude since 1973, it remains a race not worth running for many reservoirs.

CHAPTER 17 –
SUPPLIES AND SERVICE

The Industrial Revolution, nursed on coal and whale oil, evolved into the modern age of high technology on a diet of rock oil. Increasing cost and scarcity imposed limits on a coal-fired and whale oil-illuminated technology that only the discovery of easily produced petroleum could surmount.

The combination of oil and high technology has proved mutually beneficial. Modern industrial and highly technological economies grew up on oil; similarly, the petroleum industry thrives on the very technology it has helped make possible.

For example, modern helicopters hardly could exist without cheap, abundant petroleum fuels and lubricants; without helicopters, those fuels and lubricants would be neither so cheap nor so abundant. Few fruits of high technology have had as much impact on the petroleum industry as the helicopter. It is transport, skyhook, instrument platform, communications base, truck, and bus. It gets crews and equipment to places once accessible only by mule train, if at all, and gets them out again. Without

the helicopter, development of the vast North Sea oil fields would be barely under way now.

Sometimes oil-patch support technology uses some arcane discovery of little apparent value; more often, it adapts some prosaic or exotic technology developed for another purpose and makes it its own. Later, that same technology may move, altered, to yet another industry.

In other instances, engineers and scientists working directly for the petroleum industry or for one of its hundreds of support firms devise equipment and techniques for petroleum-related purposes that eventually find application elsewhere. The magnetometer is an example. This device spent more time looking for enemy submarines during World War II than it did seeking oil.

Perhaps one of petroleum's oldest supporting technologies is metallurgy. Drake relied heavily on metals of good quality for his engines, drill bits, pipe, and tubing. Of course, in his day, oilmen most likely used cast and forged iron rather than steel or the more unusual materials now common to drilling sites. Supplies and equip-

The exploration and development of the oil fields on the North Slope of Alaska depend heavily on the use of air transport. Helicopters bring in supplies and help build roads and camps. The helicopter in the foreground, for instance, is delivering loads of grass seed for a land reclamation program.

ment came from the local founder, ironmonger, or black-smith, who was more concerned with horseshoes than with oil-field equipment. These businessmen were quick to recognize a new opportunity. Advertisements offering their wares and services to the nascent petroleum industry, with appropriate testimonials, soon decorated barnsides during the early Pennsylvania oil boom.

As the petroleum industry grew, so did the metal industry. Oil-field metals had to be capable of operating under conditions of high stress, great pressure, and extreme temperatures and in infinitely varied abrasive and corrosive working environments. Moreover, because they are used in enormous quantities — a single 10,000-foot well requires 100 tons of drill pipe and leaves another 1,000 tons of casing in the ground when completed — the metal goods had to be relatively inexpensive.

At first, oilmen had to adapt equipment made originally for some other purpose and to use materials with less than ideal performance. Eventually, oilmen suggested a better design to the local blacksmith or requested a metal that would last longer. In time, the more astute smithies developed their local businesses into major corporations that still furnish equipment and supplies to the oil industry.

A vast and amorphous group of companies sprang up whose primary purposes are to supply and service the giant international petroleum industry. The firms range from small, locally owned service companies to very large corporations that rank well up in the *Fortune 500*.

Today, the oil industry's metallurgical standards are so high that other industries often turn to oil-field technology and equipment to do jobs that other technologies cannot handle. An intriguing example of this reverse adaptation occurred in San Antonio in 1968. A top house designed to contain a restaurant was being placed on a 605-foot-high tower built for a world's fair. Jacks at the top of the tower, acting through high-strength steel reinforcing rods, pulled the 650-ton top house into place. The brittle rods broke, nearly precipitating a disaster.

Pragmatic engineers at nearby Southwest Research Institute suggested using drill pipe instead of the reinforcing rods. After all, they reasoned, drill pipe supports its own weight in strings a couple of miles long, comes in easily handled 30-foot lengths, and would be much stiffer over the 500-foot lift than the steel rods. So, one length of two-and-seven-eighths-inch drill pipe replaced each two strings of rebar. The lift was successfully completed.

Another example of turning oil-field technology to an entirely different use concerns the *Glomar Explorer*. This huge ship was designed and built according to standards similar to those of a floating drill ship. The equipment on board closely resembled a huge drilling rig. However, the sole purpose of the ship was to locate and recover a sunken Russian atomic submarine, not to drill for oil. The U.S. government sponsored this successful operation.

Special metallurgical skills acquired in the development of drill bits, high-pressure blowout preventers, undersea equipment, and offshore platforms have proved particularly useful in the design and forging of jet-aircraft engine components and landing-gear assemblies.

The increasing importance of operations conducted offshore and in ever-more-remote and inhospitable regions has vastly increased the range, breadth, and depth of support needed by the oil industry. Today, there is hardly a manufacturer or service organization of any size, regardless of its primary market, that does not count among its substantial customers the petroleum industry. A recent petroleum-industry exhibition listed in its catalog more than 2,300 separate service companies under 60 major categories, including aircraft, bolts and fasteners, communications equipment, heating and air conditioning, hotel and food services, ships, boats, vessels, and air and water monitoring.

Offshore drillers require the support of many services, including supply and standby vessels, helicopters, diving services, weather forecasting, communications, ice surveillance, maintenance services, catering and supplies, underwater inspection services, buoy inspection and servicing, harbor support, and miscellaneous services that involve everything from preparing and delivering the Thanksgiving turkey to emergency medical help and entertainment. The last item is not a trivial one. Offshore and remote drilling and producing locations are prime markets for movies, projectors, television recorders and playback machines, and prerecorded TV tapes.

In the oil business, as elsewhere, necessity is the mother of invention. Consider downhole testing. If the driller goes through long barren stretches before he reaches a possibly productive zone, he can afford to set casing to test that zone. But if he works through many thin formations, setting casing for each formation is costly and time-consuming. The Johnston brothers encountered this problem while drilling near El Dorado, Arkansas, in the 1920s. Tired of setting expensive casing, they sought an alternate and cheaper way to make repeated downhole tests. They came up with a spring-controlled valve that was run above a packer.

Their first device was made of bits and pieces — a discarded boxcar spring, a poppet valve, and a packer made from leather belting. The result was a drill-stem tester that was run at the end of the drill pipe and set on a shoulder at the bottom of the hole. It worked, but it needed a valve to keep fluid out of the drill pipe while the tester was being run into the hole, as well as better ways to control the valves from the surface. The brothers tinkered and refined their makeshift design. Word of the tester spread. Before long, the Johnston brothers were out of the drilling business and into the oil-field support business. They continually upgraded their basically simple device to incorporate new technology. Originally developed as a way of determining if there were hydrocarbons in economic quantities in the hole, the tester now gives surface readouts of pressure, fluid type, flow rates, and other reservoir parameters. The company is now part of Schlumberger, itself a very large company that grew as the result of new technology applied to oil exploration, electrical analysis of the earth.

The diverse engineering and scientific disciplines required to develop hydrocarbon resources are reflected in the many professional and technical organizations routinely found at industry gatherings. The American Institute of Mining Engineers (AIME), one of the original five engineering societies, dates from 1871. AIME has always been the professional focus of the oil industry, a link to the days when petroleum was considered a mineral and its recovery a mining operation.

AIME later spawned more specialized societies under its unifying umbrella. These are the Society of Petroleum Engineers, the Metallurgical Society, the Society of Mining Engineers, and the Iron and Steel Society. The Society of Petroleum Engineers alone has 74 local chapters in 18 countries.

Less directly concerned with petroleum but equally important for their technical contributions are the American National Standards Institute, the Society of Naval Architects and Marine Engineers, the American Society of Mechanical Engineers, the American Institute of Chemical Engineers, the American Society of Civil Engineers, the Marine Technology Society, the Society of Exploration Geologists, the American Association of Petroleum Geologists, and the Institute of Electrical and Electronics Engineers. Their names are a catalog of the technologies required to find, produce, transport, process, and market hydrocarbons.

Many of these organizations are international in

181

The engine room controls on the supertanker ARCO California *monitor all the functions of the physical plant on the ship. The controls, in turn, can be monitored by a single member of the engine department.*

Severe Arctic conditions force oilmen to experiment with different methods of transporting supplies and personnel. For instance, they tried towing a hover barge by helicopter over broken ice in the Beaufort Sea, but it proved unfeasible.

182

scope, although their names are American. Petroleum technology is still largely American know-how and ingenuity. Nonetheless, these groups have counterparts worldwide that make their own contributions and serve to transfer technology among disciplines as well as nations.

An intriguing adaptation of technology developed for another purpose could be the industry's future use of the air-cushion-vehicle concept. The air-cushion vehicle, or hovercraft, flies a few feet or even inches above the surface on a cushion of pressured air. The surface can be land, swamp, water, or ice. Up to a critical height that depends on the size of the vehicle, the machine can traverse uneven terrain, rough seas, even sand dunes, as long as the obstacle height does not exceed the craft's hover height. Obviously, this principle is far different from that of conventional propellers or jet engines, which give forward motion.

At first, hovercraft vehicles were relatively small. They were used for carrying moderate loads and people across coastal lands and water. The larger hover barges have proved useful in offshore activities in Alaska where they are pulled by tug in summer. Experiments in pulling them by helicopter in the spring, when the ice breaks up, have proved unsuccessful so far.

Air-cushion vehicles carry heavy loads and are easily towed because they offer no water resistance. Some, rated for 160-ton loads, served as ferries on the ice-clogged Yukon River during construction of the Trans Alaska Pipeline System. They also were used for light core-drilling operations in tidewater areas. They can hover in places where boats cannot go. In this way, they eliminate the need to build a permanent platform for a short-term job.

With modular air-cushion barges, an operator can fly a complete drilling rig to a remote site over water or ice, and then anchor the unit in place much like a conventional drillship. Preliminary plans call for a vehicle weighing 6,000 tons, measuring 300 feet by 250 feet and hovering 6 feet above the surface on a pressure of only 1.5 pounds per square inch.

Offshore support technology can be seen most clearly at work. To the oilman, there is nothing strange

The need for underwater work at offshore oil operations has spurred the development of diving technology. This submersible system, shown at a Chevron platform, allows a diver to work at great depths without worrying about decompression problems later.

about one- and two-man submarines, robot submarines, sonar, underwater television, mixed-gas systems for deep dives, diving bells, radar, and solar-cell-powered satellite communications systems. The major difference between items produced for the petroleum industry and those produced for the military is that the oil company paints its equipment in bright colors rather than drab battleship gray. Oilmen truly ride in yellow submarines.

The submarines are not very big, but they are versatile. Although some weigh a little more than two tons and measure only 14 feet in length, they can dive deeper than 500 feet while allowing the operator to work in a shirt-sleeve environment, using robot manipulators, lights, and TV cameras for underwater exploration, inspection, and salvage operations.

Sometimes yellow submarines are used to supervise deliberate pipe freezing. From time to time, pipelines or process piping must be shut down for inspection or repair. Because the contents of these pipes are under pressure and the pipes themselves are large diameter, shutting them down also means closing down very large production facilities and draining and storing large amounts of material. Ever present is the danger that an inadvertent valve opening during shutdown could cause a catastrophe.

One British-developed solution freezes the pipe and its contents, thereby making an impermeable plug of water, brine, or the process material. External jackets filled with liquid nitrogen or other refrigerant are used to freeze the contents of pipes up to 30 inches in diameter into plugs many feet long. Upstream of the plug, the system continues to operate at full pressure; downstream, men work on the pipe in safety.

At a depth of 266 feet in the North Sea, contractors used a sea-ice plug in a 16-inch oil riser to shut off the riser while men worked on an in-line valve. The job was done in 24 hours at a considerable saving in time and reduction in hazard compared to conventional or underwater mechanical sealing methods.

Computers have become another major support technology for the petroleum industry. Oil companies were among the first to use computers for everything from planning and modeling to operation and control.

Today, computers operate pipelines and chemical plants, pilot aircraft, and log data, interpret them, and print them out in any desired form, including topographical maps and subsea contours. Computers design new types of drill bits, perform structural analysis of offshore platforms, plot ice movements as functions of weather and drilling-island shape, develop construction schedules, maintain personnel records and payrolls, and monitor the financial health of the companies. Some computer service companies add electronic-circuit design, aircraft design, geophysical research, and reservoir analysis to this list.

Computers also help oil companies rediscover old but overlooked fields by subjecting huge accumulations of geophysical records to new and better analyses. Without the computer's ability to handle millions of routine operations each second, this could never be done. Yet this task will take the largest computer years to finish.

These are a few examples from the vast reservoir of technology that the petroleum industry draws upon and

The LOOP system supervises and controls the offloading of crude oil from supertankers and the delivery to receiving pipelines. The state-of-the-art system monitors over 7,000 variables continuously.

Computers not only speed up the transmission of seismic data, they aid in their interpretation through color-enhanced seismic trace analysis as shown here in ARCO'S Plano, Texas research center.

contributes to. Since its inception, the petroleum industry, the rest of industry, and science and technology have been inextricably interdependent. They will remain so. As costs skyrocket and the industry pushes the search for oil to more and more inaccessible locales, higher levels of technology help restrain costs and, at the same time, find, produce, and transport the oil.

In the future, new instrumentation will give more information about downhole conditions. Better bits will help contain the cost of drilling by reducing the frequency of round trips of the drill string to change bits — no small matter in very deep offshore holes. One experimental design puts a number of bits down the hole, like bullets in a magazine, so that the operator can change the bit underground instead of pulling the entire string. Small, light, rugged and cheap programmable calculators and computers give closer control at the rig of mud properties, treatment rates, bit selection, and bit weight.

No technology is the exclusive property of any industrial sector. All technology results from the specific application of nature's laws, which are the same for everyone, to a particular problem. The laws of fluid dynamics that enable an airplane to fly are, at base, the same as those that allow a submarine to maneuver. The physics is the same, whether exploited by the military, the oil companies, or the industries that support them; such laws do not recognize nationality. Ingenuity and know-how, however, determine how well or poorly those laws are put to work to create the technology. Oilmen can be confident that the ingenuity and know-how that has given the petroleum industry its support technology will continue to do so.

CHAPTER 18 –
MARINE TRANSPORTATION

Each day, more than 60 million barrels of liquid petroleum are produced around the world, mostly in areas of sparse population or limited industrial development. The job of getting this oil from its point of production to where it is needed — industrial Europe, the United States, or Japan — has created a giant subindustry: petroleum transportation.

Today, two-thirds of all liquid petroleum moves by water, making the modern tanker, or crude carrier, indispensable to modern industrial society. Without a way to transport enormous volumes of petroleum over thousands of miles at low cost, the American way of life would be profoundly different.

Modern very large crude carriers (VLCCs) carrying 200,000 to 350,000 deadweight tons (dwt) and ultralarge crude carriers (ULCCs) transporting 500,000 dwt and up bear no resemblance to the earliest petroleum carriers, but the descriptive nomenclature, the designation of size in dwt, remains the same. This designation describes the vessel's carrying capacity in cargo, fuel, and stores. Since

1 ton of crude oil is about 6 barrels, a 200,000-dwt ship holds about 1.2 million barrels.

Sailing ships of the 1840s carried their liquid cargoes in wooden barrels stacked on deck or in the hold. Metal barrels eventually replaced the wooden ones, and later these gave way to built-in tanks. One legacy of those days remains: the American 42-gallon barrel is still the standard of measurement in the industry.

The first modern American tanker was the *Gluckauf*, a sail-assisted steamship with an iron hull, launched in 1886. It carried its cargo (about 3,000 dwt) in separate compartments within the hull, a configuration followed to this day. For its time, the *Gluckauf* was a large ship. Tankers did not become substantially larger until World War II, when demands of the Allied war effort pushed tanker sizes up to around 16,000 dwt.

Many heroic and tragic stories were written about tankers during World War II. German submarines made these slow-moving vessels their prime targets during the Battle of the North Atlantic, seeking to choke off fuel

Major Tanker Routes

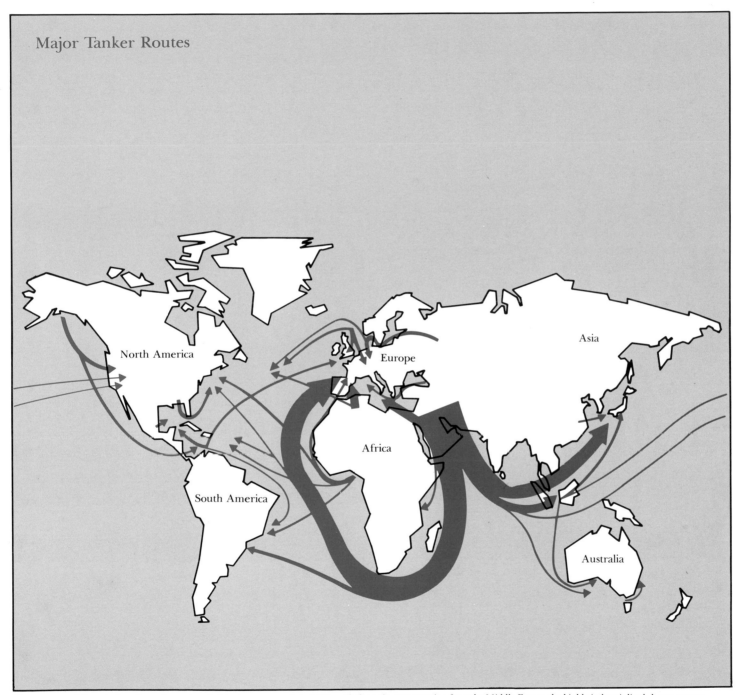

The arrows on this map show the major routes of oil tanker movements in 1982, with the main concentration from the Middle East to the highly industrialized, heavy energy-consuming areas of Europe and Japan.

supplies to the Allied war machines in Britain and the Soviet Union. Tankers laden with aviation fuel were especially vulnerable to explosion and fire when struck by torpedoes.

Even though it took more than half a century for tankers to double in size the first time, the next 25 years after that saw tonnages double, redouble, and double twice again. By 1956, 45,000-ton crude carriers were in service. In the mid-1950s, the first 100,000-tonners came down the waterways, followed shortly by ships in the 200,000- to 250,000-dwt class. In 1973, the 476,292-ton *Globtik Tokyo* was launched, the first of what has become the standard half-million-ton class. Even larger vessels are contemplated.

Two factors promoted this quantum jump in tonnage. First, the discovery of major oil deposits in the Persian Gulf greatly lengthened supply routes to Europe and North America over what they had been from the United States, which had been the world's leading petroleum exporter before World War II. Second, the surge in

worldwide petroleum demand following World War II meant that much greater volumes of oil would be moving by sea.

The new vessels of the 1960s and 1970s offered several advantages besides their much greater capacities. New hull designs and propulsion systems made them faster (up to 18 knots, compared to about 12 knots for older ships) and more fuel-efficient. New technologies in metallurgy and epoxy paints made them more durable, so they needed dry-docking less often. Though crews were slightly larger, the tonnage carried per man increased severalfold.

On the other hand, these huge ships had certain disadvantages. They could not negotiate the Suez Canal; they had to go the long way around the tip of Africa to reach Europe and America from the Persian Gulf. Even the shorter run to Japan, via the Strait of Malacca, was a tight squeeze for some of the deepest-draft vessels, which have been known to scrape their keels while transiting the shallow strait.

188

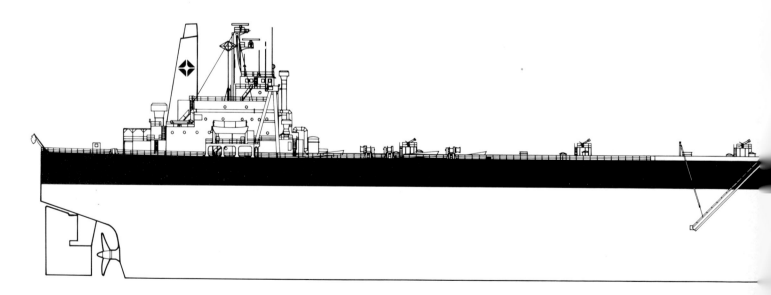

This drawing of the modern supertanker ARCO Spirit *compares her to the* Gluckhauf, *the first ship designed specifically to carry petroleum across the ocean. The steam-powered* Gluckhauf *began carrying oil from New York to Bremen in 1886. She was 300 feet long and carried 20,000 barrels of petroleum products. By contrast, the* ARCO Spirit *is 1,100 feet long and carries over 1.7 million barrels of crude on a run from Alaska to Panama.*

Another drawback was harbor access. Their deeper drafts and great beams required new ways and new facilities to handle the ships. Finally, their sheer bulk appeared threatening in the eyes of some. The very thought that one of these giant ships might break up and spill 100 million gallons of crude oil on some pristine strand was alarming.

The Suez Canal problem proved illusory; the added capacity and speed of the VLCCs and ULCCs more than made up for the longer trips. Docking and unloading were dealt with in various ways, but two proved most effective.

At times, the new supertankers were "lightered," that is, unloaded in batches by smaller tankers or barges that shuttled between the big ships and shore. Later, as the number of VLCCs and ULCCs increased, harbors and ports around the world were modified to permit supertanker docking. One such port, built at Bantry Bay, Ireland, by Gulf Oil, regularly handles many tons of petroleum each month.

This tanker offloads its shipment of crude at Cherry Point, Wash., after a three-and-a-half day passage from its loading point at Valdez, Alaska.

189

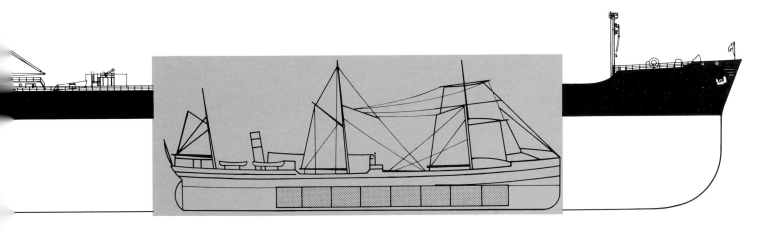

Today, about 65 ports around the world can handle vessels in the VLCC and ULCC class. These new ports, called "superports," are built with offshore mooring facilities where underwater pipelines transfer cargo from ship to ship. In the United States, however, only one such superport has gone past the talking stage. That one is offshore Louisiana in the Gulf of Mexico. In general, though, oil cargoes bound for U.S. ports must move in ships of 150,000 dwt or less, or they must be lightered to middle-sized tankers off the U.S. coast.

Opposition to superports in U.S. waters is largely an environmental concern, particularly on the East and West Coasts. The oil industry is concerned about the high costs involved as well. Indeed, the industry's own doubts about the economic value of offshore superports long delayed the establishment of the Gulf Coast superport and made it the nation's only facility of that kind.

Statistically, smaller crude carriers pose a potentially greater threat to the environment than large tankers. Smaller vessels need more trips to deliver the same amount of cargo, as well as more loading and off-loading, the most environmentally sensitive part of tanker operations. The additional trips mean more tanker traffic in and out of American ports. This increases danger to the environment.

With so much at risk in a single ship, safety has been a topic of great public interest and concern in recent years. Because owners are loathe to lose ships, cargoes, and crews, these vessels are designed and operated at the leading edge of marine technology to assure their long life and safe, clean operation. Inertial-guidance navigational aids, electronic control systems, automated power plants, and the standard and sophisticated array of communications and radar equipment allow companies to run these ships with smaller crews and with greater security than those of the very much smaller World War II tankers.

Nevertheless, the huge bulk of these ships makes them hard to manage in close quarters. They acquire tremendous momentum and take literally miles to stop at speeds of only fifteen knots or so. They are many times larger and heavier than the largest aircraft carriers. When they are full, they ignore the wind but fight the sea

190

The supertanker class of ships, which includes the Esso Malaysia *(above), is the largest in the world and demands a new and special kind of seamanship to operate.*

and the currents. But when they are empty, their great sides act like sails, so that they must be sailed as much as driven. A ship with a dead engine is almost unmanageable in heavy weather, even with the aid of tugs or support ships. These modern tankers require a new and different breed of ship handler on whom old windjammer experience is not wasted.

Ships in trouble have always foundered on lee shores in bad weather. But when those ships are carrying millions of barrels of crude oil, as were the *Torrey Canyon* in the English Channel in 1967 and the *Amoco Cadiz* off the French coast in 1978, the problem is not just another shipwreck. Spillage of tons of crude on beaches focuses public attention mightily and builds strong pressures for government action to prevent similar disasters in future.

Unfortunately, lack of international agreement on construction and safety standards has impeded reform. Many nations, like the United States, have strict and tightly policed regulations governing tanker construction and operations. But many other countries use a minimum of regulation as a lure to attract registry fees. Tankers registered in these countries fly "flags of convenience," or "flags of necessity." This means that their registry is essentially ceremonial. Ships carrying these flags rarely, if ever, call on their "home" ports, and almost never employ crews or carry cargoes from those countries, few of which have oil in any case.

Some tanker operators flying flags of necessity run their fleets with scrupulous regard for safety and the environment. They register their ships in those countries for tax reasons, not to escape responsibility for the safe and clean operation of their vessels. But other operators are less concerned. They run their fleets at the very edge of regulations and, sometimes, over the edge. These marginal shipowners concern the responsible shippers, among which are the oil companies that operate their own fleets.

Tankers of American registry (called "U.S. bottoms") must conform to numerous rules and standards that unquestionably increase operating costs but also make them safer and more reliable. Many devices contribute to the remarkable record of the U.S. tanker fleet. These include systems for keeping ballast water separate from crude

The Torrey Canyon *broke up in heavy seas off the coast of England March 23, 1967, spilling its cargo and covering 70 miles of the Cornish coast with crude oil.*

The March 16, 1978 wreck of the Amoco Cadiz, *the worst tanker spill on record, caused up to $290 million in environmental damage. It focused attention on the need for high safety standards in tanker construction and operation.*

191

compartments. Vapor deadeners, another device, minimize the danger of explosions or emissions from carrying tanks. Still another important safety measure, double-bottomed hulls, lessens the chance of ruptured oil tanks in case of collision. But controversy abounds, even about safety. For instance, some experts challenge the safety advantage of double bottoms. They believe that vapor leakage into the space between the hulls could be an explosion hazard in the event of collision.

The U.S. government's role in tanker operations goes well beyond safety and environmental concerns. The Jones Act, passed by Congress in 1920, imposes a number of restrictions on shipping in U.S. waters, for both native and foreign shippers. This legislation restricts trade between U.S. ports to vessels built in U.S. shipyards and crewed by U.S. citizens. The Jones Act is costly to U.S. fleet operators, since construction costs and pay scales in the United States are much higher than those costs abroad. On the other hand, the act insulates these operators from foreign competition. Thus, they can pass the higher costs on to their customers. The U.S. consumer ultimately pays the penalty in this energy bill.

Other federal legislation subsidizes operators who build their vessels in U.S. shipyards and put them into international service. This law has helped American shipbuilders, although the share of world tonnage built in U.S. yards has steadily declined nonetheless. In 1975, when more than 50 million tons of tanker capacity were launched worldwide, the U.S. share was less than 1 percent.

The economics of tanker operations are complicated not only by government regulation but also by the vagaries of the international petroleum trade. In the early 1970s, demand for oil and, hence, tanker capacity seemed insatiable. A worldwide shipbuilding spree ensued. But when petroleum demand fell sharply in the wake of the fourfold price increase of 1973-74, the industry suddenly found itself with huge overcapacity. In ports around the world, fleets of giant new ships lay idle, eating up money in capital and maintenance costs.

Charterers in the international market reported losses of up to $100,000 per month on 135,000-ton tankers. On the other hand, operators of U.S. flag vessels, protected by law from foreign competition in the domestic market, are earning as much as $400,000 per month with each 100,000-ton vessel.

Typically, it costs about $40,000 a day to maintain an idle VLCC. These are mainly capital costs, the burden of interest and depreciation on the investment. An operating tanker also has crew and fuel expenses, which have risen along with the cost of oil. Bunker fuel prices have risen tenfold since 1970, from less than $3 a barrel to over $30. Liability insurance is another major expense. Many nations, fearful of environmental and property damage, require operators to carry very high levels of liability coverage.

Builders, owners, and operators of large tankers have managed to insulate themselves from the worst economic effects through several financing devices. For instance, complex leasing and underwriting plans spread the risks connected with enormous construction costs. The simplest of these are equity investment arrangements, in which an owner or group of owners puts up its own money to build a ship. Another plan is called the beneficial ownership lease, or leaseback, where a company or individual in need of a ship arranges for others to put up the necessary capital and then signs a long-term lease that assures the investor-owners a suitable return. A leverage lease is one in which a third party lends most of the money to build the ship. He recovers his investment much as a lender is repaid on a home mortgage, while legal ownership remains with the borrower-builder.

Once built, a ship may go directly into service for its owner or lessor, or it may be chartered to a third party. For the most part, ownership of the world's tanker capacity, which is around 350 million dwt, is shared by major petroleum companies and large independent fleet owners. What these two groups do not own is controlled by smaller independents or national ownership.

Various chartering arrangements have developed that allow greater flexibility among owners, operators, and charterers. Chief among these are bare-boat charters, which allow for rental of vessels without crew, fuel, insurance, or other amenities. Another method is the spot charter, by which ships are rented, usually for a

Double Bottoms: Yes or No?

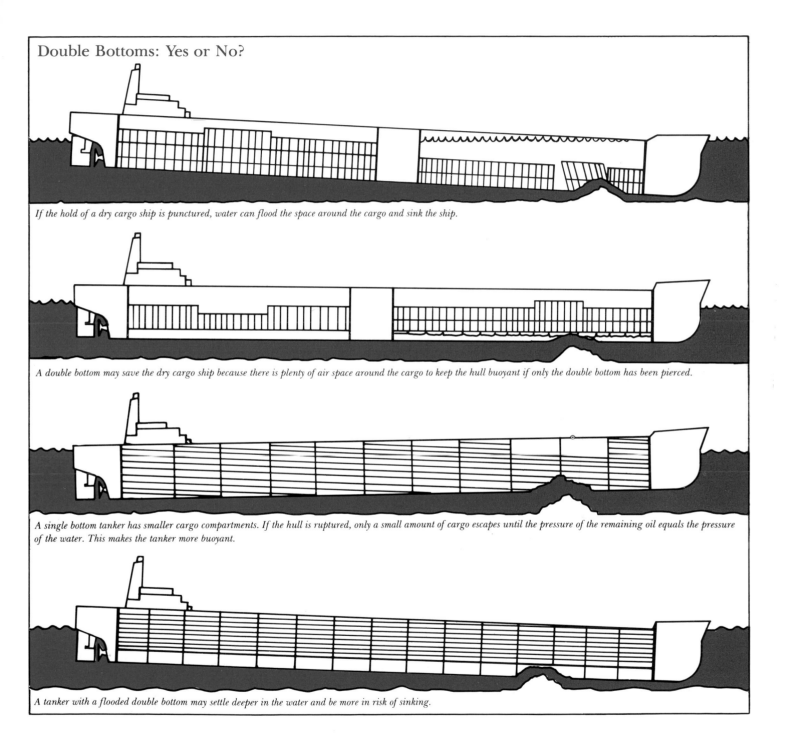

If the hold of a dry cargo ship is punctured, water can flood the space around the cargo and sink the ship.

A double bottom may save the dry cargo ship because there is plenty of air space around the cargo to keep the hull buoyant if only the double bottom has been pierced.

A single bottom tanker has smaller cargo compartments. If the hull is ruptured, only a small amount of cargo escapes until the pressure of the remaining oil equals the pressure of the water. This makes the tanker more buoyant.

A tanker with a flooded double bottom may settle deeper in the water and be more in risk of sinking.

Liquefied natural gas tankers like this one deliver world gas resources to market. Specially designed just to carry this product, these vessels are highly sophisticated. They transport the gas in liquid form under great pressure or at extremely low temperatures.

specific voyage, fully crewed, fueled, and insured. Term or time charters obtain when the owner supplies vessel and crew and bears all the fixed costs except the fuel expenses, for a certain period of time at a set price.

Much water transport of petroleum is performed by smaller vessels, including barges, which sometimes carry crude oil from a VLCC to a small refinery upriver or, at other times, carry gasoline or fuel oil from a refinery to a product terminal. The United States is fortunate in having a complex system of inland waterways that permits large petroleum cargoes to move into the interior of the country at relatively low cost.

The really big ships must transship their cargoes bound for the United States elsewhere, since U.S. ports cannot handle them. As a result, the Caribbean has become a giant way station for supertankers. Ships unload their cargoes at massive storage facilities at such islands as Curacao and Bonaire in the West Indies. From there, crude oil either is carried directly to U.S. ports or is refined first and then shipped.

As natural gas became an international commodity of increasing importance, different and extraordinarily costly ships were designed to transport it. These were the liquid petroleum gas (LPG) and the liquefied natural gas (LNG) carriers. Highly sophisticated vessels, they transport gas in liquid form under great pressure or at extremely low temperatures. They require special facilities to handle and store their cargoes. Such ports now exist in the United States at Baltimore and Boston. However, these ships may be an idea whose time has come and gone. As U.S. production of natural gas increases, the need for these tanker ports has declined.

Tomorrow's petroleum probably will be transported much as is today's, in large, efficient carriers plying essentially the same sea-lanes. An unexpected discovery of a major oil field in an out-of-the-way corner of the world could change this picture somewhat, but geologists are not optimistic about such a prospect. Ice-breaking tankers, in all probability, will be among the ships of the future. The specially strengthened but otherwise conventional tanker *Manhattan* demonstrated their feasibility as far back as 1968.

Tankers continue to be improved in both efficiency and safety. Satellite-controlled guidance systems, new communications systems, and other fruits of modern technology make ocean transportation better and safer in many ways. Despite the publicity attached to spills like that of the *Amoco Cadiz*, the total amount of petroleum lost into the world's waterways each year amounts to a small percentage of the total transported.

Of major concern to shipping operators today is the direction of future government policies. While most responsible shipowners have no direct obligation to better international standards of safety and environmental protection, a uniform code would enable U.S. vessels, now built under strict standards, to compete more effectively in world markets.

Politically motivated actions, such as closing U.S. harbors to foreign vessels or requiring a certain percentage of U.S. crewmen aboard ships operating in U.S. trade, seem dangerous and counterproductive. They could invite retaliation by foreign governments. Congress has resisted cargo preference laws in the past and should continue to do so to preserve a competitive and open market.

Finally, American regulators and lawmakers must consider the growing problem of moving petroleum through U.S. waters more efficiently. Regulatory delay, environmental confusion, and prolonged litigation cause huge overruns that cost the American taxpayer and consumer dearly.

CHAPTER 19 – LAND TRANSPORT

While ships move oil cheaply over great distances and in immense quantities, they cannot move material over land, nor are all oil-producing and using points within easy reach of waterways, natural or manmade.

The best and least expensive way to move any fluid continuously and in large amounts is through pipes. Civilizations have known this for millennia. In 1000 B.C., the Chinese used bamboo pipes to move natural gas to light their palaces and temples. The Greeks transported water over mountains, using closed pipes and the siphon principle. They even made pipes of ceramic materials that could withstand the tremendous flows and enormous pressures generated in the valleys between mountains. Earthquakes eventually wrecked these impressive public works. Their ruins are still visible throughout western Turkey, around Istanbul, and all over the old Roman Empire from Britain to Bessarabia.

These wholesale works of antiquity were complemented by retail networks that brought water to reservoirs and cisterns and distributed it to homes and institutions through pipes and troughs of wood, lead, and ceramic.

When oil became a commercial commodity with the discovery and exploitation of the Drake well, the new industry moved crude overland in unit quantities, just as producers moved other commodities. For 30 years, oil was taken to refineries or to market by horse-drawn wagon in wooden barrels. Frequently, the barrel became its own transportation vessel. Producers simply floated them singly or lashed onto rafts downriver and into dammed-up streams that served as receiving depots.

These primitive methods were not cheap. Transporting a 42-gallon barrel of oil from western Pennsylvania to New York City in 1865 cost $8.35, more than 50 percent of the sales price. Hauling a barrel only 20 miles in a horse-drawn wagon cost $3.00. Such inefficient and costly practices could not continue. Oilmen could not long ignore what was known about mass movement of fluids. They adapted this knowledge to their own use.

The first to act was Samuel Van Syckel, who in 1865 built a small wrought-iron pipeline to carry crude from Pithole City, Pennsylvania, to a railroad five miles away. The two-inch-diameter line moved 800 barrels per day for about $1 a barrel. While this was no great improvement over the teamster wagon, the vast increase in convenience and efficiency offset the temporary disadvantage and showed the way to the future.

Soon many more pipelines carried oil, but only from the field to the railroad, never to the refinery. The railroads controlled transportation outside the oil fields and handled the bulk of crude and products from the railhead to the refinery and to the final customer.

Competition in oil transportation intensified. There were turbulent times as teamsters and even railroads lost work to the pipelines. And there was retaliation, legal and extralegal. Railroads forbade pipelines to cross their rights-of-way, often forcing a line to stop short of a railroad, transship the oil by wagon, and re-enter the line on the other side. Railroads gained control of many pipe-

For centuries, the Chinese in the Szechwan Province used bamboo pipes to transport natural gas. With only minor modifications, descendents of the same people used the same technology in the 1940s, as pictured here.

lines clandestinely and suspended them, often helped by the inept business practices of the many small companies in the industry.

Stability of a sort came in the 1870s, when Standard Oil, by then the largest refiner, secretly arranged with the railroads to divide and allocate markets and to fix prices and rates. Standard Oil gained tremendously, while many smaller firms collapsed. Though Standard became the target of riots and sabotage, there was greater order in the marketplace.

Perhaps of more lasting significance, in 1874, the Pennsylvania legislature passed the first laws to give relief to the pipeline companies in their fight with the railroads. Pipeliners gained the right of eminent domain — they could use the sovereign power of the state to secure their rights-of-way — and the law compelled the railroads to offer and publish uniform rates for all customers. These changes allowed the companies to complete the first field-to-refinery-to-market pipeline system of 1874, the Columbia Conduit Pipeline, which served the Pittsburgh area. But the changes did not wholly end strife between railroads and pipelines; it continues in a limited way to this day in other areas and for other specialized products.

197

As Standard Oil continued to gather control of the industry in the 1880s and 1890s, competitors built several major pipelines. But by 1900, the Standard Oil Trust owned or controlled almost every line and had virtually 100 percent control of the petroleum-refinery, transportation, and export business in the United States. That lasted until 1904, when the Standard Oil Company's monopoly on the transportation of oil was challenged in state courts. The eventual outcome of the various court decisions allocated existing pipelines to new companies born from the dissolution of the trust in 1911. Huge new discoveries in the West spurred the construction of many lines to move crude to eastern refineries.

Pipelines are subject to many government rules and regulations because they cross state lines and transport flammable materials under pressure. Such regulation goes back to the Hepburn Act of 1906, in which Congress declared that pipelines are common carriers and must carry any shipper's material that is tendered to them for delivery. It placed them under the jurisdiction of the

Interstate Commerce Commission (ICC) and exempted only private lines that move material owned by the company to its own facilities. In 1909, the ICC acquired jurisdiction over pipeline safety, though in 1942 the commission ruled that the industry's own established safety standards were so good that separate government standards were not needed.

In 1967, the Department of Transportation (DOT) assumed responsibility for pipeline safety. In 1970, DOT published a formal safety code that closely follows industry codes. Pipelines, just like other sectors of the petroleum business, came under the stringent Environmental Protection Agency rules as well as those of the Occupational Safety and Health Administration (OSHA).

Nowhere is the hand of the regulator more evident than in those areas concerning pipeline tariffs and earnings. Pipelines pay their way by charging shippers a tariff per barrel moved. Until 1977, the ICC governed tariffs based on a 1940 ruling that limited a crude pipeline's earnings to 8 percent of the fair value of the pipeline company's property. This 8 percent effectively set an upper limit; it did not guarantee that rate of return, and companies could (and sometimes did) earn less. For product pipelines, the maximum return on fair value was set at 10 percent in 1941. For shipper-owner pipelines, where the shipper is also the owner of the line, the effective return was set at 7 percent in a 1941 consent decree that settled a Justice Department suit with shipper-owners.

Pipeline construction mushroomed during World War II. Older lines were rebuilt, and more than 10,000 miles of new lines were added in only two years. War brought not only expansion but also the new technology that made possible the famous 24-inch "Big Inch" crude pipeline and the 20-inch "Little Inch" product line. These systems proved the economy of building on a large scale, despite huge initial capital outlay, and laid the technical foundations for all subsequent large-diameter pipeline projects, including the world's largest, the 800-mile Trans Alaska Pipeline System (TAPS). Before the war, pipelines moved 250,000 barrels per day; by the end of the war capacity was 750,000 barrels per day.

From 1945 to today, pipeline mileage has increased

198

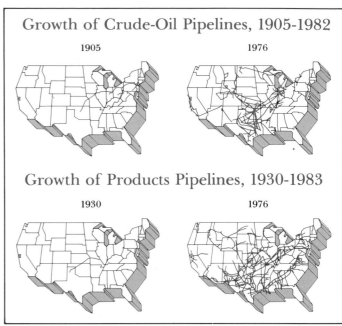

Growth of Crude-Oil Pipelines, 1905-1982
1905   1976

Growth of Products Pipelines, 1930-1983
1930   1976

Due to skyrocketing demand for crude since World War II, pipeline capacity in the United States has increased at a geometric rate.

30 percent, but since newer lines are mainly large diameter, the increase in oil moved exceeds 500 percent. In 1978, more than 230,000 miles of pipeline in the United States carried some 8 million barrels of crude and 10 million barrels of product every day. Gas pipeline totaled nearly a million miles.

Pipelines today transport many materials — crude petroleum, petroleum products, natural gas, liquid gases at ultralow temperatures, and even coal suspended in water. They serve all sectors of the oil industry and its customers — producer, refiner, and marketer — serving each sector through networks that differ more in degree than in kind.

First is the oil- and gas-gathering system, of which there are about 70,000 miles in the United States. These pipelines, usually no more than four inches in diameter, collect the oil (or gas) from individual wells or, more commonly, from small individual lease tanks (or gas separators) in the field and convey it to a central location. Sometimes a production field has parallel gathering sys-

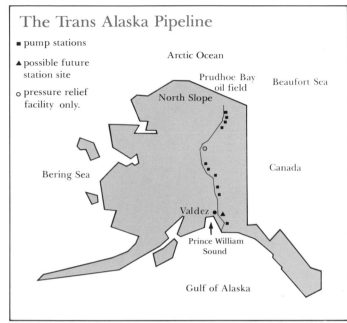

The Trans Alaska Pipeline

- ■ pump stations
- ▲ possible future station site
- ○ pressure relief facility only.

Arctic Ocean

Prudhoe Bay oil field

North Slope

Beaufort Sea

Bering Sea

Canada

Valdez

Prince William Sound

Gulf of Alaska

The 800-mile-long Trans Alaska Pipeline carries crude oil south from Prudhoe Bay on the North Slope to the ice-free Port Valdez on Prince William Sound.

tems to handle oil of differing qualities from adjacent wells.

Trunk lines are large-diameter pipes that move oil from the gathering system's central storage to major refining areas or to marine terminals onshore and offshore. Since the unit cost of moving materials through a pipeline decreases as the size of the pipe increases, the trend, particularly in trunk lines, has been to larger-diameter pipe. Trunks range up to 48 inches in diameter, twice that of the Big Inch, but have four times the capacity. The TAPS line can transport up to two million barrels per day with additional pumping capability.

At the output end of the trunk pipeline are the products distribution systems, which carry products from refineries to terminals, where they are stored until delivered to the final customers by tank trucks. Product pipelines handle 40 percent of all intermediate delivery of product from refineries; small ships and barges handle the remainder.

Pipelines cost a great deal of money to build and

This aerial photograph of a field gathering system in West Texas shows the tank batteries, which separate the gas, oil, and water, and the pipelines, through which the oil is pumped to a refinery.

operate. The Alaska Pipeline cost $9 billion to put into operation, although its $11.25 million per mile is not typical, even for a 48-inch line. Very difficult terrain and climatic conditions more than quadrupled the costs. The economic success of a pipeline venture depends at least as much on finance and management as on technology. Above all, a pipeline must have prior commitments from customers who will pay to ship material through the system in quantities sufficient for profitability. In the absence of advance shipper commitments, called "throughput agreements"or "guaranties," the line usually will not be built. Throughput agreements are the security and collateral that pipeline owners use to get the long-term financing necessary to build the line.

The structure of pipeline ownership is limited only by the ingenuity of the participating financial and legal professions. A pipeline can be a private venture built by a company that has no other interest in the petroleum industry. More commonly, it is a joint venture of several companies that also own the materials to be shipped or wish to use the material being transported. Most recent pipeline ventures have been undertaken by subsidiaries of oil companies that have large volumes of oil to transport. Of course, most interstate pipelines, private or joint-venture, are common carriers that by law must accept and ship other shippers' material at uniform, approved, and posted rates.

Joint ventures typically are organized in one of two ways — the undivided interest or the stock company. A prominent example of the undivided interest is the Trans Alaska line, in which each of the eight owners in effect has undivided sole possession of a percentage cross section of the line — as if the pipeline's interior were divided into pie-shaped wedges of differing sizes. Each owner has an undivided interest in one of those imaginary wedges, through which he schedules his own shipments, charges his own tariff, and looks after his own operating and financial arrangements.

The other joint-venture type is the stock company. Here, owners form a separate corporation in which each holds stock, charges a single uniform tariff, has one movement schedule, and has one financial entity. The owners, of course, may provide the throughput

200

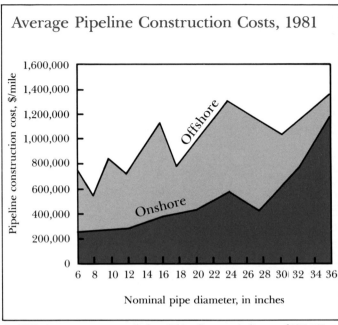

In 1972, the average cost per mile for a 36-in.-diameter pipeline was $276,163. By 1981, this cost had quadrupled to $1,144,937.

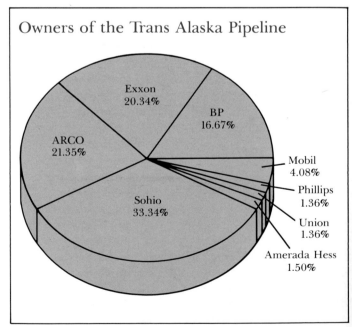

While Alyeska Pipeline Service Company holds responsibility for the design, construction, and operation of TAPS, it is a joint venture company composed of the eight member companies listed above.

agreements that enable the corporation to build the pipeline. And they receive the dividends, if the project earns any.

There are also 61,000 miles of private or single-owner pipelines in the United States. These are not open to any shipper but are built for the exclusive use of the owner, usually between two of his plants or terminals. Most such lines cover only short distances within a single state.

Unlike the early days of the Standard Oil Trust, the pipeline industry today is highly dispersed. It might even be considered fragmented; concentration has actually declined since the 1950s. In 1976, the largest interstate line carried only about 10 percent of the total barrel miles. The top four companies moved 34 percent, and the top eight, only 54 percent. Each of the eight had a highly diversified ownership.

Pipeline transportation is unique in the sense that the material moved provides the transport medium. Once in motion, the fluids push themselves through the line. A pump or compressor provides the impetus for the fluid, of course, but sometimes gravity or pressure moves the fluid just as it did in ancient times. For instance, gravity propels the oil at the downhill end of the Alaska pipeline.

To be able to operate, a pipeline must be completely filled. That takes a lot of oil and a long time. It took ten million barrels of oil and a month's time just to fill the 800-mile Alaska pipeline. This amounted to about 15 days of production from the Prudhoe Bay field, or full cargoes for twenty 62,000-ton tankers. At $35 per barrel, the inventory in the pipeline was worth $350 million.

Oil does not flow continuously from the gathering system into the trunk line. Rather, it travels in batches, or volumes of a minimum size (called a "tender") of 10,000 barrels, which the shipper or producer usually must accumulate over a number of days in suitable storage tanks. During the accumulation process, the producer strips out gases, water, and sediments and carefully measures the clear, gas-free oil for volume, temperature, sediment, and, most important, API gravity (density), in order to meet the standards set by the line. Once measured, the oil travels through the gathering system into the trunk line, usually by the batch, consigned to various shippers.

Alternatively, rather than physically measuring the oil in the gathering-system tanks, a producer or shipper uses the Lease Automatic Custody Transfer system. This is a means of automatically metering, measuring, and testing the oil as it goes from the well lease into the custody of the pipeline system.

Sometimes, a pipeline transports dissimilar batches of oil that belong to different owners and are not allowed to mix. Therefore, scheduling is critical, so that shipper and receiver know exactly what is going into the line and at what time. This enables the line to operate at optimum efficiency. Today, this task is considerably easier, thanks to computers and satellite communications, that monitor every movement.

If oils in adjacent batches are similar, they are run through the pipeline without separation, and the mixture where they meet, the interface, will be split between the shipments on delivery. In instances where two materials must not mix, the pipeline operator inserts a rubber ball or plug into the line to separate batches.

Such operations obviously require much record keeping, the major documents of which are the run ticket and the delivery ticket. The run ticket is the receipt and record for the lease operator, crude producer, crude shipper, and consignee, as well as for state and federal agencies. It shows the amount and grade of oil purchased and moved from the well to final destination. The delivery ticket does the same at the end of the pipeline movement. The two should match — and they do, to a remarkably high degree.

Once it begins flowing into the line, the oil is the responsibility of the pipeline operator, specifically that of the dispatcher, who from a single console watches and controls the flow over hundreds of miles. The dispatcher monitors pressure, gravity, and flow; starts and stops pumps; opens and closes valves; detects malfunctions; diverts flow; and shuts down the system in an orderly fashion. Microwaves, telephone land line, and satellite communications are his main aids. Ultrasonic flow meters on the Alaska line, for example, can detect a leak of less than 100 barrels a day or a 0.005 percent loss from the average daily flow. Such precise and continuous measurement assures not only that the consignee gets what the shipper has sent but also that prompt and effective control of events that threaten losses for the owners is maintained.

Product pipelines operate in much the same way as crude lines, with an important difference. While crudes may be intermingled to some extent, petroleum products have rigid quality specifications that the line operator must maintain. He must not mix gasoline and fuel oil, for example, as they travel through the same line. Two similar gasolines may move in the same batch with some mixing, and the same is true of similar kerosenes. But for sensitive discrete batches, plugs must be used. If direct interface is permitted, the operator normally diverts the flow at the destination in time to assure that none of the lower-grade product mixes with higher-grade material.

Planning is vital in the design of a pipeline. Once the line is built, its route cannot be changed. There must be assured supply and demand at appropriate locations. There must be a need for the line despite the presence of parallel or competitive transportation systems. And the regulatory environment must allow a rate of return sufficient to make the investment attractive. Many existing pipelines have been in operation for 50 to 60 years.

Not the least problem in building a pipeline is the right-of-way. Acquiring a narrow strip of land hundreds of miles long from hundreds or thousands of individual owners, not all of whom are eager to cooperate, is difficult at best. In most states, pipeline companies have the right of eminent domain. Thus, landowners cannot prevent the line from crossing their land, although agreement is not always easily reached. Eminent domain, however, does not affect Indian lands; Indian tribes have the absolute right to reject a pipeline route.

Today, before any major project is permitted to go forward, an Environmental Impact Statement must be filed and approved by regulatory authorities. This can take weeks, months, or even years. Environmental considerations were mainly responsible for delaying completion of the Alaska pipeline by four to five years and for contributing to a substantial increase in its cost because of inflation during the delay. And environmental regula-

202

At a Carson, Calif. crude farm, an employee checks the run ticket to make sure the company has received exactly what was ordered.

tions are only a few of the thousands of regulations the line must meet before it can begin operating.

The specific pipeline design depends on the type and quantity of material to be shipped. Pipe diameter, alloys, and wall thickness must be specified and ordered months before construction begins. So must pumps, valves, instruments, and controls. Terrain and flow rates determine the number of pumps required; if more capacity is needed later, additional pumps can be added.

Suitable tankage is needed for a variety of purposes. Some systems develop large tank farms to accommodate their shippers and their operational needs. The Alaska line even has its own refineries to make fuel for its turbine-driven pumps.

A pipeline-construction site, known as a "spread," is a dramatic scene of ordered, scheduled, and controlled activity. Large earth-moving machinery opens the ground. Automatic traveling rigs weld lengths of pipe into sections, wrap, as well as bend, and lift them into the trench. Automatic and manual devices inspect the welds,

Rusting equipment and pipes stacked up when construction of TAPS was delayed until the environmental impact of the line could be determined.

In Alaska, the pipeline is built both above and beneath the ground. Here, a worker adds insulation to the pipes to protect against Arctic temperatures.

using many techniques, including X rays. When welded together, individual lengths of pipe as stiff and unyielding as columns become as supple as a rubber hose, bending and curving to match the terrain.

Trenching and burial are the usual modes of construction, but portions of the Trans Alaska Pipeline were placed above ground on special heat-resistant supports to protect the delicate permafrost. Sometimes lines cross waterways on above-water spans. Sometimes they are sunk to the bed of the waterway, weighted to keep them in place.

Cathodic protection against corrosion is essential. A light electric charge passes through the pipe, which inhibits the chemical process of oxidization between pipe and surrounding soil. Without such protection, tiny natural electric currents eventually make pinholes in even the thickest, best-protected pipe.

Marine pipelines that move oil from offshore wells to onshore terminals present special problems. These lines, of which there are some 7,000 miles throughout the

Except in Alaska, most pipelines are buried. Pipe ends are measured carefully and cut so that ends can be aligned easily. Tie-in welds are made in the ditch.

world, now move more than 10 percent of U.S. crude oil. They are exceedingly expensive, costing up to $3 million a mile in such severe locations as the North Sea, because they must be built to withstand great external pressure from the weight of overlying water and must hold fast against the scouring and wrenching actions of ocean currents. Typically, these lines are laid by a barge into a water-scoured trench. Natural water movement then covers the pipe with soil, or special ships dump gravel, sand, and rock on top of the pipe.

When a pipeline is completed, it must be tested. Usually the line is filled with water and held at above-normal expected operating pressures for 24 hours. The water is injected into the line behind a pig, a spherical device equipped with sensors that measure curvature, dents, and other internal features of the pipe and then transmits the findings to the test crew.

Once the pipeline is in operation, maintenance becomes the primary concern of its operator. Corrosion must be watched, and pumps must be inspected and replaced from time to time. Pipes can be damaged by hunters' guns or farmers' plows.

Instruments help tell the operator when things go wrong, assisted by spotter aircraft with sophisticated remote sensing devices. Still, an inspector actually walking the line on a regular schedule is best for early detection of problems.

During the course of operation, deposits build up on the inside wall of the pipe. Some crudes leave substantial coatings of wax as they cool, reducing the pipe's diameter. Because North Slope crude is quite paraffinic, wax was a major prestartup concern on the Alaska line. Salts and other foreign materials also accumulate. Special pigs similar to those used for testing but equipped with scrapers and brushes are run through the line periodically to clean it and remove the deposits. Pigs enter and leave pipes through pig traps, or locks, that allow the line to continue to operate while it is under pressure.

From small beginnings, pipelines have become the leading factor in the transport of petroleum crude and products. From 1938 to 1976, tonnage moved per year has increased sevenfold, from 140 million to more than a billion tons. Despite the advent of special carriers for liq-

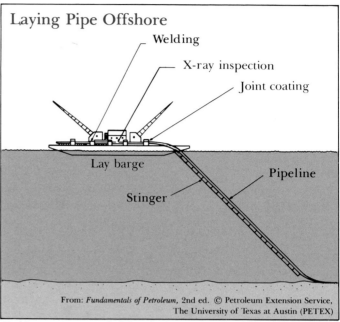

Laying Pipe Offshore

Welding

X-ray inspection

Joint coating

Lay barge

Pipeline

Stinger

From: *Fundamentals of Petroleum*, 2nd ed. © Petroleum Extension Service, The University of Texas at Austin (PETEX)

Used in swampy areas and offshore, lay barges are equipped for welding pipe joints on board and then laying pipe out over the stern as the vessel moves forward.

uid gases, pipeline systems are virtually the only mode of transportation from producing field to ultimate consumer, large or small, in the transportation of gas.

About the only form of fossil energy that does not depend heavily on pipelines for transportation is coal. But that may change. Pulverized coal suspended in a mudlike solution of water today moves through a pipeline to fuel a power station in the Four Corners area of the West. Some people estimate that 20 million tons of coal per year could move by slurry pipeline from the Powder River Basin of Wyoming without significantly affecting the water table in that area. Railroads still thrive on coal transport, however, and attempts to build competitive coal pipelines have already brought legal battles reminiscent of the early days of petroleum pipelines.

The pattern of future pipeline growth depends on the terms of the energy supply-demand equation. Some producing areas will not be able to fill their pipelines. Pipelines that serve these areas will have to be abandoned or turned to other uses. One example is the Four Corners

204

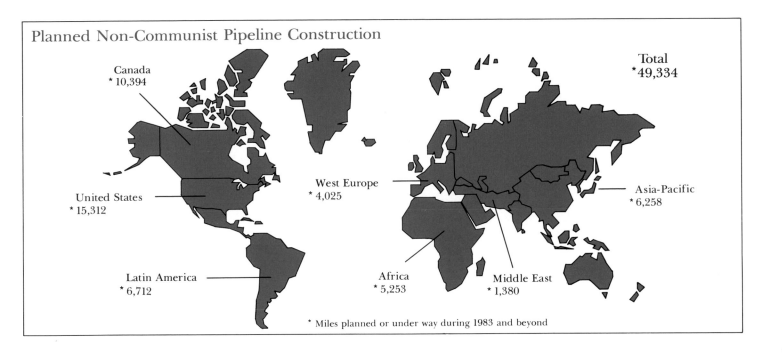

Planned Non-Communist Pipeline Construction

Canada
* 10,394

United States
* 15,312

Latin America
* 6,712

West Europe
* 4,025

Africa
* 5,253

Middle East
* 1,380

Asia-Pacific
* 6,258

Total
*49,334

* Miles planned or under way during 1983 and beyond

Pipeline, which once transported crude from New Mexico to California. It was recently reversed, bringing surplus oil from the West Coast to the midcontinent area. Such a reversal is a major engineering project in itself because pipelines are designed to accommodate flow in one direction only, with thicker pipe walls near the exits of a pump station, for example. Merely reversing the pumps does not do the job.

Nonetheless, over the next 20 years, the United States probably will not need more than one or two new large oil pipeline systems to match new demands with existing supplies. The trend, rather, will be toward more reversals and conversions, to bring oil from ports to the interior, and to move gas.

Whatever the material and regardless of the direction, the largely hidden pipeline systems in the United States will certainly continue to predominate in bringing energy to the nation's consumers. They function quietly and smoothly without clogging the surface channels of most commercial freight.

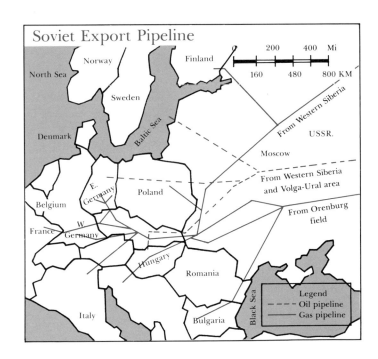

Soviet Export Pipeline

200 400 Mi

160 480 800 KM

Norway

Finland

North Sea

Sweden

Baltic Sea

From Western Siberia

USSR.

Denmark

Moscow

From Western Siberia
and Volga-Ural area

E. Germany

Poland

Belgium

From Orenburg
field

France

W. Germany

Hungary

Romania

Italy

Bulgaria

Black Sea

Legend
- - - - Oil pipeline
——— Gas pipeline

CHAPTER 20 –
COMMUNICATIONS

While communications technology is vital to petroleum operations, the industry began developing its own communications applications only in fairly recent times. From its earliest days, the industry has adapted communications techniques developed for other purposes.

From a communications standpoint, the oil industry obviously was born at the right time. When Drake drilled his well, Samuel Morse's telegraph — the first step away from centuries of communicating by carrier pigeon, smoke, and bonfire signals; flashing mirrors; post; and semaphore — was already more than 15 years old. Newspapers were quick to recognize the virtues of the "lightning wire." In 1848, even small communities from the Atlantic seaboard to the Mississippi received news of the Mexican War by telegraph. Railroads strung wire for signaling, dispatching, and traffic control. By 1856, Western Union came into existence, with other networks soon to follow.

Thus, the news of petroleum's momentous debut traveled rapidly. It set off a tidal wave of exploration throughout the United States and abroad. Similarly, many a fortune was made because advance word of a new strike or a refinery investment was acquired by men who had easy access to a telegraph office as well as a better-than-usual grasp of how to use this information to financial advantage.

Curiously, at the end of the Civil War, a breakdown in communications may have helped spur the development of American oil. Unaware of Lee's surrender at Appomattox, the Confederate raider *Shenandoah* continued to destroy the American whaling fleet in the Pacific for five months after the war. Her victims were no better informed. The U.S. whaling industry, which provided the high-grade illuminating oil that competed with petroleum during the infancy of the oil industry, never recovered.

Although the oil industry continues to borrow techniques developed for other uses, it has proved an innovator in the imaginative use of communications. Its purposes, as varied as the industry itself, range from re-

mote monitoring and control of untended production sites to use of video transmissions by satellite for face-to-face conferences of widely dispersed specialists and managers.

Information is essential to the petroleum industry. A weather report sent to a tanker captain 12 hours ahead of a storm is much more valuable than 3 hours' notice. Detailed geophysical data about a promising region help the company that receives them first. Efficient operation of a pipeline centers on instant-by-instant information about operating conditions at every point along its untended length.

The range of communications equipment the petroleum industry uses is hierarchical and interdependent. It starts with portable hand-held radios and rises through mobile radios to closed-circuit television transmission over radio channels, microwave, and satellite and land lines. The telephone and teleprinter and even the telegraph have not lost their value in the flood of new technology. Radio waves, after all, are open to intercep-

A meteorologist at the Kuparuk oil field launches a weather balloon to measure local conditions for pilots. The data also are transmitted to the National Weather Service where they help make up the worldwide forecasts.

tion and decoding by anyone with the right equipment, while land-wire communications are relatively secure from both prying and atmospherics.

The oil industry operates the world's largest transportation system. Its pipelines girdle the earth and contain thousands of pumping stations, most of them unmanned. Radio transmissions over hundreds and sometimes thousands of miles turn them on, shut them down, regulate, and monitor them remotely. The industry's computers communicate with one another by satellite and ask for human intervention only when something goes off limits. A communications web knits together fleets comprising 500,000-ton ultralarge crude carriers as well as tiny coasters and barges and their support infrastructure, such as tugs, tenders, and terminals. Schedules, routes, loadings, orders, company information, weather data, and messages from home all travel from shore station or ship to satellite, then to other ships or shore and back again.

Through the Marisat system, the marine satellite communications system operated by Comsat, ships at sea now have all the communications facilities of shore stations, including telephone, high-speed teleprinters and facsimile printers, well-log recorders, and even visual terminals for information transfer or television. Totally interactive computer terminals enable shipboard people to exchange engineering, operational, personnel, and financial data with shore-based facilities on a real-time basis. No ship or offshore drilling rig or platform so equipped, regardless of location, is less than seconds from information of any type originating anywhere in the world.

Navigation satellites make it possible for vessels to pinpoint within feet their position anywhere on the globe in any kind of weather. On shore, dedicated microwave systems carry company information from place to place and help monitor and operate pipeline systems running through remote territory. Secure telephone lines link computers and permit the instantaneous transfer of all sorts of information from any location to any other. Processing systems automatically gather data from production, transportation, and marketing sites. They perform the necessary calculations and produce at appropriate

Scientists at Kennedy Space Center encapsulate the Marisat spacecraft, launched in 1976 as the first satellite in a worldwide communication system that became a major navigational aid for tankers.

Offshore Early Warning System

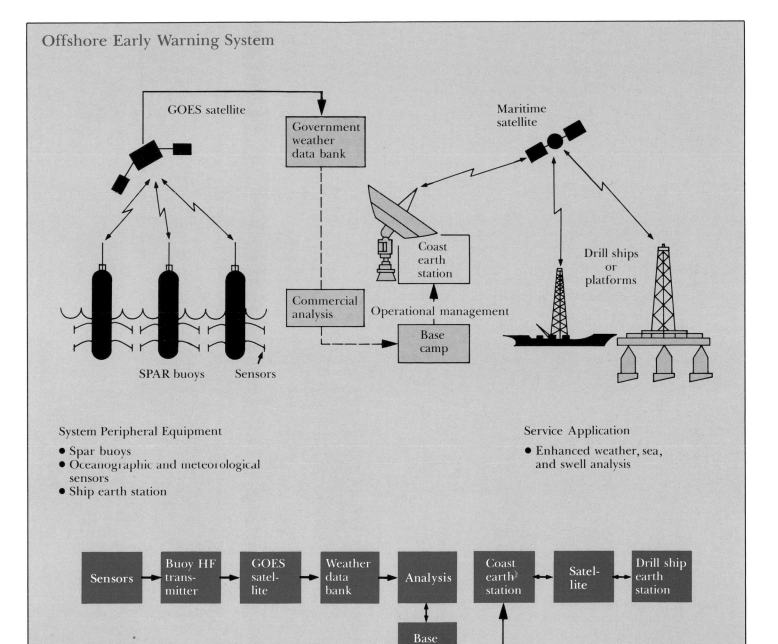

GOES satellite

Government weather data bank

Maritime satellite

Coast earth station

Commercial analysis

Operational management

Base camp

Drill ships or platforms

SPAR buoys   Sensors

System Peripheral Equipment

- Spar buoys
- Oceanographic and meteorological sensors
- Ship earth station

Service Application

- Enhanced weather, sea, and swell analysis

Sensors	Buoy HF trans-mitter	GOES satel-lite	Weather data bank	Analysis	Coast earth station	Satel-lite	Drill ship earth station

Base camp

In areas of extreme weather, drill ships and platforms need advance warning of threatening conditions. This system collects data from floating sensors at sea, transmits the information via satellite to an analysis center, and from there, via another satellite, to ships and rigs.

209

locations comprehensive and timely reports that company management needs to operate the business profitably.

Today, a major portion of all international communications, from telephone to high-speed data processing, and about 65 percent of all transoceanic communications go by satellite. Without satellites such as Marisat, ships sailing in Alaskan waters could be out of communication 22 hours out of every 24.

Meteorological satellites differ from communications and navigational vehicles only in that they carry optical sensing equipment — cameras — for photographing the earth. They transmit a constant stream of pictures to receiving stations that interpret the data. For as little as $1,000, drilling platforms, ships, and remote locations can be equipped with antennas, receivers, and photocopiers that help provide a real-time picture of the weather and other conditions of interest in their vicinity.

There are two types of weather satellites. One travels a polar orbit every 100 minutes at a 450-mile altitude and can distinguish an object as small as a tanker. They show storms and sea state as well as clouds and ocean currents. This last capability is especially useful to ships, which conserve fuel by running with the current. Knowing the ocean current is important to the control of oil spills as well.

The second type of satellite, the geostationary, orbits 22,300 miles above the equator at a velocity identical to the earth's rate of rotation. It remains in one place relative to the surface of the earth. The geostationary satellites currently in place cover the globe.

Since communications and navigational satellites are also geostationary, no point on the earth ever need be out of communication with any other. These satellites allow moving objects, such as ships and drilling rigs, to have a permanent point of reference from which to plot their positions. Receiving systems for these position-determining satellites are now so small that a worker can carry one and be informed of his precise location while in the field.

Modern communications technology made possible the Trans Alaska Pipeline System. This 48-inch pipe could not operate safely without assured, secure, contin-

210

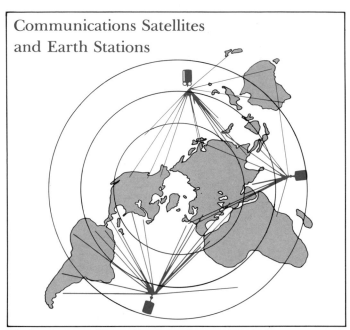

Communications Satellites and Earth Stations

The three Intelsat satellites positioned on station over the globe form communications links to every land mass on earth.

GOES-1 DPT 298 1645Z 25 OCT. 75

Meteorological satellites regularly monitor and transmit back to earth pictures of weather systems around the world.

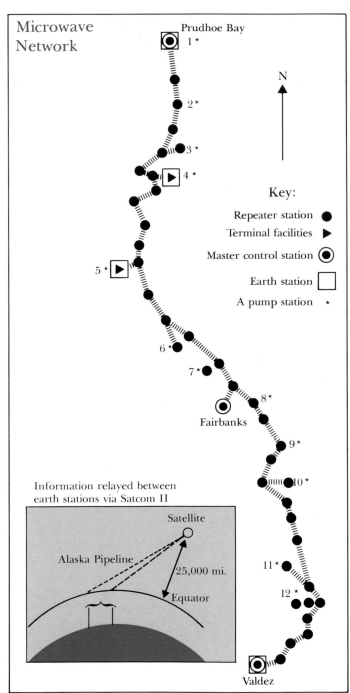

Microwave
Network

Prudhoe Bay
1 *

N

Key:

Repeater station ●

Terminal facilities ▶

Master control station ◉

Earth station ▢

A pump station *

2 *

3 *

4 *

5 *

6 *

7 *

8 *

Fairbanks

9 *

10 *

11 *

12 *

Valdez

Information relayed between
earth stations via Satcom II

Satellite

Alaska Pipeline

25,000 mi.

Equator

In Alaska, communications are transmitted both by microwave and by satellite on a route that parallels TAPS. This system enables operators to monitor and control the complex workings of the 800-mile pipeline.

uous, and instantaneous communication with the source at Prudhoe Bay, the terminal at Valdez, the pump stations between, and, perhaps more important, every valve, heat pipe, and potential trouble spot along the way. Most of these elements of the system are unattended.

TAPS employs three communications systems, satellite, microwave, and radio. The basic system is the satellite relay, which transmits operating data from relay towers at each pump station to the central computer at Valdez and returns instructions to the stations. The microwave system provides point-to-point backup and voice communication. The radio provides local contact between employees in the field and the pump stations and serves as a second backup. At Arctic latitudes, however, atmospheric interference is severe at normal radio frequencies, which limits both range and dependability.

Satellite communications have more uses than monitoring and controlling equipment or transmitting data. Satellites provide clear channels without the fading and interference that, until recently, were common in long-distance communications. These still plague conventional earthbound radio.

Conventional radio, of course, remains extremely useful over relatively short distances. It is the backbone of mobile radio, telephones, and walkie-talkies. With radio, a base station readily talks to people in the field on foot or in vehicles over an area of several hundred square miles. Such sets permit communications between tanker captains and crews handling ground tackle, hawsers, or loading lines a thousand feet away at the bow. Tug crews speak to their charges during docking and sailing maneuvers, and geological crews coordinate their seismic shots. Fire chiefs coordinate damage control. Salesmen learn of good prospects from the head office, and the head office summons executives to emergency meetings.

Satellites are already beginning to make such meetings unnecessary, at least in the sense of gathering all parties in a single location. Not only has the cost of travel and executive time escalated, but ever-faster-breaking problems multiply. On occasion, there is not enough time to bring together the people needed to solve problems before the problems get out of hand. Now communications technology can help.

211

Armed with highly sophisticated information gathered by the most modern exploratory tools, analyzed by the newest computers, and transmitted by the latest communications equipment, bidders gather at a lease sale where millions of dollars ride on the accurate interpretation of these data.

For example, in 1973 Atlantic Richfield began to build a closed-circuit television network that is unique in the petroleum industry. Completed in September, 1982 it enables its employees to confer "in person" between major operational centers in the United States without leaving their offices. Using satellite links to seven regional centers as far apart as Alaska, Texas, and Washington, D.C., the system allows as many as ten managers, geologists, tanker captains, economists, or whoever else is needed to talk with each other from different sites by means of a large 3- by 8-foot screen. In addition to providing face-to-face conversation, the system can transmit documents, computer-to-computer data, and electronic mail.

Atlantic Richfield plans to expand the system far beyond the initial seven sites. It already has a data and voice communications system that uses the same satellite as is used by the telecommunication system. This system links the company's 120 offices to each other as well as to exploration teams in the field through small portable computers. These connections enable seismic teams to send their data, previously recorded on magnetic tape in the field for later processing, directly to a central computer. The computer can process the data immediately and return the results to the crew.

Noncommunication sometimes is as important and as critical as communication. Any business wants those who should have information to get that information as quickly and accurately as possible. In some cases, it is important to withhold information from those who should not have it. The technology that eases information access makes security more difficult as espionage technology keeps pace.

The oilman prizes geological data. Whether he derives those data from preliminary surveys or as the result of test wells, he will lease more land, pay a higher price, abandon an area, or go into a new one on the basis of such data. Naturally, competitors would like nothing more than to have a peek at each other's data. Modern technology, in principle, makes those data readily available to anyone with the equipment to pluck them out of the air.

Since much of today's exploration occurs in remote areas where there may be no land lines, all communication is by wireless. Data from seismic shots go by radio to a van and then to a central computer. Computed data return by radio. Satellites pinpoint field crews, making it rather difficult for companies to conceal their areas of operation. Sudden increases in radio traffic can tip off a competitor that something big is afoot.

Corporations have code books just as complex and secret as those of governments, and the codes are changed just as often. Even so, there is no code generated that cannot be broken. Therefore, oilmen try to employ a new code so close to the event being kept secret that their competition lacks the time to break it.

A lease sale is a form of poker. Each player has his cards — the geological data that each has generated. Shaping up the hand is the job of the analysts and their computers at headquarters. They put the raw data into a form that tells the company how much to stake on a lease. Everyone in the bidding wants to know how much the others are willing to bid and on what information those bids are based. At such times, the people involved in the bidding arrange to have headquarters transmit the critical data to them at the last minute and in coded form.

Once the word is out, of course, there is no stopping others from stampeding in at whatever price they can join the game. In that sense, the oil business has not changed at all. In Drake's day, and for a long time afterward, they had to wait until a telegrapher translated the Morse code and a messenger brought the telegram. Today, the people who need to know have a communications terminal right at hand. It makes a difference of only minutes or hours, perhaps,but the message is just as crucial to the oilman today as it was to the pioneers of the nineteenth century.

CHAPTER 21 –
REFINING

214

While exploration and production symbolize the glamour and excitement of the petroleum industry, both are pointless if the industry does not convert crude oil into salable products. The revenues for everything the industry does, including exploration, come from the eventual sale of products produced by the refineries, not from the oil as it emerges from the ground.

As of this writing, in the United States alone, 47 separate companies own more than 200 refineries with a total capacity of almost 16 million barrels per day. These range from "teakettles" with a few thousand barrels' capacity to monsters that process half a million barrels a day. These facilities turn out products ranging from gasoline, kerosene, and jet fuel to solvents, lubricating oils, greases, waxes, petrochemicals, asphalt, and coke. All are derived from crude petroleum, although they were not always so derived nor will they always be. The world will depend on liquid fuels and feedstocks far into the foreseeable future, in fact, far beyond present estimates of remaining petroleum reserves.

When there is no more petroleum, refineries will still be needed. They will employ many of today's techniques to process oils from shale, coal, and tar sands. The major difference will be that the feedstocks will undergo additional steps, at additional costs, to produce many of the same products. Indeed, the industry will have come full circle, as plants that produced oil from coal were the precursors of petroleum refining.

Neither crude petroleum nor diamonds are very useful in the raw state. But just as the maximum value of a fine diamond is realized after it is cut and shaped, the maximum value of a barrel of crude petroleum is attained after it is refined. Unlike diamonds, petroleum is nearly infinite in the variety and quantity of things it can become.

The refiner's task is made immeasurably easier by the fact that petroleum consists of virtually only hydrocarbons — molecules composed solely of atoms of carbon and hydrogen — and not also of carbohydrates, like sugars, starches, and cellulose. The relative simplicity of

Early refineries like Monitor Refinery along Oil Creek boiled only small batches of crude oil over open fires. They produced only 15 to 300 barrels a day using a closed cast iron kettle and copper tubing as a condenser, but they wasted half the crude they processed.

215

In sharp contrast, modern refineries like this one in Philadelphia process 100,000 or more barrels of crude each day. They are an intricate network of storage tanks, pipes, pumps, furnaces, reactors and other equipment, manufacturing hundreds of different products, all prepared uniformly according to exact specifications.

the petroleum mixture allows the refiner to separate one group of hydrocarbons from another by simple means. Further, depending on his refinery configuration, he can vary his mix of products to correspond to market demands, within limits, regardless of the composition of the crude feedstock. The refiner can lighten or darken the barrel at will, depending on demand and the profitability of various end products.

In summer, when gasoline demand is high, the refiner runs a whiter barrel. In winter, when demand for heating oil increases, he runs a blacker barrel, producing more fuel oil and less gasoline from the same feedstock.

Refineries have always made gasoline. Once a nuisance, gasoline gradually replaced lamp oil and since the advent of the automobile has become the premier petroleum product. The problem has been what to do with what is left over after the gasoline has been removed, since there is a market for only so much lamp oil, kerosene, diesel fuel, heating oil, solvents, lubricating oil, waxes, and bunker C, the last a fuel manufactured especially for the diesel engines of ships. Satisfying the primary gasoline demand sometimes causes wasteful surpluses of other products.

The solution has been and continues to be the development of new processes that allow the refiner to convert the residue from one process into the feed for another. Now, through a series of steps ranging from simple distillation of crude petroleum to complex and expensive catalytic processes, a refiner theoretically can produce 100 percent of one product from any feedstock.

What is possible, however, is not necessarily economic. In practical terms, the refiner can lighten the barrel only so much. Today, the normal yield is almost 50 percent gasoline, but new technology and plant upgrading may increase that to as much as 75 percent, depending on what the refiner does with the distillate. That flexibility also means that the refiner can make less gasoline and more of the other light fractions, such as diesel fuel and jet fuel. As people drive less in more fuel-efficient cars, as diesel-engined vehicles become more popular and as utilities increasingly switch back to coal, this added flexibility will help reduce crude-oil needs and imports.

216

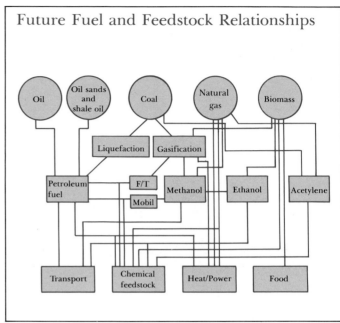

As petroleum increases in value, refiners must make the most of every barrel. Thus, processing has become more complex, with many intermediate stages.

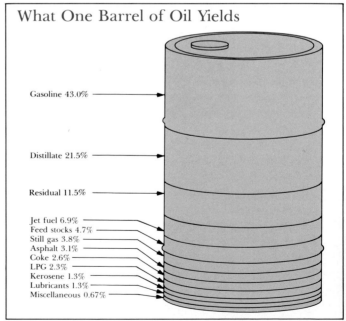

When the value of crude oil first became obvious, kerosene was the main refinery product and gasoline was considered waste. Now, gasoline is the main output and kerosene is insignificant.

Major Processing Steps in a Typical Gasoline Refinery

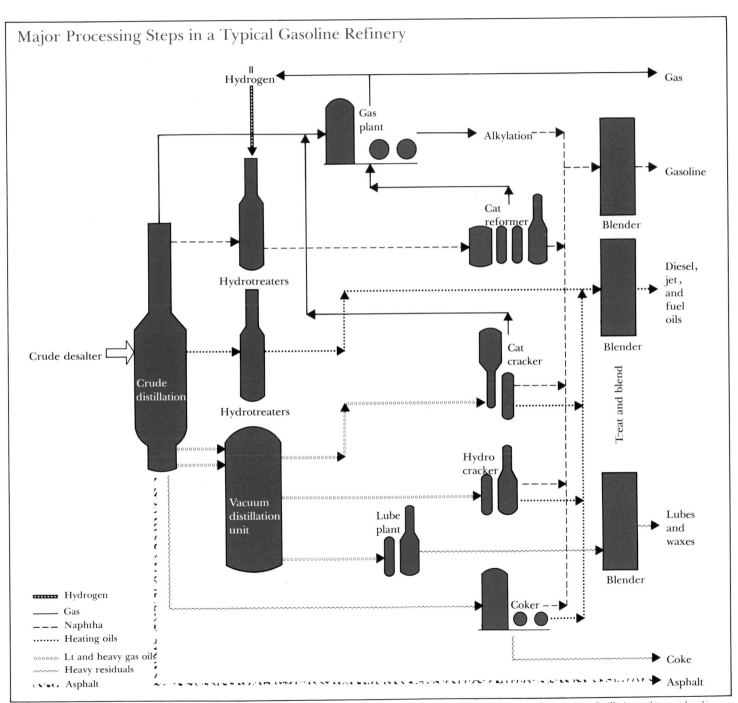

Hydrogen

Gas

Gas plant

Alkylation

Gasoline

Hydrotreaters

Cat reformer

Blender

Crude desalter

Crude distillation

Hydrotreaters

Cat cracker

Diesel, jet, and fuel oils

Blender

Treat and blend

Vacuum distillation unit

Hydro cracker

Lube plant

Lubes and waxes

Blender

Coker

Coke

Asphalt

▪▪▪▪▪ Hydrogen
——— Gas
– – – Naphtha
········· Heating oils
ᵒᵒᵒᵒᵒᵒ Lt and heavy gas oils
∿∿∿∿ Heavy residuals
⌒⌒⌒⌒ Asphalt

217

Processes concerned with reducing the yield of heavy fuel oil and increasing the yield of higher value middle-distillate products—such as vacuum distillation, coking, visbreaking and hydrocracking—have become increasingly important because of the dramatic rise in crude oil prices at the end of the 1970s.

The extent to which refining technology has changed the complexion of the barrel is astonishing. In 1935, refiners produced nearly 27 barrels of residual fuel oil along with 44 barrels of gasoline from 100 barrels of crude oil. From the same 100 barrels today, the yield is only 6 barrels of residual oil, 51 barrels of better-quality gasoline and 7 barrels of jet fuel, which was unknown in 1935. The remainder consists of a wide range of products from coke to gaseous petrochemical feedstocks and fuel for the refinery itself.

Hydrocarbon molecules come in a number of forms: straight chains (paraffinic), branched chains (isomeric paraffins), ringed structures (aromatics, cycloparaffins); and in chains where fewer than the maximum number of hydrogen atoms are present and at least some of the carbon atoms have two or more bonds between them (olefins, a type of compound called unsaturated, even polyunsaturated), though the latter do not exist in the natural crude oil. Physically, these hydrocarbons vary from colorless gases to water-white liquids to solids such as tar or asphalt. Coke, the ultimate end product, is essentially pure carbon. Pure hydrogen gas also can be an end product, though it is generally reused in the refinery's various processes.

All hydrocarbons are structurally versatile. By appropriate processes, the refiner can disassemble any hydrocarbon and turn it into any other. The refiner can string methanes end to end to make molecules of any desired length. They can break down waxes, tars, and asphalts into ethane, gasoline, and fuel oil.

Some hydrocarbons have identical or nearly identical properties, such as boiling point, that make them difficult or impossible to separate from one another by distillation. Yet they differ greatly in some other property, such as reactivity, solubility, or melting point, which makes them easy to separate by another method. Thus, the combination of properties for each hydrocarbon is unique, allowing the refiner, through successive steps, to separate them all from one another.

These steps, or processes, include distillation, alkylation, thermal cracking, catalytic cracking, reforming, isomerization, and hydrocracking, as well as catalytic desulfurization and hydrotreating. Used separately or in combination, they change a sulfurous crude oil into pure compounds and tailored product mixtures.

Refinery processes are classified in four broad categories:

1. Physical processes.
2. Breakdown processes.
3. Change processes.
4. Buildup processes.

Physical processes are the oldest and most simple. They do not change the physical structure of the crude-petroleum molecules, but through physical operations, separate it into its components. These operations include distillation, solvent extraction, and crystallization. In distillation, heat separates high-boiling components from low-boiling ones. In solvent extraction, a solvent removes a more soluble component from a less soluble one. In crystallization, cooling causes a high-melting-point component to solidify and separate from the remaining liquid ones.

Solvent extraction and crystallization apply mainly to lubricating oils. The two processes often work together. For example, the solvent removes the wax from the lube stock, and then crystallization recovers the solvent from the wax.

After treatment to remove gross impurities, such as inorganic salts and some trace metals, the first refining step for most crudes is distillation, generally done in three stages, one at atmospheric pressure and two under successively higher vacuum. All three steps take place in pipe stills that look like an assembly of vertical pipes.

The first-stage products are straight-run gasoline, naphtha, kerosene, and diesel fuel. The gasoline is called "straight-run" because it is a natural component of the crude. It undergoes further processing that removes gases and blends it into a suitable market product. In the early days, straight-run gasoline was *the* gasoline and went directly from the still to the gasoline station.

The heavy bottoms, or the part left over, go to a vacuum second stage for further distillation under reduced pressure, or they are blended and sold as residual fuel oil. The boiling points of bottoms at atmospheric pressure are so high that the compounds decompose into elemental carbon, or coke, and hydrogen before they boil

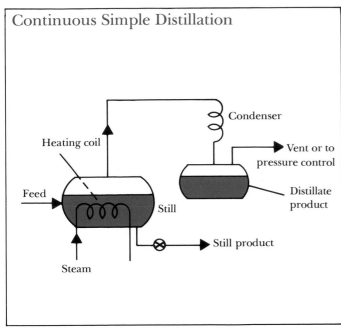

Continuous Simple Distillation

Heating coil

Condenser

Feed

Vent or to pressure control

Still

Distillate product

Steam

Still product

Since each hydrocarbon compound boils or vaporizes at a different temperature, the distillation process separates components with different boiling points.

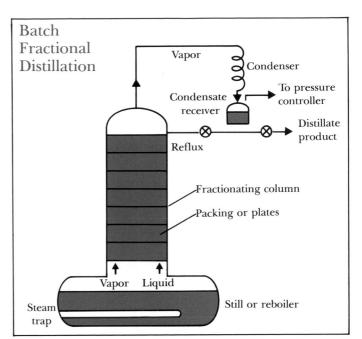

Batch Fractional Distillation

Vapor

Condenser

Condensate receiver

To pressure controller

Distillate product

Reflux

Fractionating column

Packing or plates

Vapor   Liquid

Steam trap

Still or reboiler

By the batch fractional distillation system, the separation obtainable is much greater than that by simple distillation.

Marathon's crude distillation unit at Robinson, Ill., processes 110,000 barrels a day. Since distillation alone rarely yields the proportion and quantity of products that consumers need, the refiner turns to other conversion methods.

and turn to vapor. The refiner reduces their boiling point by operating the equipment at reduced pressure or partial vacuum. The heavy liquids then vaporize without charring or coking.

Products of this stage are "gas oil," that becomes feed for later processes, and heavier oils destined to become lubricants or asphalt. Bottoms sometimes undergo yet another vacuum distillation at a still lower temperature to yield more fuel for cracking processes. The final bottom product is asphalt, a low-value material that yet has many uses, including road paving and roofing.

Breakdown processes then take over. These encompass all types of cracking processes that use heat, pressure, and catalysis to crack large hydrocarbon molecules into smaller ones. Under heat and pressure, the molecules literally break apart. As a rule, they do not change their shape, only their size.

The oldest of these operations is thermal cracking, now generally used in the United States only on asphaltic vacuum bottom fractions to make coke and to increase the yield of light products. In a thermal cracker, the charge is heated to crack it to lighter products and coke.

In a coking unit, a specific type of thermal cracking unit, gas, gasoline, and distillates leave the top of the drum and the coke stays behind. When coke fills the drum, high-pressure water jets clean it out. The refiner alternately fills and cleans several drums. He runs the fractionater continuously. In this way, coke, the ultimate irreducible refinery product, is sold for fuel and a variety of other uses including metallurgical applications.

The most widely used breakdown process today is catalytic cracking, which efficiently converts gas oils from the pipe stills and other units into high-octane gasoline and other lighter distillates. Catalytic crackers, nicknamed "cat crackers," operate at high temperature near atmospheric pressure. They use catalysts rather than pressure to promote the desired reactions. Catalysts themselves are not used up in the reaction but are recycled and used again.

Cat cracking goes back to the late 1930s. Then it was discovered that when vaporized oils pass over pellets of activated clay, known as zeolites, at temperatures of 850 to 900 degrees Fahrenheit, the oils change to gasoline at much higher yields than those from thermal cracking. The cat-produced gasoline also had higher octane ratings and ran cleaner in engines. The discovery revolutionized the refinery industry.

In time, the zeolite loses efficiency as a catalyst because of clogs in microscopic pores. The refinery regenerates the catalyst by burning off the coke with air very carefully and with thorough purging between cycles. If the catalyst is based on a precious metal, like platinum, palladium, or rhenium, trace metals in the crude as well as carbon buildup eventually deactivate it. In these cases, the catalyst cannot be regenerated, but must be replaced.

Cat cracking has improved over the years. In the continuous cracker, the catalyst continually cycles between the reactor and a regenerating kiln. In the fluid cat cracker, the catalyst, a fine powder, behaves as a fluid through which the vaporized oil bubbles.

The ultimate in cracking is hydrocracking. Here, hydrogen in the presence of a platinum-based catalyst reduces heavy gas oils to gasoline, jet, and diesel fuels (the

220

After being processed in a coke calcinator like this one at Cherry Point, Wash., green-delayed petroleum coke is used in the manufacture of aluminum.

In the cat cracker, the heavier hydrocarbon compounds are broken down to lighter compounds such as gasoline and jet fuel. Using catalysts rather than pressure, cat crackers increase yields.

petroleum industry, not the jewelry trade, is the largest user of platinum). The refiner controls the process to vary product type in response to market demands. Hydrocracking also converts already cracked gas oils to lighter products more effectively than do cat crackers.

Hydrocrackers are expensive because their high operating pressures (1,000 to 3,000 psi) and great size require heavy construction. But the fixed-bed catalyst can operate for a year or two before the unit must shut down for decoking. Such a shutdown, during which the refinery also performs major maintenance and reconstruction, is called a "turnaround."

Change processes reform hydrocarbons, altering the shape of the molecules by changing them from straight-line hydrocarbons to ring-shaped ones. This process is the most common way of making high-octane unleaded gasolines today. Change processes take over when, after cracking has arranged the product distribution to match the market, some hydrocarbon streams remain whose molecules are the right size but the wrong configuration. For example, heavy straight-run gasolines or naphthas have too low octane numbers. Streams like these must undergo upgrading.

The most important gasoline-upgrading process is catalytic reforming, which increases octane in a hydrogen atmosphere over a platinum-rhenium catalyst. Prior to reforming, impurities must be removed because the naturally occurring sulfur and nitrogen compounds in many straight-run gasolines poison the catalyst, which is very expensive. So, the refiner first hydrotreats the charge over a cheaper catalyst to remove the impurities.

Metals must also be removed before the feed can go to the reformer. Another catalytic process removes the heavy metal contaminants, such as nickel and vanadium, and allows upgrading of even the heaviest crude fractions, where the contaminants tend to concentrate.

Another change process is isomerization, which converts normal paraffins to branched chains. The process can convert a low-octane normal pentane to a high-octane isopentane. Nowadays, refiners do not often use isomerization because the value added in the product does not justify the investment and operating costs. However, as the trend upward toward unleaded gasoline ac-

221

Lower-octane gasolines produced in the distillation tower, coker, and hydrocracker are processed in the magnaformer, where their molecular structure is rearranged to make high-octane, low-lead gasoline.

celerates and octane requirements stay high, the process may again become important.

Buildup processes are used to form larger, higher octane molecules with low volatility by combining smaller, lower octane molecules of high volatility. These smaller molecules would have limited use except for buildup processes because they are too volatile to be added to gasoline. One important buildup process is alkylation. Undesirable products of cat cracking are small-molecule hydrocarbons of relatively low product value. Alkylation occurs when, in the presence of such catalysts as sulfuric or hydrofluoric acid, these molecules unite to form compounds in the gasoline range with very high octane numbers. Alkylates are important constituents of unleaded high-octane gasoline or aviation gasolines.

Another buildup process is polymerization, which links short-chain molecules. Refineries seldom use the process now because polymer gasolines have low-octane ratings, and yields are only 60 percent that of alkylation. Polymerization, however, is extremely important in the petrochemical industry.

A number of additional operations support these primary processes as well as meet environmental requirements. The refinery gas plant also extracts from by-product gases alkylation feed and heavier components for later blending into gasolines. This plant also recovers liquefied petroleum gas (LPG) and petrochemical feedstocks. The remaining "dry" gas is then treated to remove sulfur compounds. The hydrogen sulfide recovered in this way is converted to elemental sulfur and sold, while the gas becomes fuel in the refinery.

Even straight-run gasoline may need more processing. If made from sulfurous crude, it may contain sulfur compounds called mercaptans. Several processes sweeten the product by exposing the gasoline to water-solutions of chemicals that react with mercaptans.

The gasoline must also be stabilized. This involves a closely controlled distillation that removes enough lighter hydrocarbons to give the fuel the desired volatility. Too much volatility produces vapor lock in an engine; in cold weather, too little volatility leads to difficulty in starting an engine. Refinery products must be blended to yield a final product of the desired viscosity, volatility, and octane number. A refiner can manufacture a fuel oil that is a mixture of a heavy residue and a much lighter fraction that together yield a product with suitable viscosity and burning characteristics.

Refineries make waxes and lubricants, although the wax is often an unwanted by-product. Lubricants account for less than 2 percent of yield. Strictly speaking, the refinery does not make the final product but rather makes the refined stock from which the lubricants, particularly greases, are later formulated and blended.

The manufacturing of lubricating oil is a special operation based on the heavy gas oils produced from vacuum distillation at the very beginning of the refinery process. Lube stocks have poor flow characteristics and contain many impurities. Through solvent treating, dewaxing, and hydrofinishing, the refiner changes these unpromising feeds into high-quality lubricants.

Most ways to remove the impurities involve a countercurrent-extraction operation in which a solvent and the lube stock flow against one another in a tower.

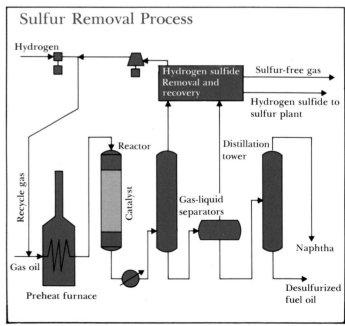

Sulfur Removal Process

The refiner must remove sulfur and other impurities that diminish product performance or pollute the environment. Once removed by the process illustrated above, the sulfur is sold to chemical manufacturers.

222

Once removed, the impurities go back to the refinery for processing into gasoline, solvents, or feedstocks, proving that one product's contaminant is another's feedstock.

Most crude oils contain some wax, and some contain a great deal of wax. Wax in lube stocks causes high pour points — the oil will not flow at normal temperatures — and the refiner must remove it. He does so through an extraction process in which a solvent dilutes the oil. Then the mixture is chilled to crystallize the wax, which is then filtered out. The wax goes on to further processing, while the lube goes to a finishing step, usually hydrofinishing and percolation filtration.

The preferred finishing step today is hydrofinishing, in which hydrogen in the presence of a catalyst removes impurities. The process is similar to the hydro-desulfurization that prepares reforming feedstocks.

In percolation filtration, mainly a color-stabilizing process, the lube stock percolates through a tower filled with tons of fuller's earth and bauxite.

Another type of lube oil stock, called "bright stock," comes from select crude oil in the vacuum-tower bottoms. The refiner treats these in the same way as the solvent-treated stocks, after he removes asphalt in the step called "propane deasphalting." The heavy reduced crude goes to a tower where it mixes with the propane, dissolves in it, and goes out with it, leaving the asphalt behind. This bright stock then separates from the solvent to yield a clean lube stock. The asphalt is also recovered.

The use of gasoline illustrates the changing market and the emergence of conservation. Demand has declined. Gasoline must meet engine octane requirements, and it must do so without lead compounds. Lead, an environmental hazard, poisons the catalytic converters that reduce emissions from automobile exhaust gases. The refining industry has met this challenge through new investment in technology and processing equipment. That investment is not small. To build a modern 200,000-barrel-per-day refinery today from the ground up costs more than one billion dollars.

While gasoline is by far the industry's largest-volume product, others are also significant. An efficient 100,000-barrel-per-stream-day (bpsd) plant produces about 50,000 barrels of gasoline, 2,600 barrels of

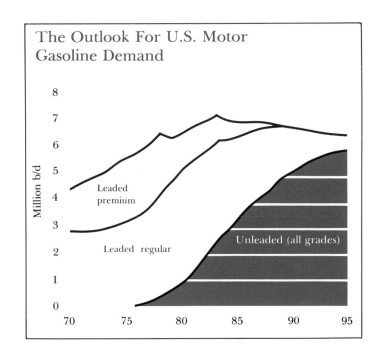

The Outlook For U.S. Motor Gasoline Demand

Million b/d

Leaded premium

Leaded regular

Unleaded (all grades)

223

liquefied petroleum gas (LPG), 6,300 barrels of jet fuel, 3,500 barrels of kerosene, nearly 21,000 barrels of heating oil and diesel fuel, and coke, asphalt, lube stocks, and even 85 tons of sulfur.

High energy costs affect refineries just as they do every other energy-consuming activity. When energy costs escalated in 1973, refineries had to adopt the same levels of energy conservation as those of other energy users. Today, refinery fuel and loss can make up as much as 40 percent of the cost of refinery operation. Efficiency is costly in terms of investment. But in energy conservation, as well as environmental control measures, today's refineries lead the way.

CHAPTER 22 – PETROCHEMICALS

Dorsey Hager's classic work on the petroleum industry, written in 1939, covered petrochemicals in a single paragraph. Hager acknowledged what is today an immense and complex area of chemistry with the prediction that "chemical research in the field of petroleum has such unlimited possibilities that predictions which would today seem fantastic may in short time prove realities."

On the eve of a war that changed so much in the world, even that prescient author could not know how a few years would alter the very concept of petroleum. Oil fueled the war machine but, in addition, increasingly clothed it, provided its munitions, packaged and protected it, gave it tires to ride on, and produced new materials to replace those which were no longer available or had found more valuable uses. With the outbreak of World War II, the petrochemical industry came into its own. Before very long, people everywhere were echoing what Mendeleyev, the famous Russian chemist, had said in 1872 after visiting Pennsylvania: "This material is far too precious to be burned. When burning oil, we burn money; it should be used as a chemical base material."

The history of the petrochemical industry does not begin with oil, however. The first petrochemicals came from materials derived from hydrocarbons and other compounds that occur naturally in coal. A large coal-based chemical industry grew up in Europe in the last half of the nineteenth century. It was largely a European development, another step in the Industrial Revolution, which itself was built on coal energy rather than oil. Europeans in general and Germans in particular developed their coal-based industry, which produced everything from photographic film to liquid fuels and household cooking gas. Coal-tar dyes developed in Germany yielded brilliant new hues that quickly changed the fashion industry.

In the United States, a parallel but smaller coal-chemicals industry grew up, often based on European discoveries and frequently owned by European parent companies. But the United States had plenty of oil; its economy, therefore, became more dependent on petro-

225

ARCO's Lyondell plant at Channelview, Tex., can produce 1.3 billion pounds of ethylene annually, plus propylene, mixed butylenes, butadiene, aromatics, hydrogen, and a number of fuel byproducts. It saves energy by using a special process that recycles materials that previously would have been flared.

leum resources than on coal. By the late 1950s, petroleum had displaced coal everywhere in the nation as both an energy source and a chemical raw material.

The U.S. chemical industry was based largely on inorganic materials — minerals that do not contain carbon. American chemists borrowed technology from Europe to manufacture acids and alkalis and to treat ores to recover metals and make paper, soap, and explosives. The variety of chemicals made was not great, and the industry was not large. Eventually, the growth of the coal-tar-chemical industry led to a greater variety of chemicals: notably dyes for the textiles industry and, of greater significance for the future, the beginnings of the science of organic chemistry.

Organic chemistry is the exploration and exploitation of the carbon atom in its infinite permutations and combinations. The carbon atom is almost indiscriminate in the company it keeps, bonding to virtually any other element, as well as itself, and able to form molecules that are long chains, rings, cubes, pyramids, branched and unbranched, simple and complex. Each such compound is unique. Each has chemical and physical properties that distinguish it from all other compounds in nature and that may or may not make it useful to man.

The chemistry of hydrocarbons, basically that of oil and gas, although coal derivatives could be included, forms a huge body of knowledge and application. Yet it occupies only a tiny corner of the total realm of organic chemistry. Hydrocarbons consist of only two elements, carbon and hydrogen, numbers one and six in the periodic table of elements. The varieties of individual chemical compounds that exist and can be synthesized from these two elements are too numerous to count. If other elements, such as oxygen, nitrogen, sulfur, phosphorus, lead, and chlorine, are added, the bounds of mere hydrocarbon chemistry are extended. More than half a million carbon-containing compounds have been identified. Many, many more are certainly possible.

All life is based on the carbon atom and its infinite combinations. It is organic in the sense that it is central and vital. Petroleum, a mixture of hydrocarbons, is itself derived from once-living things, just like coal. In a sense, when petroleum is used as a raw material and certain

operations are performed on it to build complex molecules from simple ones, the path nature took to reduce the complex molecules of life to the relatively simple ones of oil and gas is reversed. In fact, organic chemists build molecules that never existed in nature to do jobs that would not need doing if man had not devised them.

The petroleum and chemical industries have been close allies practically since the Drake well in 1859. In 1861, the Civil War created a demand for lubricants and illuminating oil, formerly obtained almost entirely from whales and plant materials. The fledgling petroleum industry supplied these. Petroleum spirits substituted for turpentine, a natural product of southern pine forests that was understandably in short supply in the North. Petroleum-based products have been substituting for and often surpassing natural products ever since.

These examples do not, however, represent an early beginning of the petrochemical industry. Lamp oil, lubricants, and spirits are not petrochemical products but items distilled, refined, and purified from natural crude oil. The petroleum industry itself, however, used many chemicals. In the latter half of the nineteenth century, expansion of petroleum products increased demand for sulfuric acid, which is used to treat various oil stocks. Continual advances in the technology of petroleum production and refining gave rise to a distinct chemical sector based on servicing this industry.

Despite the close relationship between the petroleum and chemical industries, only in the early twentieth century did manufacturers seriously consider petroleum itself as a feedstock for chemicals. Even the critical demand for chemicals generated by World War I failed to promote the use of petroleum feedstocks or the development of a petroleum industry. In the 1920s, the first commercial production of a true petrochemical — a new material deliberately synthesized from natural petroleum fractions — took place. Standard Oil of New Jersey, at its Bayway refinery, produced isopropyl alcohol through a new process that involved hydration of propylene, which was in turn obtained from cracked refinery gases.

Isopropyl alcohol, an important component of medicines, cosmetics, and pharmaceuticals, is a chemical

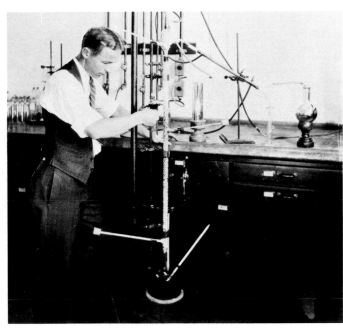

"*Shorty*" *Holm, shown in his laboratory, was part of a 1929 Chevron research team credited with developing synthetic alcohols from petroleum.*

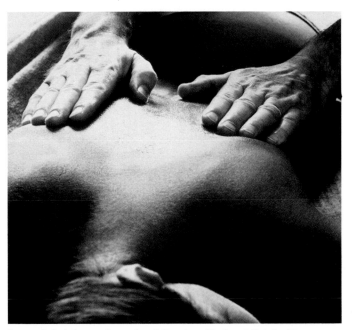

Isopropyl alcohol, manufactured at olefins plants like Lyondell, is familiar to athletes everywhere as rubbing alcohol.

raw material in its own right. It was previously made from acetone, which in turn was made by bacterial fermentation of natural carbohydrates such as sugars and cellulose. Many chemicals continue to be made by fermentation, usually for applications involving food, drink, and medicines. The same chemicals are manufactured synthetically from petroleum feedstocks. The enormous quantities of these materials required to satisfy society's needs can be obtained at reasonable cost only through synthesis.

The modern petrochemical industry owes its beginnings to the development of the internal-combustion engine. Improved engines could not operate well on the early straight-run gasolines of relatively low octane. By 1912, refineries were using thermal cracking processes to produce higher performance fuels for automotive and aircraft engines. This produced a lot of gas as by-product, which was at first burned to supply energy in the refinery. When the value of propylene and butylene gases was recognized, the gases were recovered and converted to chemicals as well as to higher-octane polymer gasoline.

In the 1930s, additional isomerization processes were developed to convert an ever larger portion of cracking gases into useful isomers and polymers. Most of this went to make better gasoline, but some isomers also became feedstocks for the limited petrochemical processes then appended to refineries almost as afterthoughts.

Commercialization of catalytic-cracking processes in the mid-1930s enabled refiners to lighten crude oil significantly as demand for higher-performance gasolines rose. But these processes also produced ever-greater quantities of refinery gases, far in excess of the needs of a petrochemical industry whose total output in 1939 was only 2.5 billion pounds.

World War II changed that. It brought a huge demand for textiles, explosives, elastomers, plastics, drugs, and synthetic rubber. From those beginnings, petrochemicals have grown to the point that about 5 percent of all crude and natural gas now consumed in the United

States is used as feedstock by chemical companies to make billions of dollars worth of products each year. The value added in the process is immense.

Today, the petrochemical industry comprises about 40 percent of the total chemical industry in terms of the value of products shipped. Capital expenditures for petrochemicals were $5 billion annually on the average from 1977-81, reflecting the industry's large investment per unit of product.

The industry is extremely competitive, both domestically and internationally. In the United States alone, about 300 plants make petrochemicals of one sort or another. Petrochemical companies outnumber oil companies. No manufacturer has a monopoly on feedstocks, processes, or products. As soon as one maker brings out a new and cheaper process to make a new product that enjoys great demand, his competitors rush in with new plants and slightly different processes. In a few years, only the most efficient manufacturers remain. Then a new product or process comes along. The pattern has repeated itself time and again for virtually every large-volume petrochemical company, both domestically and internationally. Obsolescence resulting from improved technology moves at an extremely rapid pace.

Until the end of the 1970s, competition held U.S. petrochemical prices below world prices and gave the U.S. industry an edge in international trade. The U.S. balance of trade suffered because large amounts of foreign petroleum was imported and paid for with dollars. Exports of petrochemicals contributed to a positive balance of payments in the industry, reaching $9 billion in 1981 and falling to $6 billion in 1983. Of course, as oil-rich countries elsewhere expand their operations from production to refining and eventually to chemical feedstocks and even chemicals, as in Saudia Arabia today, where 19 petrochemical plants were built by 1983, the United States is facing a declining export market and also the possibility of having to increase imports, because other countries manufacture at lower cost.

Historically, the petroleum industry has been divided into two broad categories. The first separates feedstocks and converts them to chemical building blocks and intermediates. This function naturally falls to the

Olefins manufacturing plants like this one at Bay Port, Tex., have been important in the production of propylene oxide, a major ingredient in urethane plastics.

petroleum companies and is usually carried out in plants within or next to refineries.

The second category converts intermediates to finished products. Chemical companies usually fill this role, although oil companies and chemical companies occasionally carry their operations a considerable distance downstream toward the manufacture of the ultimate consumer product, or upstream to feedstocks. Thus, some oil companies manufacture synthetic rubber and plastics, as well as the styrene, butadiene, ethylene, and propylene that are the feedstocks and intermediates. Some have gone even further and acquired companies to make consumer products, such as plastic bottles or film.

In the other direction, some large chemical companies have integrated backward, or upstream. These companies have acquired or built refineries, producing properties, and even oil companies to assure their own sources for at least some feedstocks and intermediates.

While the petrochemical industry can produce nearly innumerable chemicals and finished products, i

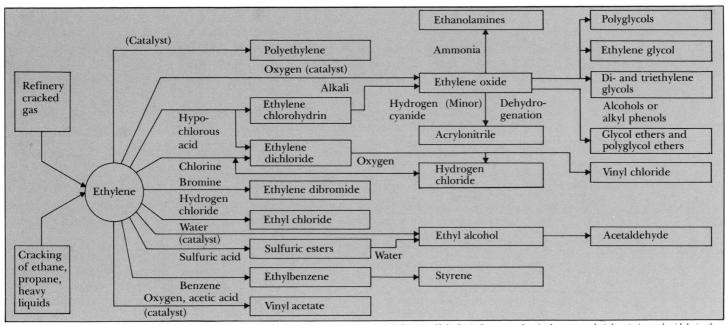

Olefins are the basic building blocks of petrochemicals, particularly of propylene and ethylene. Ethylene, itself the basis for many chemical compounds (above), is used widely in the manufacture of most plastics, synthetic fibers, and anti-freeze.

rests on a relative handful of basic building blocks. One basic feedstock is naphtha — a generic term for a simple range of petroleum distillates that are somewhere between gasoline and kerosene, and thus not a specific compound. It is obtained from petroleum crude in the course of normal refinery operations. The other basic feedstocks are natural gas itself, largely methane, and the individual refinery gases: ethane, propane, and butane.

In the next tier are the primary intermediates. Of these, the most important are ethylene, propylene, butadiene, benzene, paraxylene, ammonia, and methanol. Except for ammonia and methanol, all are normal refinery by-products whose production varies according to the basic product mix of the refinery. But because demand for these intermediates long ago exceeded the ability of refineries to supply them as by-products, many specific processes have been developed to convert the naphthas and gaseous feedstocks into intermediate chemicals with high efficiency. Though externally indistinguishable from refineries, facilities using these processes are chemical plants.

These tend to be very large plants. For ethylene, by far the most important of the chemical intermediates, accounting for some 25 percent of total output for eight key petrochemical intermediates, plant size has grown over the years from an average capacity of 50,000 tons per year to 500,000 to 600,000 tons per year. Economies of scale have prompted similar increases in the size of other basic plants, such as those for ammonia and methanol.

Building these big plants requires a considerable commitment by the manufacturer, who must be assured of both feedstock supply and market. And though it seems that almost any feedstock can be used to produce any intermediate, given the versatility of hydrocarbons, it is true only in the sense that a plant can be designed to use one of several feedstocks. Engineers build any plant around a specific process that, in turn, requires a particular feedstock to produce the desired products in an economic mix. Once the plant is built, the feedstock can vary only within narrow limits. For some chemicals, there is little choice of feedstock, even at the design stage. Only certain routes from raw material to product are economically feasible, regardless of how many actual routes there may be.

Six basic processes convert petroleum hydrocarbons into raw materials for the chemical industry. These are:

1. Dehydrogenation — removal of hydrogen to produce unsaturated hydrocarbons, the olefins and aromatics;

2. Oxidation — addition of oxygen to the molecule to produce a range of oxygen-containing compounds, a process that has only recently become a serious contender;

3. Chlorination — addition of chlorine to make chlorinated compounds;

4. Sulfuration — addition of sulfur, relatively limited applications;

5. Carbonization — removal of all hydrogen to leave elemental carbon;

6. Hydrogen generation — removal of all carbon to leave elemental hydrogen.

Dehydrogenation of aliphatic or saturated hydrocarbons to yield unsaturated olefins (ethylene, propylene, butylenes, and butadiene) is the most widely used of these processes. There are several ways to strip hydrogen. In one, under relatively mild conditions with catalysis, hydrogen is removed from the molecule without changing its basic structure. For example, normal or straight-chain butane heated to 650 degrees Celsius in the presence of a catalyst becomes normal butadiene.

Another reaction used with heavy fractions and without a catalyst removes hydrogen under such conditions that the molecular chain breaks. Thus, wax cracks at about 500 degrees Celsius to yield a mixture of olefins with different chain lengths. The processor so arranges conditions that his plant produces only large-olefin (or higher-olefin) molecules. He then separates these into fractions, the higher olefins going into detergents and the lower ones, through oxidation, becoming plasticizer alcohols.

The more important reactions are those that produce the lower olefins: ethylene, propylene, butylenes, and butadiene (C_2 through C_4). At temperatures in the range of 650 to 850 degrees Celsius, various low-boiling or gaseous feedstocks are cracked in the presence of steam, then rapidly cooled or quenched to stop the reaction. A mixture of C_2 to C_4 olefins results. The proportions depend on the nature of the feedstock and the reaction conditions. With naphtha as feed, for instance, the ethylene (C_2) yield varies from 14 percent to over 30 percent as the temperature goes from 650 to 850 degrees Celsius. The ethylene yield increases at the expense of propylene (C_3) and butylenes (C_4), although the amount of butadiene (also C_4) changes very little.

Since ethylene is the more valuable olefin, the trend in processing has been to higher temperatures and faster quenching to maximize the yield of this intermediate.

In the United States, lighter gaseous fractions have been used as feedstocks, although the current trend is also toward naphtha, in particular ethane and propane. These give much higher yields of ethylene, 75 to 80 percent from ethane, with corresponding lower yields of the other lower olefins. The lower yields matter little since the ethylene is wanted, and other refinery processes make enough of the others.

Regardless of yield, the product stream from an olefin unit contains a wide mixture of compounds that are both liquids and gases at normal temperatures and pressures. The liquid components have high-octane numbers and can be blended into gasoline. The gaseous components are complex mixtures that contain not only the desired olefins but also hydrogen, methane, and other hydrocarbon gases. Distillation of the gaseous fractions under refrigeration and pressure splits these into the C_1 through C_4 fractions. Special distillation units capable of extremely fine fractionation then separate the ethylene and propylene from the C_2 and C_3 fractions. But that is not good enough for the C_4 fractions, which are more complex. There are four C_4 olefin isomers, compared to only two each for C_2 and C_3. The C_4 fractions require distillation in combination with selective extraction by appropriate solvents to yield the pure products of butadiene, isobutylene, normal butylenes, and butane.

These processes actually differ little from those a refinery uses to produce gasoline and other products.

In blender silos, like these at Port Arthur, Tex., batches of products are separated by baffle plates or tubes and then recombined to make a continuous and uniform product.

Trends in Feedstock Type
(% of Ethylene produced from each feed)

■ Natural Gas Liquids

■ Naptha   ☐ Gas Oil

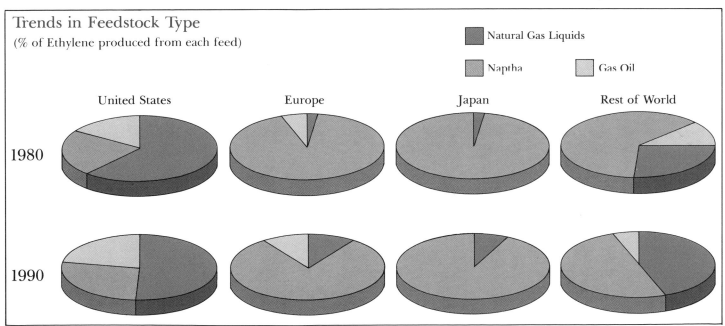

United States   Europe   Japan   Rest of World

1980

1990

Throughout the 1970s, natural gas liquids were the dominant olefins feedstock, accounting for almost 65 percent of all industry ethylene production. As this supply continues to decline, new sources must be found. More low-octane naphthas will be available to the chemical industry because gasoline demand is lessening. Industry substitution of coal and gas has created an excess of heavy fuel oil, which, after further refining, will be an additional feedstock for the chemical industry.

But where refinery products are generally blends, the final products of the chemical plant are pure compounds.

While the olefins as a group are the most important intermediates, the ring-structure compounds, or aromatics, such as benzene, toluene, and xylene, are not far behind. Petroleum is now the source of more than 95 percent of all aromatic chemicals. The remainder still comes from coal tar. Until World War II, coal was the only source of such chemicals, except for a period of World War I when toluene was extracted from a naphtha stream to make trinitrotoluene (TNT).

Crude oil contains 1 to 2½·percent benzene, toluene, and xylene, referred to in the industry as BTX, plus another 7 percent or so of naphthene precursors. The BTX rings are unsaturated, like the olefins, while the naphthenes are saturated rings that can be stripped of hydrogen to yield aromatics. In a dehydrogenation process similar to that used for olefins, these naphthenes are converted to BTX through a catalytic-reforming process first commercialized in 1940. Yields can be very high and can be varied at will to meet changes in market demands. These include high-octane gasoline as well as petrochemical intermediates. The use of unleaded gasoline, in fact, has greatly increased refinery demand for aromatics. However, the increasing use of naphthas for olefin production also increases output of aromatics from that source and helps meet petrochemical demand.

Benzene is the simplest of the three aromatics. It is a six-membered carbon ring with one hydrogen atom attached to each carbon (C_6H_6). Toluene has an additional carbon attached to one of the ring carbons and three hydrogens attached to that carbon. In fact, it is a benzene ring with a methyl group attached and is also called methylbenzene. Xylene has two methyl groups attached to two of the ring carbons. Since there are three different ways two methyl groups can attach to a ring of six members, there are three different xylenes. These, named according to the positions of the methyl groups in relation to one another, are orthoxylene, metaxylene, and paraxylene, usually designated as o-, m- and p-xylene.

Unfortunately, the proportions in which these aromatic compounds are extracted from petroleum are not the same as the demand for them. Of the three, benzene is the most important, but toluene is two to three times more abundant, as are the xylenes. Of the xylenes, only the ortho and para forms have value as intermediates, yet the meta form is twice as abundant as the other two. Therefore, the industry has had to develop processes to bring supply in line with market demand. Hydrodealkylation is one process developed to do this. It removes a methyl group from xylene and turns it into toluene or turns toluene into benzene.

Another multistage process changes the natural mixture of xylenes into one more in line with chemical production needs. Distillation removes o-xylene from the mix. Then crystallization removes the solid p-xylene from the remaining mixture. The now m-xylene-rich mixture runs through a catalyst bed at 700 to 900 degrees Fahrenheit. This isomerizes the xylenes to a new mixture that has the same proportions of the various xylenes as the original. Then this mix is recycled through the entire process.

Generally, benzene is the aromatic in greatest demand, since its derivatives form the major portion of the aromatic division of petrochemicals. Processes that use toluene or xylene as starting materials can also start with benzene, though an additional step is required.

Between them, the olefins and aromatics account for 90 percent of the carbon-based petrochemical intermediates. The other 10 percent is methanol, which is made through an oxidation process. Until 1923, this important intermediate was produced through destructive distillation of wood, from which it got the name "wood alcohol." Today, the raw material is hydrogen and carbon monoxide, a mixture called synthesis gas, obtained through the steam reforming of natural gas over a catalyst. In this reforming process, the hydrogen reacts with steam (water) in the presence of a chromium oxide/zinc oxide catalyst at a temperature of about 650 degrees Celsius. The carbon in the hydrocarbon oxidizes and becomes carbon monoxide (CO) and carbon dioxide (CO^2). The synthesis gas therefore contains a mixture of pure hydrogen and carbon oxides. Proper control of processing conditions adjusts the ratios of hydrogen and carbon oxides to those needed to make methanol.

Passing the mixture over a chromium oxide/zinc

oxide catalyst at high temperature and pressure, more than 4,000 psi, converts the synthesis gas to methanol. Newer processes that use a copper catalyst operate at much lower pressures, only 700 psi. These less severe conditions mean lower capital and operating costs and cheaper product, a saving passed along the line to all the products for which methanol is a feedstock.

Unfortunately, severe conditions such as high temperature and pressure are normal in making the organic chemical molecule rearrange or recombine to yield useful products. Suitable catalysts that help the reactions along have gone far to reduce the costs and increase the abundance of both petroleum-refinery products and petrochemicals. It simply costs less to build, operate, and maintain a plant that uses low temperatures and pressures. That saving shows up in the product.

Polyethylene offers a typical example of how new technology brings substances formerly inaccessible to the public within the reach of everyone. Polyethylene is a polymer of ethylene, a long string of ethylene monomer molecules strung end to end to make a useful plastic solid from a gas. But in the early days, ethylene could not be made to polymerize (join its molecules together) at reasonable temperature and pressure. It took very high pressure, to 28,000 psi, in costly equipment. The resulting products were correspondingly expensive.

In time, researchers developed a low-pressure process in which the ethylene, when dissolved in a hydrocarbon solvent in the presence of catalysts, polymerizes at relatively low pressures, as low as 280 psi. Production costs for the low-pressure polyethylene are lower, and the product is cheaper. But the newer product has not driven the older from the market. The two types of polyethylene differ somewhat in their molecular structure: the high-pressure product has more side chains, and the other is more linear. These differences give the two somewhat different properties and thus different applications.

Ammonia, the petrochemical produced in the greatest quantities, is not really an organic chemical at all but a chemical derived from petroleum. It contains only nitrogen and hydrogen. It is, however, essential to life and, in nature, is a product of living organisms.

The process for nitrogen fixation is relatively old,

Polypropylene pellets like these are the finished product of an olefins plant but the raw material for the next user, frequently a molder who makes automobile parts.

233

having been patented in 1908. The process is also relatively simple; the conditions under which it occurs are not. To make ammonia, three hydrogen atoms must be grafted to one nitrogen atom. But nitrogen, which by volume makes up over 78 percent of the atmosphere, is relatively inert and resists joining up with other elements. In nature, nitrogen fixation takes place through bacterial action and through the action of lightning passing through the air. The tremendous energies involved in a lightning bolt temporarily knock the nitrogen atoms loose from one another and free them to join other atoms, usually oxygen in the atmosphere. These nitrogen oxides then become accessible to life-forms.

The process of making ammonia in a petrochemical plant starts with synthesis gas, the same feedstock as that used for making methanol. But this time the manufacturer converts the carbon monoxide to inert carbon dioxide and removes it, leaving only the hydrogen. Or there may be a methanol plant nearby that uses the carbon content of the synthesis gas for its own feed. The purified hydrogen then mixes with nitrogen obtained from the air, and the mixture passes over a catalyst at 700 degrees Fahrenheit and about 3,700 psi.

For a long time, plants with such temperatures and pressures could not be built very large — a 100-ton-per-day ammonia plant was a big one. As a result, it took many plants to supply the world's growing need for fertilizers, the chief use for ammonia. But in recent years, new technology in control systems and compressor design has made possible the construction of larger plants. The first giant plants of 900 tons or so per day began to appear in the middle 1960s. Today, the normal plant runs from 1,000 to 1,500 and even 2,000 tons per day. These plants are built wherever in the world there are both hydrocarbons to make synthesis gas and a market for ammonia. Thus petrochemicals not only clothe the world but also feed it. As a simple rule of thumb, each pound of ammonia fertilizer produces an additional 10 pounds of grain. Much of the world's population depends on ammonia for its continued supply of food.

Carbon itself, manufactured as carbon black, an extremely finely divided form of pure carbon that many call soot, is a petrochemical created in huge plants. Natural gas or liquid feedstocks such as heavy residues from cracking operations are subjected to high temperatures, 1,300 to 1,650 degrees Celsius, in the absence of oxygen. The hydrocarbon completely decomposes into elemental carbon and hydrogen. The process is much like what occurs in coking furnaces, except that the product is an extemely fine dust instead of solid coke. This carbon finds a market in tires, plastics, inks, phonograph records, and products made of pure carbon, such as high-temperature furnaces, nuclear-reactor components, and synthetic diamonds.

As these chemical compounds, simple and complex, progress through the manufacturing chain from crude petroleum to finished sweater, telephone, or tire, they undergo additional processing, some of it chemical, some mechanical. In a further chemical reaction, the intermediates styrene and butadiene polymerize to form a synthetic rubber. In still further operations, that rubber is mixed with carbon black, plasticizers, antioxidants, vulcanizing agents, and cords of synthetic fiber to make tires

and other end products.

Hexamethylenediamine ($C_6H_{16}N_2$) and adipic acid ($C_6H_{10}O_4$), both of which are made from cyclohexane, which itself is obtained from benzene, react together to make nylon. The process was discovered in 1935, and the first product was named nylon 66. Now many nylons, each with different properties, have many applications ranging from textile fibers to high-strength molding plastics. All modern polyester fabrics are made from petroleum and natural gas.

As these very few examples of downstream processing indicate, the petrochemical industry is immensely diversified. This brief description of intermediates barely scratches the surface.

Thousands of end products too numerous to mention specifically flow from these intermediates. Even ammonia, which is largely its own end product, has extensive markets in fibers and plastics, in explosives, and even in livestock feeds. Polyethylene and polypropylene are the main end products of ethylene and propylene

Some Petrochemicals and Their Uses

Feedstock	Styrene	Paraxylene	Metaxylene	Propylene
Petrochemical	Polystyrene	Terephthalic acid (PTA), dimethyl terephthalate (DMT)	Isophthalic acid (IPA)	Polypropylene
Typical Final Products	Cups and glasses, phonograph records, radio and TV cabinets, furniture, luggage, telephones, ice chests, lighting fixtures.	Polyester fibers used to make wearing apparel, tire cords, polyester film used to make electronic recording tape, photographic film, cooking pouches, specialty packaging items.	Fiberglass-reenforced auto bodies, surf boards, snowmobile housings, outboard motor covers, cooling fans, vaulting poles, many paints and coatings.	Appliance parts, bottles, safety helmets, disposable syringes, battery cases, construction materials, feed bags, carpet backing.

Products manufactured as a result of the petrochemical industry touch almost every aspect of modern life—from space exploration, communications, and construction to home appliances, clothing, and coffee breaks.

Automobile tires have been made from synthetic rubber since butadiene, a derivative of butane, was discovered during World War II.

monomers. But ethylene also is the base for additional intermediates such as ethylene oxide, ethylene dichloride, styrene, various olefins and alcohols, ethyl alcohol, and vinyl acetate. These products in turn become plastics, fibers, antifreeze, dye, explosives, rubbers, and so on.

Similarly, propylene becomes acetone (the basis for a class of transparent plastics often used in place of glass), propylene oxide, isopropyl alcohol, cumene, and butyraldehyde. Butadiene becomes rubber, nylon, and neoprene. The aromatics, mainly benzene, become styrene, cumene, aniline for dyes, chlorobenzene, and maleic anhydride, which in turn become fibers, plastics, detergents, and insecticides.

The list of useful products that flow from petroleum has no end. The ones already developed number in the thousands; tens of thousands more are known but have not yet found a use or need. Without doubt, many thousands more of even more complex combinations of carbon with other elements will in time be discovered and used. When that occurs, the petroleum industry will be ready with the intermediates and the processes that will bring the new products from laboratory to market at an economical cost.

235

CHAPTER 23 –
NATURAL GAS

Before World War II, the easiest way to find a producing oil field at night was to head for the glowing horizon. As one drew closer, the pervasive orange light gave way to clearly distinguishable flames, sometimes so many of them that it seemed the world was afire. The flames, burning jets leaping and dancing in the blackness, silhouetted the derricks. The fires were burning the unwanted and unusable gaseous portion of the hydrocarbons produced in the wells.

For many years and in most places, the gas was vented to the air and set afire at the vent-pipe exit to prevent dangerous concentrations. Producing oilfields were known as much by daytime smoke and nighttime flame as they were by their forests of derricks, christmas trees, and pumping units.

In those days, as the oil surfaced, relieved of the pressure that compressed it in the underground strata, the lighter hydrocarbons (the gas) came out of solution like gas fizzing out of soda water. In the field, the operator controlled the fizzing by use of an oil-and-gas separator.

By that means, he sent gas-free oil into the distribution network, from small gathering lines to field storage tanks, to larger lines and terminals and on to trucks, tank cars, ships, or refineries.

In some parts of the world, an oil field is still identified by its smoke and flames. But this phenomenon is increasingly rare. Natural gas has become too valuable to vent and burn and waste. Because it is the cleanest-burning fuel available, natural gas now justifies the effort and expense of bringing it to market. Indeed, today, far from being a pesky by-product of the production of oil, natural gas merits massive exploration and production efforts in its own right.

People in the United States used gas long before they found much need for liquid hydrocarbons. The first well drilled specifically for hydrocarbons was a gas well. In 1821, 38 years before the Drake well, citizens of Fredonia, New York, drilled a 27-foot gas well to supply 30 burners. In 1854, the first gas well in Pennsylvania was drilled at Erie. In 1873, a five-mile pipeline supplied gas

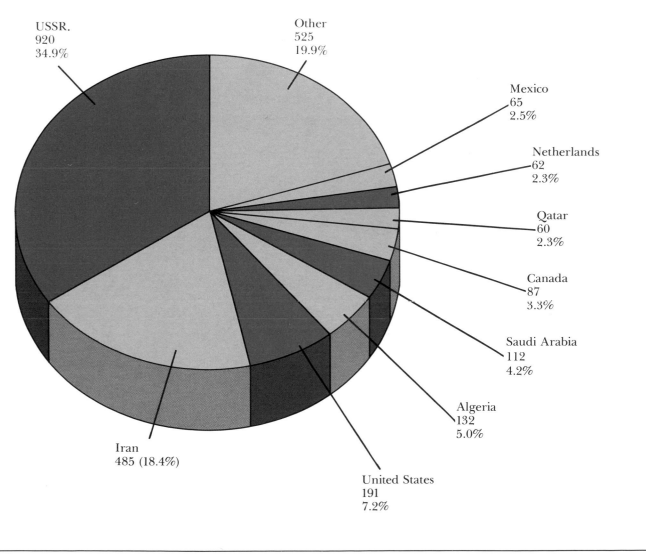

Estimate of Proved Reserves of World Natural Gas

World Total: 2,639.0
(trillion cubic feet)

USSR.
920
34.9%

Other
525
19.9%

Mexico
65
2.5%

Netherlands
62
2.3%

Qatar
60
2.3%

Canada
87
3.3%

Saudi Arabia
112
4.2%

Algeria
132
5.0%

United States
191
7.2%

Iran
485 (18.4%)

237

The world's proved reserves of natural gas total 2,639 trillion cubic feet, with the Soviet Union having nearly five times the amount as has the United States.

to Titusville, Pennsylvania, and the town of Fairview, Pennsylvania, was lighted by gas.

These early applications were isolated incidents, made possible by location and availability. Most gas in the gaslight era was coal gas manufactured locally from easily transported coal and distributed from the gasworks through a local network of small pipelines.

Fredonia and other examples notwithstanding, the gas industry started with gas derived from coal, not from wells. The gas industry grew because many utility companies made and distributed their own gas. Thus, at the same time that oilmen were flaring and burning vast quantities of gas in oil fields not all that remote from population centers municipal utilites were selling gas they made from coal. Widespread use of natural gas occurred only after World War II, when oil companies realized the value of the natural gas they were burning up and built a pipeline system to distribute it.

Today the petroleum industry is very much a gas-and-oil industry, though organizational differences remain. For example, while the gas industry shares much of its technology with the petroleum industry, it has its own technical and industry associations, such as the Interstate Natural Gas Association, the Southern Gas Association, the National Gas Producers Association, the Natural Gas Supply Association, and the American Gas Association. Indeed, though major oil companies are also major gas companies, as distinct branches of the hydrocarbon family they compete vigorously for markets and uses.

In the early days, oilmen sought petroleum for lamp oil and lubricants. The gas that separated from the oil, commonly called "casinghead gas," contained methane and many heavier hydrocarbons. Some of these heavier liquids dropped out as a result of compression and cooling. But often these liquids, including natural gasoline and natural-gas liquids, were so "wild" that the technology of the time could not handle or use them. Some casinghead gas was used as fuel in the field, but most of it was vented and flared into pits.

Times change. The internal-combustion engine created a market for gasoline that natural production could not fill. That insatiable machine gave rise to the modern, complex, and sophisticated refining industry. But natural gas and gas liquids remained stepchildren, good only for local use where expensive transportation was not needed. Often, if oil was sought and only gas was found, the discovery was shut in, particularly when the gas was far from places where it could be used.

Pipeline transmission of natural gas on a limited scale began relatively early. A 120-mile high-pressure line began carrying gas from Greentown, Indiana to Chicago, Illinois, in 1891. Two 8-inch wrought-iron lines operated at 525 pounds per square inch, a considerable pressure for those days. The first true long-distance line appeared in the 1930s, when a 1,000-mile, 24-inch line began transporting gas from Texas to Chicago.

The natural-gas era really began with the war years, when domestic oil was needed elsewhere and domestic gas was plentiful and cheap, certainly cheaper than the coal gas then in use for home cooking and heating. With the growth of the petrochemical industry, gas and the lighter liquids were upgraded to far more valuable materials.

Since 1940, the U.S. gas industry has built a million-mile, $54 billion network of pipelines and facilities that, combined with environmental concerns for clean-burning fuels and the need to conserve the energy that had been flared, has made it possible for gas to replace oil and coal for both domestic and industrial use in many parts of the United States. The 54 billion cubic feet of gas produced daily (equivalent to 10 million barrels of oil) heats 55 percent of U.S. homes and accounts for 27 percent of all U.S. domestic energy consumption.

The geology of gas differs little from that of liquid hydrocarbons. The physical geology that forms suitable traps differs only in that the trap must be even tighter to hold the gas. The techniques of exploration, drilling, completion, and even production are much the same. In fact, gas separation is a major petroleum-production operation. However, since gas moves much more easily than liquids through a formation, gas wells tend to be more widely spaced than oil wells.

Natural gas is designated as associated or nonassociated with liquid hydrocarbons. Associated gas comes primarily from oil deposits where it is either dissolved in the oil at reservoir temperature and pressure or exists as a

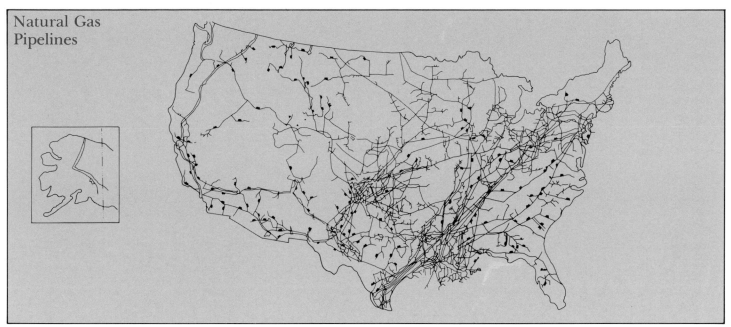

Natural Gas Pipelines

This map shows the United States criss-crossed by a maze of natural-gas pipelines, the most elaborate and efficient delivery system in the world.

cap above the oil. The gas comes out of solution when the well taps the reservoir and the pressure drops. Still more gas leaves the oil in processing after production.

Nonassociated gas may be of nonhydrocarbon origin. Theories and experiments indicate that methane can arise from sources and forces other than those that produce the more complex hydrocarbons of petroleum. Some may have been associated with petroleum at one time, but the connection has been broken.

Nonassociated gas is found under great pressure and at great depth, dissolved in water or brine, or so intimately embedded in tight sands as to be exceedingly difficult to recover. Further, as oilmen drill ever-deeper wells, the hydrocarbons found are more likely to be nonassociated dry gas than oil. Apparently, wherever hydrocarbons exist at temperatures exceeding 200 degrees Celsius, generally deeper than 15,000 feet, heat and pressure convert all the hydrocarbons to gas. The greater the depth, the greater the chance for gas and the less chance for oil.

Associated or nonassociated, oilmen find gas where they find oil. Overall, 40 percent of the world's known recoverable reserves of hydrocarbons are gas and gas liquids. In many parts of the world, including the Western Hemisphere, the proportions of oil and gas are about 50-50. Russia, however, has twice as much gas as oil, while the Middle East has four times as much oil as gas.

These are conventional sources. In addition, oilmen believe that reserves many times as large exist in less conventional deposits that are more difficult to exploit.

In the western United States, huge deposits of deep, tight sands contain enormous quantities of gas — between 50 trillion and 400 trillion cubic feet — but no oil. This stone is so impermeable, however, that in some areas the industry has not yet devised recovery methods that are economic at current price levels.

Along the Gulf Coast, geopressured sands—beds of brine-dissolved gas and gas-containing sand—exist at much higher pressure than their depth warrants. Because of the pressure, vast amounts of gas are dissolved in every cubic foot of brine. But development of these areas has been slow not only because of the pressure but also because of the huge quantity of salt brine produced (upwards of 20,000 barrels per day in some tests).

Estimates of geopressured reserves range widely from 1 quadrillion to 50 quadrillion cubic feet of gas. On the low side of the estimate, this equals all known reserves of recoverable gas from conventional sources, and on the high side, it is nearly 50 times as much.

Nonassociated gas typically consists of 85 to 95 percent methane and is often referred to as "dry gas." The other gas components are usually nonhydrocarbons, such as nitrogen, carbon dioxide, hydrogen sulfide, and helium. The nitrogen reduces the heating value of the gas, but usually can be handled in small amounts without any problems. Carbon dioxide and hydrogen sulfide are called "acid gas components" and must be removed to meet sales specifications for gas. In particular, hydrogen sulfide is a lethal substance and must be removed for safety reasons.

A few fields produce gas which is nearly pure carbon dioxide. This gas is useful as an injectant for enhanced oil recovery. Trace amounts of helium are found in many gases. When sufficient quantities are present, helium is separated and recovered as a distinct product.

As the principal component in most nonassociated natural gas, methane remains in its gaseous form under most conditions of pressure and temperature. In this form, it is referred to as dry gas because it contains no appreciable hydrocarbon liquids. However, some nonassociated gas and all associated gas contain significant amounts of heavier hydrocarbons, such as ethane, propane, butane, and pentane in addition to methane. These heavier hydrocarbon fractions are economically recovered from the gas as liquids. Therefore, these gases which are rich in recoverable liquids often are termed "wet gases."

The least volatile components, such as the pentanes and heavier fractions, are easily liquefied and are sold as natural gasoline or are added to the crude oil. Propane and butane, which are more volatile, are recovered by refrigerating the gas to subambient temperatures at elevated pressure. When liquefied, these components are

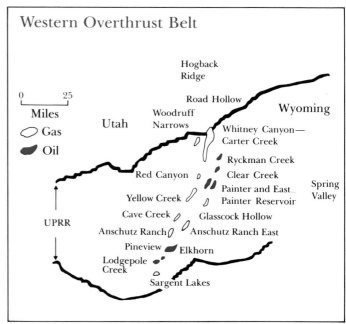

Along the Rocky Mountains lie the major producing areas of the Western Overthrust Belt, a series of geologic formations rich in oil and natural gas deposits.

termed "liquefied petroleum gas" (LPG). They must be stored and transported in tanks under pressure at ambient temperatures. LPG is commonly used for heating and cooking in areas where methane gas is not available from a gas-company distribution system.

Through the use of very low-temperature processes, the most volatile recoverable hydrocarbon, ethane, can be extracted from the gas. With the recovery of ethane, essentially all propane and heavier components are simultaneously recovered. As a group, these components are termed "natural gas liquids" (NGL) and are usually sold as a mixture at high pressure (500-900 psi) to keep them in liquid form. Then they can be separated into ethane, LPG, and natural gasoline and can be sold as individual products.

While it is possible to transport natural gas in small volumes from remote locations in pressurized tank trucks, the costs involved dictate other methods whenever possible. Gas normally travels overland in large-diameter pipelines at pressures varying from about

240

500 psi to 2,000 psi. The general range is between 900 psi and 1,200 psi.

For long-distance transport over water, the gas is liquefied at the source or at the port. As liquefied natural gas (LNG), it then is loaded onto LNG tankers for shipment to an overseas port. There the LNG is regasified and fed into the normal pipeline or distribution system. In some instances, insulated pipelines transport LNG economically for short distances.

Sea transport of liquefied gas requires special ships with pressurized containers or refrigeration plants and heavily insulated tanks. Overland, gas pipelines are normally designed for much higher pressures than those of oil pipelines. They must meet even higher safety standards, because the failure of a pressurized gas line threatens far more damage than that of an oil line. If the gas moves overland as liquid, the line must function at still higher pressures. To handle this, facilities must be refrigerated and insulated.

From Fredonia's first gas well to today's extensive

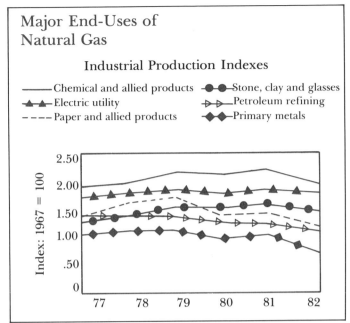

Major End-Uses of Natural Gas

Industrial Production Indexes

———— Chemical and allied products   ●—●—Stone, clay and glasses
▲—▲—Electric utility   ▷—▷—Petroleum refining
- - - - Paper and allied products   ◆—◆—Primary metals

Higher prices and inflation caused consumption of natural gas liquids by major industrial users to decline in 1982 for the third year in a row.

241

gas industry, the United States has been the world's major consumer of natural gas. As the importance of gas in the nation's energy supply has increased, so has its price. In 1954, the U.S. Supreme Court established wellhead price controls for gas shipped across state lines. Intrastate gas remained free of federal control until 1978, with foreseeable results.

Exploration for natural gas has begun to decline, though until 1973 gas remained in ample and cheap supply. Still, no energy source was immune to the OPEC price hikes of the 1970s. At the onset of the OPEC crisis, the price of unregulated intrastate gas rose rapidly compared to federally regulated interstate gas, as the free market was influenced by world energy prices. Producers naturally chose to sell their product for the best price, which was at home in the intrastate market.

As gas supplies declined, a further result was a swing back to oil and coal among some industrial users. Some plants even shut down as local shortages developed in cold weather.

This Cities Service plant at Lake Charles, La., is a typical natural gas liquids fractionation center, where liquid streams are separated into natural gasoline, normal butane, isobutane, propane, and ethane.

242

In 1980, this LNG tanker played an important role for the El Paso Natural Gas Company by transporting LNG from Algeria to regasification plants on the East Coast. The 1983 glut in the domestic supply of LNG made its importation unnecessary, so the tanker was sold.

Elsewhere in the world, the need to buttress oil supplies with alternatives led to ingenious and costly programs to move gas from its sources to places where it was needed. Most ambitious among these are vast international pipelines carrying gas from Iran to Russia, from Russia to Europe, from Africa to Europe, from the North Sea to Europe, and, biggest of them all, in a proposed project that dwarfs the TAPS oil line, from Alaska's Prudhoe Bay field across Canada to the United States, more than 1,400 miles.

Nowadays, huge refrigerated ships sail the seas with supertanker-sized cargoes of liquefied natural gas (LNG), pure methane liquefied in billion-dollar shoreside plants and carried in insulated ships at −59 degrees Fahrenheit. These LNG tankers carry gas where pipelines cannot go — from North Africa, Indonesia, and Alaska to Europe, Japan, and North America. At destination ports, terminals off-load the gas into insulated and refrigerated storage facilities. Vaporizers gasify the cold liquid and feed it into the local distribution network, just as though it came from a gas utility.

LNG facilities and ships are so costly, however, that suggestions have been proposed to convert the gas to methyl alcohol at the production site and ship it in that form. Conversion would be expensive. Much energy would be lost, and the alcohol would have lower heating value, but shipment in standard tankers in that form would be much cheaper. In any case, as more U.S. gas comes into production, importation of liquefied gas to the United States may be less attractive.

In 1978, the U.S. government increased regulated prices. Along with the prospect of decontrol, this led to a spurt of new gas exploration, drilling, and production. In 1979, independent and major producers spent $2 billion to drill more than 600 wells 15,000 feet or more deep, compared to the earlier figure of $1 billion for 400 wells a year. Partial decontrol has benefited the United States by increasing reserves. By 1981, the volume of gas added to the nation's reserves exceeded production for the first time in 14 years.

The partial decontrol of 1978 has proved counterproductive as well. Natural-gas prices, increased by legislative decision and outrageous contract arrangements between producers and pipeliners rather than by market forces, have risen in a way that harmed the consumer as well as the natural-gas industry. Currently, there is a considerable surplus of gas, just as there is a surplus of oil. The oil surplus has forced down the prices of gasoline and home heating oil. But the gas surplus has not had the same results. Prior contract arrangements have prevented the market from responding to the new supply-and-demand conditions.

Natural-gas prices are rising, yet gas exploration and drilling have been set aside in favor of oil exploration. Industrial users are switching from natural gas to oil and even to coal. Utilities have no choice but to increase prices to their remaining customers, residential users. Many argue that this situation calls for more regulation.

However, deregulation promises to correct many of these difficulties. The average price of gas will go up at a reduced rate if producers, pipelines, and industrial users are permitted to make their own decisions, based on the economics of the situation. Common sense necessitates increased production and utilization of natural gas as a cost-competitive alternative to imported oil. Increased use of gas reduces dependence on imported energy and leaves liquid hydrocarbons to their essential uses in the transportation and chemical industries.

243

CHAPTER 24 –
PETROLEUM AND THE ENVIRONMENT

A tanker breaks up in stormy seas; an offshore well blows out; a refinery accident spills oil or a pipeline breaks.

Such events occur. They cannot be ignored. They do cause damage. Oilmen understand the significant potential for environmental damage as a result of petroleum exploration, drilling, and transport.

People consider petroleum an environmental hazard for largely aesthetic reasons, even though petroleum is a natural constituent of the earth, one that nature's creatures lived with long before the modern age and its oil spills. Hydrocarbons do contain impurities, such as sulfur and nitrogen. But in nature, these are not impurities, just variations in composition.

The perception of damage centers around the fact that crude petroleum floats and is easily visible when an accident occurs. People see that crude petroleum is black, sticky, smelly, and persistent and that it leaves an even stickier residue that disappears slowly. They are particularly annoyed when this residue covers their beaches.

What some people do not recognize is that all petroleum products, including the blackest and stickiest of tars, are useful and desirable in their proper places. However, the bather who steps on a tar ball considers the environment befouled even though the black smear does no real harm, either to him or to the beach. Invariably, the bather blames the petroleum industry. He attributes the tar ball to ships cleaning their tanks, to wrecks, or to offshore drilling operations. The tar might come from any of these, but also it might be natural, the product of seepage from an offshore reservoir that may have been leaking for thousands of years. The beaches near Santa Barbara, California, for instance, were tar-collection sites for the Chumash Indians who used the tar to waterproof their canoes. The tar came from just such an underground seepage. Even today, the Santa Barbara Channel floor is riddled with thousands of natural seeps that emit large amounts of hydrocarbons into the environment.

There is, in fact, a lot of oil in the sea, some six million metric tons in any given year, according to esti-

The Chumash Indians, who lived in the area of Santa Barbara, Calif., used the tar that collected on the beaches there to caulk their canoes and baskets (pictured above), making them waterproof.

mates of the National Academy of Sciences. Of that, 8 to 10 percent comes from natural seepages. Another 10 percent comes from atmospheric sources, which may or may not be man-caused. Runoff from the land contributes 31 percent, while waste from industries and municipalities produces another 13 percent. Marine transportation, which includes the occasional massive disaster, contributes the largest portion, 35 percent. Offshore petroleum production operations, which also include infrequent but enormous disasters, add a comparatively tiny 1 to 2 percent. This last percent says a lot for the petroleum industry's stringent controls on production platforms worldwide.

   Although petroleum products are not usually very active chemically, many products and components of crude petroleum are toxic, particularly in the large amounts typical of a major spill. Ingested or inhaled crude oil has the potential to injure and kill people, birds, fish, and other animals. Vegetation is also susceptible. Thus concern about petroleum-caused damage to the environment is more than just aesthetic. It covers health, safety, and long-term ecological and environmental well-being, concerns that go far beyond appearances.

   People believe that if a contaminating source is man-made, it can be stopped, controlled, and cleaned up, as in tanker spills or the discharge of oily ballast. If the source is a shipwreck, they know that someday the oil spilled exhausts itself. In addition to man-made cleanup techniques, scientists have learned that in time oxidation, bacterial action, wind, and water destroy all traces of the deposited petroleum and leave beaches in their previous condition. But when the cause is a natural seep, such as those off Santa Barbara, cleanup is harder to implement. Paradoxically, people believe that if the cause is natural there is no cure.

   In 1978, for example, a research ship of the U.S. National Oceanic and Atmospheric Administration found an oil-rich layer of water about 650 feet deep in the North Atlantic. This layer was 800 miles long and contained at least a million tons (more than 7.3 million barrels) of crude oil. This single layer, a mile wide and 328 feet thick, contained nearly twice the oil previously estimated to be leaking into the ocean from natural seeps

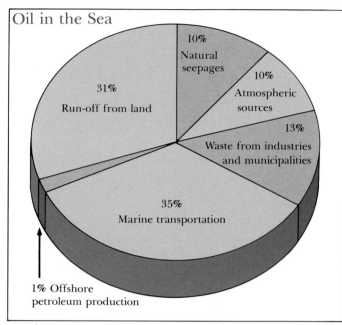

Oil in the Sea

10% Natural seepages

10% Atmospheric sources

13% Waste from industries and municipalities

31% Run-off from land

35% Marine transportation

1% Offshore petroleum production

Although tanker spills get much of the blame for pollution in the ocean, this chart shows that oil there comes from many sources.

every year. But the oil concentration was deemed natural and, thus, too low to be recoverable.

   Scientists tried to find out what happened to this Atlantic layer. When the ship returned to the area a year later, it could not find the oil. They found only microorganisms that live on hydrocarbons. This led to the assumption that the millions of barrels of petroleum had been recycled through living organisms to the biological material from which it had come. From this, scientists understood even more about the work of oxidation, bacterial action, and wind and water in the cleanup of oil spills.

   In fact, not until early in the 1980s did anyone think that natural oil seeps could be controlled by man-made techniques. In a unique effort, Atlantic Richfield Company capped the largest of the seeps on the floor of the Santa Barbara Channel and greatly reduced the amount of natural pollutants in the area.

   Development of technology to protect the environment is as old as the industry itself, though an early ob-

246

In September 1982, ARCO put a lid on one of nature's sources of pollution by placing two steel pyramid-shaped caps, 100 feet square and 24 feet high, over large seeps in the Santa Barbara Channel. These capture about 10 tons of reactive hydrocarbons a day, improving air quality on land.

server of California's Signal Hill, the East Texas field, or the Pennsylvania boom days might not have thought so. The blackened, oil-soaked earth and the forests of derricks were sights only an oilman could love.

But some measures were practiced to protect the environment, and some people did have environmental protection as a goal without calling it that. Long before anyone connected oil production with harm to the environment or thought of threats to human health and safety beyond the obvious, people preferred a pleasant aesthetic environment to an unpleasant one. Further, the brine produced with oil has been an acknowledged pollutant since prehistory. It ruined good farmland and rendered bodies of water unusable by man or beast. Safety, as well as value, mandated control of the oil, as many a disastrous fire emphasized.

So from early times in the history of the petroleum industry, oilmen have spent the time and money needed to keep their operations clean and safe. Indeed, their leases soon demanded it, and government began to regulate it. In 1863, Pennsylvania, then the only major oil state, passed the first oil industry antipollution law. Its purpose was to keep tar and distillery refuse out of creeks in producing districts.

In those days, the primary concern was land and water, not the quality of the air. Damage to land and water was visible and demonstrably costly, where air pollution appeared transient, dependent on the weather, and not obviously harmful.

Early environmental technology was primitive and hardly comprehensive. The main technique was the building of earth pits to impound oil-laden water and brine. Crude skimmers recovered the oil from the pits. But before the blowout preventer, when an uncontrolled gusher was a welcome sight, environmental protection was not a high-priority issue. In oil transportation and manufacturing, it was even less so. This attitude paralleled attitudes in all industries during that period. The land was vast; resources were unlimited. Wetlands were seen as worthless swamps, and deserts occupied wide-open spaces. For oilmen, it was easier to move on than to conserve, keep it clean, or clean it up.

There were transportation accidents that resulted in oil spills in those days, but given the size of the systems, they were insignificant. Until World War II, pipelines were small in diameter and capacity. Therefore, ruptures caused relatively small losses. Road and rail transport was and still is inherently limited in the size of possible losses. Until well after World War II, the largest ocean tankers were less than 20,000 tons.

The methods used to protect the environment before World War II were more comprehensive and varied than those used previously. With safety and economic reasons as goals, for instance, oilmen used impoundment reservoirs to surround refinery storage tanks to contain spills. Floating tank roofs were employed to limit evaporation and simultaneously reduce air pollution. Floating tank roofs were used long before air pollution became an issue. Other methods employed impounding and skimming process water, as well as the use of high stacks to carry away noxious fumes from plants.

Until the 1940s, the petroleum industry operated in geographical areas that quickly recovered from environmental injuries. The industry followed standards that were common to all industries. There was no offshore-production industry to speak of, and the scale of operations was small. In short, the need for sophisticated environmental protection was limited.

Today it is a different story. Given the use of pipelines with four-foot diameters, the increase in ship capacity, and the expansion in offshore drilling and production, the opportunity for large-scale damage has increased significantly. But corresponding advances in the technology of cleaning up spills have increased as well. The magnitude of modern operations makes the consequences of an accident far more serious than they used to be. And the cost of the spilled oil is a factor that needs to be considered in addition to the environmental harm a spill might cause.

Consider TAPS and the North Slope production wells that supply it as examples of how environmental concerns are handled in modern times. The North Slope wells and the 48-inch pipeline that crosses Alaska exist in the middle of a delicate ecological system that is well protected and respected by the oil industry. In the planning, construction, and operation of TAPS, oilmen gave special

249

The first floating roof tank was built in 1923 by Chicago Bridge and Iron Company to conserve vapors and minimize fire hazards. When oil is withdrawn, the roof lowers, preventing an empty space in the tank where fumes can collect. Today, an added benefit is the lessening of air pollution.

consideration to environmental concerns. Indeed, discussion of environmental issues stretched out construction by years and increased costs by billions of dollars.

In TAPS, which is so cleanly operated that neither the sight nor smell of oil is anywhere evident, the first line of defense is good basic design and sound operating practices to minimize the risks of leaks, breakages, and spills. TAPS also has special environmentally mandated features intended to protect the fragile environment from the heat of the oil as it comes from the ground and flows through the pipe. Where pipelines elsewhere need no insulation and are buried without concern for their temperature, TAPS is heavily insulated and carried above ground on special heat-conducting supports so that it will not thaw the "permafrost," the permanently frozen subsoil. Further, all wells and facilities at Prudhoe Bay and along the way are built on insulated and sometimes actively refrigerated pads. At points where the caribou migrate, the line is either buried or specially elevated to make crossings hospitable.

Supervisors, supported by sophisticated instruments, constantly monitor the TAPS operation. The probable flow direction of a potential spill at any point in the system has been plotted for almost any combination of weather and terrain. The system maintains containment and clean-up equipment ready for use at strategic places along the route. Should a problem occur, the system is prepared to shut down the affected part automatically and limit the maximum possible spill. Damage-control forces, guided by a computerized contingency plan, then erect protective and containment barriers in advance of the oil flow. The precautions, unfortunately, have had to prove themselves in practice. Fortunately, they have worked as intended.

Containing spilled oil, whether on land or on water, requires nothing very sophisticated. On land, a crew bulldozes a barrier or digs trenches to guide and confine the flow or deposits sandbags or special inflatable barriers. On water, containment booms of wood or other inflatable materials are floated ahead of and around the spreading oil. In all cases, the exact equipment and techniques used differ somewhat, especially on water, depending on the type of oil spilled, the location, the state of

250

TAPS was built with such care for protection of the environment that today the pipeline peacefully coexists with the surrounding ecosystem.

the weather, and the advice of ecologists about which techniques will best minimize the spill's impact. Sometimes, the best option is to leave bad enough alone and let sun, wind, and rain do the job.

Gasoline, of course, requires somewhat different handling from that of a heavy crude or fuel oil. Gasoline and other light fractions evaporate rapidly, leaving little residue. But they also spread rapidly, particularly on water, and are serious fire or explosion hazards. Although heavier fractions are easier to contain, they also form stable emulsions with water and sink, which makes them hard to recover.

Containment and recovery of a spill on quiet inland waters is relatively effective. Because oil floats, a boom spread around the spill contains it. In another containment method, boats tow a series of booms and skimmers to sweep up the oil and haul it to another vessel for collection. Towed booms include a skirt that extends a few feet below the surface to prevent oil from escaping under the boom.

In 1979, workers in Port Masters, Tex., trying to overcome the effects of a Mexican oil spill, clean tar and oil from water contained by a boom.

Gulf Oil devised this Bay Skimmer to clean up after oil spills. It can collect oil at a rate of 550 gallons a minute.

Many techniques for retrieving the oil are available. The technique used depends on the nature of the material, the location, and the weather. One method employs a giant vacuum cleaner to suck the oil from the surface into separator tanks. The tanks settle out the water, clean it, and return it to its source while retaining the oil. Another method uses rotating sponges that soak up the oil from the water's surface. The oil recovery occurs when the sponges are squeezed on board a collection vessel. In other instances, oil-absorbent materials mop oil from the water surface or beach. Straw, fuller's earth, clay, detergents, and various chemicals sweep the oil, break up the slicks, and disperse the oil into the sea, where natural forces eventually biodegrade it or sink it to the bottom. On occasion, even fire is used to help dispose of a spill.

The techniques for handling spills of any size in calm waters are very effective. Much of the oil is contained and recovered, whatever the cause of the spill. A large amount evaporates. Cleanup in the open sea, however, is far more difficult, as several famous incidents attest. Con-

Workers raking up oil-soaked straw along the shore line of Santa Barbara, Calif., are protecting the beach from an offshore oil well break that spilled 21,000 gallons a day in 1969.

tainment and recovery are possible when the weather is calm, but no method yet devised is successful in heavy seas. There is little to do but boom off the mouths of harbors and bays and clean the beaches when the storm has passed.

Happily, the ocean environment is proving far tougher and more resilient than scientists thought, even in higher latitudes, where low temperatures inhibit dispersal and the chemical and biological breakdown of hydrocarbons. During World War II, thousands of ships carrying millions of tons of oil were sunk all over the world, many of them in coastal waters of the United States. Oil and wreckage littered beaches on the East Coast. If ever permanent environmental damage was to result from spilled oil, it would have been evident after a constant onslaught such as that.

Yet a subsequent study by the Massachusetts Institute of Technology Sea Grant Program entitled *Impact of Oil Spillage from World War II Tanker Sinkings* found no trace of all that oil and no detectable permanent harmful effects of the spills. Divers found the tanks of sunken vessels swept clean. Only occasionally, when a sealed tank corrodes through, does some oil surface in relatively small amounts. However, World War II ships were small compared to today's VLCCs and ULCCs, so the amounts of oil in question may no longer be comparable.

Oilmen now try to recover oil from sunken ships. They do so partly to prevent even short-term damage to the environment but also because oil is now literally sunken treasure, worth some cost and effort to recover. In 1979, for example, nearly a thousand barrels of gasoline were recovered from a wreck 480 feet down in the Tyrrhenian Sea, off the west coast of Italy. This is believed to be the deepest such salvage operation ever. Safety was also a factor. No one wanted the gasoline to surface just as a ship was passing over. The eight-day operation required TV cameras, diving bells, and the latest in deepwater diving technology.

Some of the earlier apprehensions about long-range damage from such massive spills as the *Torrey Canyon* and the *Amoco Cadiz* are also proving unfounded. True, releasing such enormous amounts of oil on a small part of the ocean and adjacent land does wreak havoc. Some-

times even the heroic attempts to clean up make the situation worse by causing the oil to sink close inshore (where it does maximum damage to bottom life and is less susceptible to wind and wave action) and by bulldozing out of existence age-old marshes. But the evidence indicates that the local environment recovers from a massive, one-time attack, though it does take some time. In the aftermath of the wreck of the supertanker *Metula* in the Straits of Magellan in September 1974, ecologists estimate recovery will take 25 to 100 years.

Environmental protection actually begins when a well is completed. The blowout preventer keeps the lid on both land and underwater wells. Its original purpose was economic, but its effect has turned out to be environmental. If it fails, the result is costly to the environment, particularly offshore, where capping a wild well is far more difficult than it is on land.

Pollution was once a routine result of tanker operations. Empty tankers carry water as ballast. They also clean their tanks of the residue of one type of petroleum or product before loading another. It was once common practice to discharge oily ballast and tank washings into the sea before entering port. As late as 1962, industry experts estimated that these routine dumpings added a million tons of oil to the oceans per year. Eventually, the practice was abandoned as aesthetically and ecologically poor and economically unsound. Nations began to legislate against discharges within territorial waters, and this also helped shippers give up the practice. New technology and higher oil prices now make it feasible to clean tanks and discharge ballast in port to special handling facilities and to recover the oil.

Perhaps one of the most advanced facilities of this type is located at the TAPS terminal in Valdez. There 33.6 million gallons of oily ballast can be treated each day. The clean water is returned to the harbor, and the oil is retained. Another antipollution practice is the load-on-top technique. A tank collects onboard tank washings. The oil and water separate, and the clean water is decanted overboard, leaving the oil and a small amount of water. The next cargo is then loaded on top of this collected residue, which never gets into the sea. Now used by most of the world's tanker fleets, this method has effec-

The thousands of gallons of oil spilled by American tankers like this one sunk by the Nazis during World War II have left no permanent environmental damage on the East Coast.

At modern refineries, every precaution is taken to clean waste waters. For one of five water-treatment steps at ARCO's Cherry Point refinery, the water flows through a rock bed (pictured here) where bacteria consume the dissolved oil.

tively reduced the amount of oil lost to the sea, even as the tonnage transported has increased enormously.

The average person is less concerned with production and tanker operations than with a neighboring refinery that misbehaves. Environmental engineering began with refineries, which were the targets of the first antipollution legislation. Although offshore and tanker operations make more headlines, 54 percent of the oil industry's environmental investment is in manufacturing, where the number and complexity of operations and variety of products create many opportunities for mundane and exotic pollution of land, air, and water.

To check these opportunities, in the years between 1971 and 1982, the industry spent $30 billion on environmental protection, increasing from $1 billion in 1971 to almost $5 billion in 1982. Environmental protection is the cost of a mature industry's doing business. It is no longer a series of crash programs designed to upgrade existing facilities to new standards. Instead, it is constructing new facilities with the protection features built in. In the future, these expenditures will reflect the state of the economy as well as other factors, such as the cost of energy and inflation. Of the billion-dollar increase between 1979 and 1980, for instance, 80 percent reflects operating expenses, which themselves reflect the huge increase in energy costs during that period.

As industrial processes go, refineries and petrochemical plants are relatively clean. The processes use much water, but mostly for cooling. As concern not only for water quality but for its availability has increased, refineries have reduced their cooling-water demands from a typical ten barrels per barrel of oil to a barrel or less. In some locations, refineries have achieved total recirculation through air cooling and closed systems and discharge no water.

Refineries also produce small streams contaminated with various chemicals such as hydrogen sulfide, ammonia, cyanides, metal salts, and phenols. These streams are collected, the contaminants are removed by distillation and the treated water is used in other plant services before being subjected to a waste-treatment system. Sometimes the treatment plant is a municipal facility that can handle the low-level contaminants of refinery streams, or the refinery may have its own plant. Such facilities contain treatment tanks to which are added various chemicals that react with the impurities and cause them to flocculate or drop out of solution so they can be filtered out. The treated water then goes to holding ponds, where it is oxygenated and where bacterial action removes dissolved chemicals. No technology is too exotic or commonplace. Scientists have found that a certain species of water plant, for instance, thrives in phenol-contaminated water. The plant hastens the purification process.

Rain sometimes is a problem at refineries, depending on the age of the installation and the climate. In general, refineries try to segregate storm runoff in impoundment basins to avoid overwhelming the process-treatment system. Later, this water is run through the normal treatment facility at an acceptable rate.

These days, air quality concerns managers of refineries even more than does water quality, if investment is a measure. According to the American Petroleum Institute, the industry spent nearly twice as much on air quality as on water quality in 1980. In 1969, these expenditures were nearly equal. By 1971, air was already starting to pull ahead. Expenditures to preserve air quality will probably exceed those for water and solid wastes in the future because refineries can be built so tightly that no liquid or solid pollutants can leak out to threaten water supplies. But no process has yet been devised that can contain gaseous fumes entirely. Thus, the industry will continue to increase its expenditures to protect air quality.

In manufacturing, the greatest problem is sulfur, which is a natural constituent of most petroleum. During processing, it forms various compounds, none of them pleasant, though some are valuable. Atlantic Richfield Company announced in March 1983 that it had successfully developed a chemical agent, HRD-276, that reduces sulfur oxide emissions from refineries at a much lower cost than that of other procedures in use at the time. At the Philadelphia test site, emissions of sulfur oxides were reduced by more than 60 percent.

Another problem is evaporation, which is unavoidable where large quantities of volatile compounds are stored and processed. In the 1930s, the industry built tanks with floating roofs to contain evaporation because the losses were significant and costly. Nowadays, air-quality regulations not only mandate such roofs but also specify the seals between roofs and tank walls. Refineries and terminals have complex vapor-collection systems that cover every point where products might be exposed to the atmosphere. These systems also connect tankage vents, collect vapors during loading operations, compress them, and return them to storage or process. The Valdez terminal has one of the most elaborate vapor-collection systems in operation; it includes inert-gas blanketing of all storage tanks.

Because many small leaks are actually worse than one large one, no source is too small to seek out and seal. In pumps, which leak around the shaft, expensive mechanical seals have proved much better than packing to keep them tight. The value of the material conserved is but a fraction of the cost of the seals. Valves and relief valves are other sources of small leaks that improved materials and technology have helped close off.

At this water treatment facility, the clarifier (pictured here) removes suspended solids from incoming river water.

Another device that helps in the battle against air pollution is the electrostatic precipitator. Along with the catalyst regenerator, it removes carbon monoxide from the flue gases. Mixed with natural gas, the carbon monoxide is then burned to make steam, thus conserving other fuels.

Aside from evaporation, the biggest source of air pollution is combustion. Since refineries generally burn only the clean fuels, gas and oil, efficient operation should produce little pollution, only water vapor and carbon dioxide. The fuels contain sulfur and other elements, however, and the processes make nitrogen oxides. Refineries extract considerable sulfur from petroleum. In addition, they use scrubbers to remove sulfur from combustion gases and from the tail gases of the sulfur-recovery plant itself. The capital investment for a tail-gas scrubber is often more than that for the sulfur plant. A newer method of removing sulfur from combustion gases is the use of HRD-276, a dry, granular powder that captures the sulfur oxides and releases them into existing refinery equipment, where they are ultimately converted into pure sulfur, a dry substance with value as a commercial chemical. With this substance, a modern refinery can keep a considerable amount of sulfur dioxide from entering the atmosphere.

About the only sources of particulate emission in a refinery are the catalytic crackers, which use a very fine dust as working material. Electrostatic precipitators — expensive, complex, and much less reliable than the cracking units — collect the dust before it can enter the air. Typically, one of these machines captures 90 to 95 percent of the dust passing through. In many areas, no dust is allowed. Catching the other 5 percent in additional precipitators, scrubbers, and bag filters is far more costly than catching the first 95 percent.

Of course, precipitators, scrubbers, various processes, and boiler-water treatment with lime produce solid wastes that must also be disposed of carefully. This presents no great difficulty. The materials are generally inert and are more nuisance than hazard. Water treatment with lime produces a sludge that is easily and safely incorporated in land fills. Oily wastes are also safely tilled into the ground in selected sites for soil biodegradation. Powdery spent catalysts are more trouble because they must not be allowed to blow on the wind. Many contain valuable metals and are returned to the manufacturers for recycling.

Environmental protection, never secondary, has become a most serious consideration of the oil industry.

Environmental engineers and scientists rate the salaries, position, and regard that their discipline's impact on the annual budget now warrants. In many companies, a top executive heads the environmental effort and is as much a maker of corporate policy as are the heads of manufacturing, exploration, or finance. After all, work frequently cannot proceed unless the environmental department concurs, and it cannot concur without a hand in the planning. These workers devise the methods and equipment that enable the industry to provide products at a reasonable cost while maintaining the environment.

CHAPTER 25 –
WHO OWNS THE BUSINESS

In oil, success depends more on luck than it does in most other enterprises. As in any other business, however, continued success and growth hang on financial and managerial acumen and adequate financing. Developers must have enough money to get started. In addition, they must generate enough money to keep going and to grow. Especially in these days, even the largest personal fortunes are a drop in the ocean compared to the industry's financial needs. It would not be good business for oilmen to drill only with their own capital. They have to spread their risk.

Even Drake used money supplied by backers "back East" to pay his crew, buy his tools and supplies, and secure the land on which he drilled. Since there was no oil industry then, that seed money came from other businesses and private capital. One of the company founders and backers was a physician.

Drake himself died broke, but in the drilling flurry that followed, many became wealthy. A few with special skills and luck succeeded on their own assets, although most had to borrow or find money, usually by selling off shares in their enterprises. Not the wholly private company but the stock company with many shareholders distinguished the early oil industry and continues to do so.

John D. Rockefeller, Sr., founded an enduring American fortune. A careful accountant, he did it through financial and managerial skill, not luck and eccentricity. Even though the industry was very nearly a Rockefeller monopoly at the turn of the century, Rockefeller's personal holding was but a fraction of the assets he controlled. A complex web of interlocking operating and holding companies held shares in one another and in firms on lower tiers. Former owners of these companies also held shares, as did their families and friends. Ownership in the Rockefeller empire was actually widespread. Rockfeller controlled his empire through his personal relationships with others, who were also owners, not through overwhelming direct ownership.

The Standard Oil Trust was considerably greater than the sum of its parts. Its mass generated a sort of

The Shift in Oil Ownership, Product Sales *

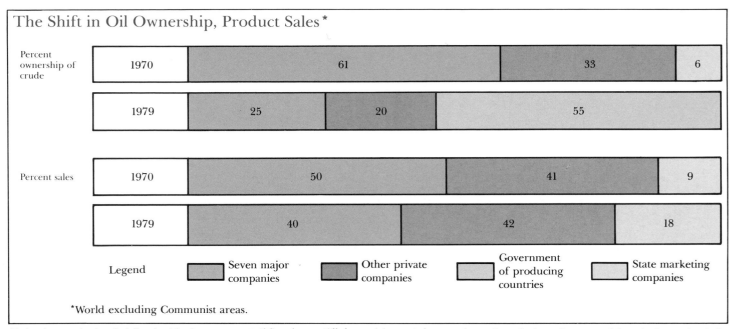

Percent ownership of crude			
1970	61	33	6
1979	25	20	55

Percent sales			
1970	50	41	9
1979	40	42	18

Legend
- Seven major companies
- Other private companies
- Government of producing countries
- State marketing companies

*World excluding Communist areas.

National companies that deal directly with other government oil firms have rapidly been gaining control over crude supplies and sales at the expense of the major international oil companies. This trend was exacerbated by the Iran-Iraq War of 1980.

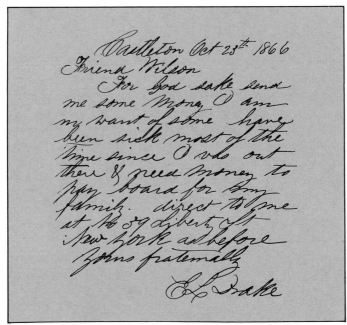

In 1863, Drake left Titusville and went to New York where, speculating on the stock market, he lost most of the money he had made from his well. He wrote to his oil industry friends, begging for money.

World Oil Supply
(Millions of barrels per day)

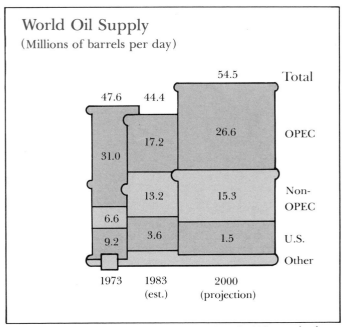

	1973	1983 (est.)	2000 (projection)
Total	47.6	44.4	54.5
OPEC	31.0	17.2	26.6
Non-OPEC	6.6	13.2	15.3
U.S.	9.2	3.6	1.5
Other			

Conservation measures and worldwide recession in the early 1980s caused a sharp decline in demand for oil. This, in turn, influenced OPEC to curtail its production, a situation which will probably change by the year 2000.

gravity that held it together. Standard Oil of New Jersey occupied its center, much as the sun holds together the solar system. Dissolution of the Standard Oil Trust in 1911 diminished neither the Rockefeller fortune nor the Rockefeller holdings in the successor companies. But the center of gravity was gone. The components began to drift away as independent companies. The industry acquired an appearance similar to what it has today.

But there was, and still is, room for individuals with friends, faith, perseverance, some money, and a lot of luck to become independently wealthy — or to go broke despite success — in the oil business. Dad Joiner helped found some of the largest modern fortunes when he discovered the East Texas field. He himself profited little because he had to sell all his shares to get the money to keep going. Independent drillers with good lines of credit occasionally still bring in a well and keep it all for themselves. But, in the main, the oil business — from the smallest entrepreneur to the largest corporation — is one of partnerships and shareholders, bankers and backers.

Private oil-industry ownership is confined largely to North America, Western Europe, and Oceania. The United States is the only oil-producing country in the world that has no national government-owned oil company or any significant government interest in privately owned companies. Attempts to set up a national oil corporation during World War II came to nothing. But the government certainly is interested in the business.

In response to the energy crises in the 1970s, the government established the U.S. Synthetic Fuels Corporation. It might have become a government-owned energy company itself, but the world energy situation changed so drastically after 1980 that the Synfuels Corporation was hardly established before it faded from daily view. Congress did fund the new entity with $15 billion. The Great Plains coal-gasification plant in Beulah, North Dakota, for instance, is being constructed with the help of $2 billion in federal loan guarantees. Yet, in 1983, its sponsors estimated that, owing to lower gas prices, it would take 16 years for the company to repay its construction costs alone, compared to a projection of eight years when construction began, and that in its first decade it would lose $773 million rather than earn an

earlier anticipated $1.2 billion. No wonder interest in synthetic-fuels projects like this has lessened. The organizational structure for a government-owned energy company still exists, however, and should the price of oil increase dramatically again, government interest in synfuels may revive.

The U.S. government owns substantial hydrocarbon reserves through ownership of land and offshore tracts, of course. But unlike the practice in most other nations, U.S. custom and law leaves to the landowner or his assigns the ownership of subsurface minerals, including oil. Over generations, mineral rights have become severed from the surface ownership in many cases, but the principle remains intact. This applies to the government as much as to private individuals and corporations, though the government has a right of eminent domain.

The same rules once applied in most other countries and still do in some, notably Canada, Australia, Japan, and the British Isles. But nationalism and concern for local nonrenewable resources have caused most countries to arrogate to their governments ownership of all minerals. Many private oil and mining companies were early victims of these policies. By the 1980s, outright ownership of overseas oil reserves by private American companies was limited. Companies now hold precarious government concessions or are contractors to foreign national companies that permit them to find, produce, and lift oil.

Where there is still private ownership of the companies themselves, it exists alongside companies wholly or partly government-owned, everywhere except in the United States. In West Germany, 25 percent of oil production is government-owned; in Austria, 100 percent. The British government owns about 50 percent of British Petroleum. British Petroleum, in turn, owns a controlling interest in Standard Oil of Ohio. Sohio owns part of the Prudhoe Bay Unit on the North Slope in Alaska. As a result, the British government has a substantial interest in American petroleum reserves.

The French government, through Compagnie Francais des Petroles, owns 35 percent of the stock in the worldwide Total group. Through Société Nationale Elf Aquitaine, it owns 70 percent of Elf Aquitaine, a diversified mineral and energy company. Even in Canada,

Petro-Canada, a $1.5 billion government corporation, has interests in Panarctic Oils, Ltd., and the Syncrude Oil Sands Consortium. Canadian government ownership of its industry undoubtedly will increase.

Given the national ownership in whole or in part of virtually all foreign oil companies and reserves, it follows that the day of the individual entrepreneur staking his fortune and that of his friends on the sands of Arabia is past. Budding oilmen are still a flourishing breed, but one largely confined to the continental United States, where the stakes are much higher than they were even in the early 1960s.

In practical terms, the search for and exploitation of oil and gas has become so costly, so risky, and so entwined in government regulation and participation that it is now a business only for the established and well-financed company. Even a wealthy person, if he is realistic, can consider entering the business only through participation in syndicates or by buying shares in existing companies.

There are more than 12,000 oil-and-gas companies in the United States. Most of these are very small independents founded by individual entrepreneurs with a little help from their friends. Many of these smaller companies started out as drilling contractors who, with their own capital and bank loans, bought a drilling rig and other equipment. They then contracted with established companies to drill and rework wells.

In time, when these contractors got to know the people in the business and the territory, they yearned to drill for their own oil. They had the equipment, they knew the business, and they also knew that they could drill a hole in known geology for $60,000 or less. If it was successful, the well cost another $100,000 or more to complete. But in that case, the additional financing usually was not hard to get.

Typically, the small operator acquires a lease or perhaps takes over some abandoned wells and reworks them into strippers that produce a very few barrels a day. One such company, the Double S Oil Operators, eventually had five strippers that produced a total of 50 barrels per day — minuscule production by any standards but worth $45,000 a month in gross revenues. To expand

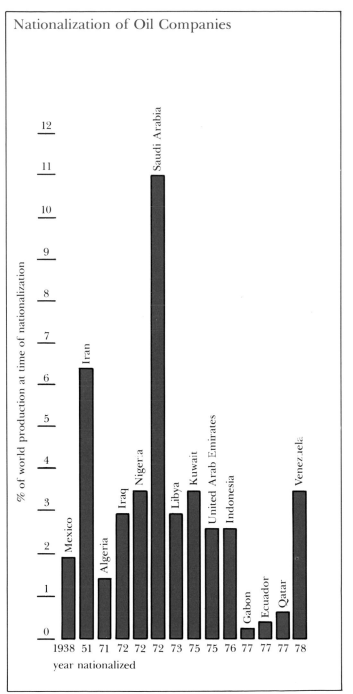

Nationalization of Oil Companies

Nationalization of production was the first step most oil-producing countries took to establish their independence from the multinational majors. That phase was virtually completed by 1979.

from this small base, the company borrowed money from a bank and bought a used drilling rig from an estate for $30,000. The oilmen, David Bullock and Richard Scott, raised another $40,000 from investors to buy a lease and to finance the drilling of an entirely new well. The investors were not big-city financiers, but, as in Titusville, just local people — the scoutmaster, his son, and his secretary, among others. Successful at 2,600 feet, the new well cost $110,000 to bring in, produced 50 barrels per day and gave the company reserves worth $1.5 million. Double S and its investors were lucky. The odds were 17 to 1 that they would lose their entire investment and the company too.

Of course, present success on so small a base is no guarantee of future success. The revenues from one well and five strippers cannot support a significantly expanded exploration and drilling program. To increase operations, a company has to spread its risk by going to new investors. There are thousands of such companies, some smaller, most a little larger. Their numbers and

262

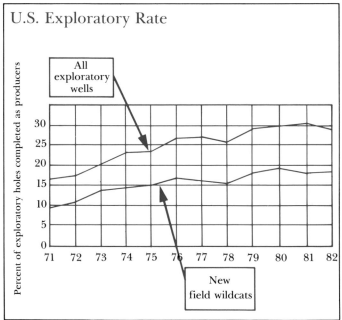

1981 was a peak year for oil exploration in the United States. Although the number of wells drilled increased, the number actually completed declined slightly. Wildcats were more successful than wells in general.

their relative positions change daily with their success or lack of it in finding oil. The one overriding premise in the business, regardless of the size and relative success of any company, is that no company can rest. Once a well comes in, it is only a matter of time until it stops producing. To stay in business, let alone grow, oilmen must find more oil. Not even the largest companies can do so continually and entirely from self-generated funds. In recent years, the independents as a group have spent 105 percent of their wellhead revenues on drilling. They are putting back into the ground more than they are taking out.

The small independents are small operators. Yet they account for about 30 percent of the oil produced in the United States. In 1979, independents drilled 91 percent of all wildcat wells in the United States and 86 percent of all wells. They also account for much of the new oil being found in smaller fields. These companies, however, do not have the financial strength to undertake exploration and production in those parts of the world where most large deposits are likely to be found in the future. Even the small independent who can raise several times the cost of an average well, about $400,000, quails at the prospect of an offshore or Arctic well, which costs several million dollars. These remain the exclusive realm of the majors and the large integrated independents.

The modern majors began as independents. Many of the top eight companies still bear the mark of a single man or family. They grew through the business savvy of individuals but gained their financial strength not only from self-generated funds but also through mergers, purchases and new stock issues, which widened ownership.

Although the eight largest companies account for 55 percent of all refining capacity, no individual owns even 1 percent of any of them. Rather, their owners number in the millions. In the section that follows, the percentages of ownership cited for each company have probably changed since this writing. However, they truly illustrate the general relationship of owners to company. This relationship has not changed substantially.

Atlantic Richfield, often regarded as the Eighth Sister, has about 250,000 owners. ARCO's records, like those of other major companies, show the largest amount of its

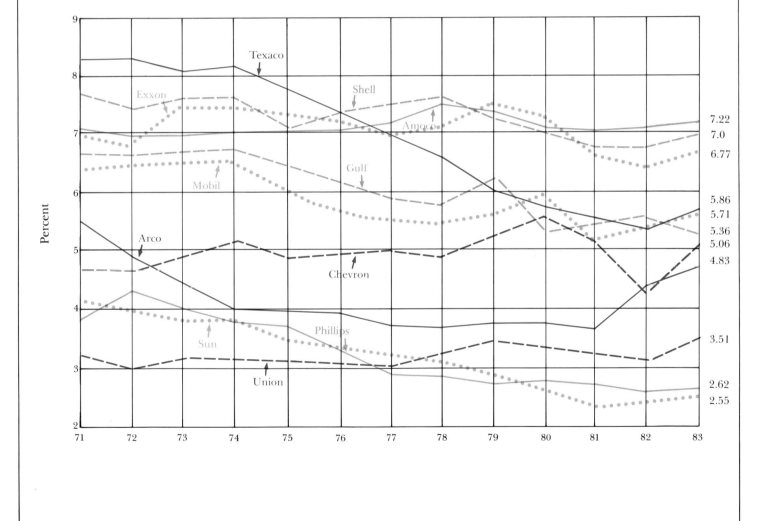

Percent Share of the Market

Amoco retained its leadership as the nation's number one gasoline marketer for the third consecutive year, followed by Shell and Exxon. Seven of the top eleven companies illustrated here show increased sales over 1982, indicating for the first year since 1978 a slight increase in gasoline consumption.

263

shares held by a depository. This depository holds almost half the company's stock for many different banks, which, in turn, hold their shares for many thousands of individuals who are the actual owners. Another large holder is the employees' savings plan, which owns a share of just over 3 percent. Though Atlantic Refining Company, the predecessor of modern Atlantic Richfield, was once a part of Standard Oil, Rockefeller interests are not among the top 70 owners. But this former Pennsylvania company continues to show the lingering effects of regional investment, since the Pennsylvania-based Mellon family interests own a share of about 1 percent.

Gulf Oil Company also grew with large infusions of Mellon money. At one time, that Pittsburgh family owned 70 percent of Gulf. That ownership, held in a relatively few hands, became more diverse as shares passed to the heirs of the original holders and large blocks went to various Mellon foundations. Finally, total Mellon ownership of Gulf Oil was upwards of 11 percent, spread over many family members and several foundations. The Mellon group owned the largest block of shares, but the top 35 shareholders together owned almost 35 percent of the company. None of these top shareholders was an individual. Rather, the holders were banks, trust companies, pension funds, and insurance companies. More than 350,000 individual holders owned the balance of the company. When Chevron took over Gulf in 1984, this pattern of ownership was absorbed, and, in a way, diluted even further.

Exxon, the survivor of the Standard Oil Company, John D. Rockefeller's flagship, is the world's largest oil firm. Yet Rockefeller family interests, which include the Rockefeller Foundation and Rockefeller University, own less than 2 percent of Exxon. Altogether, the actual descendants of John D. Rockefeller own about 1 percent. The largest stockholder is the company's own thrift plan, which is to say, the employees themselves as a group. The top 60 holders, out of a total 693,000, own about 31 percent of the company. Not one is an individual.

The remaining majors — Mobil, Shell Oil, Texaco, and Standard of California — reflect the same pattern of dispersed ownership. Some of these companies are strongly identified with their top managers, many of whom are articulate spokesmen for their companies and industry. For many people, perhaps, the company and the man merge. But these men, though part owners, are not, as in the old days, *the* owners.

Mobil has nearly 270,000 shareholders. The largest shareholder, with just over 6 percent, again is the employees' savings plan. Mobil was once Standard Oil Company of New York, so it is not surprising that Rockefeller family interests own a part of Mobil that is just under 2 percent. No individual is among the top 50 shareholders who, together, own 34 percent of the company.

At Texaco, a company long identified with the autocratic rule of Gus Long, the top 40 shareholders are groups or institutions that own 20 percent of the company. The employees' savings plan is by far the largest owner, with a share just under 6 percent. Again, out of 430,000 shareholders, no individual owns as much as 0.2 percent of the company.

The major owner of Shell Oil is the Royal Dutch/Shell Transport group, the Anglo-Dutch international company that is one of the Seven Sisters. The parent group owns 70 percent of Shell Oil, which has 35,000 other shareholders. At the same time, the parent companies themselves have about a million owners between them.

Standard Oil of California has over 260,000 shareholders, the top 50 of which own 32 percent of the company. Again, the employee stock plan is the largest holder, with about 10 percent. And again, the former Standard Oil component counts a Rockefeller interest of just over 2 percent among its owners.

Even the independent oil companies, many of which are far more closely identified with their founders than are the majors, show widely dispersed ownership. In recent years, few oil companies have been so closely identified with one man as Occidental Petroleum Company has been with Armand Hammer — and rightly so. Hammer took a failing company and built it into a major competitor. He greatly benefited himself, but he benefited the other 162,000 owners far more. Hammer and his family together own only a little more than 1 percent of Occidental. The single largest owner is the employees' thrift plan,

Andrew Mellon, the founder of Gulf, became involved in oil just after Spindletop when Lucas's associates had to turn to Mellon for funds to complete their well.

Dean McGee (left), a geological genius, joined in 1937 with Robert Kerr (right), a man who attracted talent, to found their modern corporation.

J.P. Getty spudded his first well near Tulsa, Okla. Then he acquired the Tidewater Oil Company. After World War II, he won a 50 percent concession in the Middle East and soon became the richest man in the world.

Harry Ford Sinclair was a prolific deal maker. He started Sinclair Oil Corporation in 1916 with an initial capitalization of $50 million, making his firm one of the country's largest on the day it opened.

which holds a little less than 3 percent.

The nineteenth largest U.S. oil firm, Kerr-McGee, founded by former Oklahoma Senator Robert Kerr and oilman Dean A. McGee, has 19,000 owners. McGee and Kerr family interests are first and third with just over and just under 3 percent, respectively.

In recent years, oil-industry profits have appeared to the public as being too large. Total industry profits for 1980, for example, the year decontrol began, were around $25 billion. In absolute terms, the numbers are indeed enormous, but in relative terms, the numbers are much more modest; in terms of need, they are too small.

In absolute terms, the petroleum industry is itself huge. Combined assets of the ten largest domestic companies alone exceed $200 billion. Worldwide, oil is the first trillion-dollar industry in sales. Aside from capital investment, the value of the industry's products and raw materials have increased tenfold in less than ten years. The industry is as affected by these increased costs as are any of its customers. Higher energy costs raise equipment costs and increase the cost of doing business at every level.

Higher costs mean higher prices. At first, the industry benefited from higher prices because it sold its inventory, which was developed in times of lower prices, at the new higher prices. But that temporary bulge in profit was soon offset by the much higher cost of the oil the industry had to find to replace its older stock. Those new, higher costs were absolute; they were not relative to the current selling price of oil. They reflected the fact that it took more constant dollars to find and produce oil.

The situation resembles a store whose stock includes goods bought at last year's much lower prices. If the market price of those goods increases, but the shopkeeper continues to sell below market at a price based on what he originally paid, he will not have the money to restock when his old inventory is depleted. However, if he sells the old goods at current market prices, his markup appears excessive and his profits obscene. If he deludes himself into believing that he actually has a sudden 'windfall' profit and spends it accordingly, again he will have no money to restock at the new, higher prices.

As older stocks deplete and higher-cost oil moves into the market, these profit levels — in both absolute and percentage terms — are falling. The problem for the future is to figure out where the money will come from to develop supplies and alternate sources of energy.

In 1980, the cost of an average well was $398,691, up from $250,000 in 1978 and an average of $95,000 in 1968. For an offshore well, the cost today easily exceeds $10 million. In other terms, it takes $2,000 in investment to produce one barrel per day of low-cost oil. Thus, a 50-barrel-per-day shallow onshore well in a proven area comes in at about $100,000.

The investment for medium-cost oil runs about $8,000 per barrel per day. A 1,000-barrel-per-day North Sea well costs around $8 million. High-cost oil from heavy crudes and tar sands requires an investment of $20,000 per barrel per day.

These figures help oilmen and financial people calculate how much money they need to produce a given amount of oil from a given reservoir. The numbers mount quickly. To develop medium-cost reserves that would augment U.S. production by one million barrels per day takes new capital investment of $8 billion. As long as demand continues, the industry must invest that amount, adjusted upward again and again for inflation and increasing costs.

Production is only part of the cost. New refineries cost well over one billion dollars. New transportation networks must be built and older ones repaired or replaced. Like any other industry, oil must constantly update and renew its plant.

Smaller independents obtain the additional funds they need for expansion by selling more of their equity, setting up drilling partnerships, and borrowing from financial institutions. Congressionally defined tax incentives have made investment in drilling ventures attractive to individuals.

The large integrated independents and the majors, however, with millions of shares of stock already outstanding, can hardly raise the additional funds they need by selling still more of themselves. Indeed, some of the projects they undertake are so huge that the investment required exceeds their net worth, let alone any ability to pay for them out of operating income. So, despite their

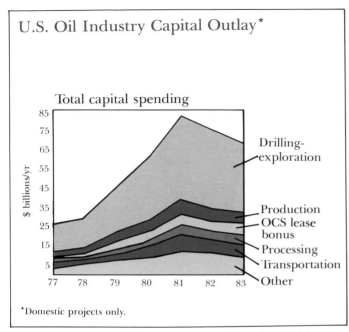

U.S. Oil Industry Capital Outlay*

Total capital spending

$ billions/yr

85
75
65
55
45
35
25
15
5

77   78   79   80   81   82   83

Drilling-exploration

Production
OCS lease bonus
Processing
Transportation
Other

*Domestic projects only.

In response to declining oil prices, lower profits, weak demand, and high interest rates U.S. oil companies in the 1980s invested less in capital and exploration.

huge assets and great financial strength, even the largest companies are forced to turn to the capital markets, to one another in joint ventures, and even to government.

Joint ventures in projects financially too large or politically too risky for any one company are common in the oil industry. They involve making the most of the complementary strengths of different companies. One company may have technology or products that another does not have, and the second may have a marketing organization the first lacks. A joint-venture company combines the attributes of each to the benefit of all.

Many foreign operations of majors and independents are joint ventures. Perhaps the best known of these is Aramco, a joint venture of Socal, Texaco, Standard Oil of New Jersey, and Mobil. Aramco's physical assets were taken over by Saudi Arabia in 1976, however, even though the original four companies still own all of its stock. It thus represents a variation on the usual joint-venture theme. CalTex, a joint venture of Socal and Texaco, also operated in the Middle East and still operates

in Asia. It demonstrates the classic melding of the marketing channels of one company with the supply position of another.

After World War II, many companies established new petrochemical ventures as joint enterprises. Many of these grew out of the work of the government's wartime synthetic-rubber plants.

Domestically, TAPS is a huge joint venture made necessary by the immense costs of the project. Despite the size of the Prudhoe Bay reservoir and the value of the oil in place, even before the 1973 runup, the cost to develop the region exceeded the total value of some of the discovery companies. The logical solution was to combine the financial strengths of many firms, not only to draw on their current revenues but also to increase their borrowing power — and, of course, to spread the risk.

The last and first source of financing for major projects remains borrowing. Borrowing allows the industry to close the gap between operating income and capital requirements for projects that must not be deferred until they can be managed with self-generated funds. The industry's reserves are the security. Revenues gained from production and marketing of these reserves over the years ahead go first to repayment of the billions of dollars in principal and interest borrowed to develop them. Industrywide, this figure can only increase.

Oil-industry borrowings do not differ in form or source from those of other industries. Long-term funds obtained at fixed rates finance long-term projects that have predictable payout periods. Fundings of this kind are not bank loans but bonds and debentures. Investment banking houses underwrite the issues and sell them to the general public. This route is open only to the largest and best-rated companies in any industry.

Short-term borrowings frequently are straightforward bank loans for equipment, with the equipment as security. Smaller operators usually have no other resource, though bankers rate them, too, on the underlying value of their reserves.

The petroleum industry is no octopus run by and for the benefit of a greedy few. It has made some people and families immensely wealthy and continues to do so. But the successful ones were, and are, people who risked a lot.

267

268

They are greatly outnumbered by those who risked and lost.

Until 1981, the average person believed that being in the oil business was nothing less than a license to make unbelievable profits. Indeed, from 1979 to 1981, oil companies reported enormous profit gains over those of previous years, with increases as great as 200 percent. For the very large companies, the dollar size of those profits was beyond the imagination of the average person. Major companies took out full-page advertisements to show that the industry spends almost as much as it takes in and that its long-term profitability is actually less than that of other manufacturing industries. In general, return on equity for the top 25 oil companies averaged 12.5 percent during this period, while for all industry the average was 13.3 percent.

The industry reached its high point during the Iranian crisis, when OPEC jacked up the price of oil to $30 a barrel or more. Spot prices on the Rotterdam market hit $50 a barrel. Everyone wanted to get on the bandwagon.

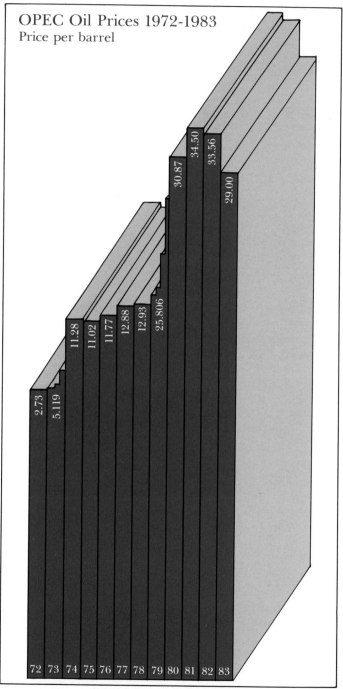

In March, 1983, a worldwide oil glut forced OPEC countries to retreat from their long-standing policy of raising oil prices as a response to political events and inflation and as a means to allocate limited supply. Prices fell abruptly.

Thousands of individuals rushed into the business. In the United States, there were not enough drilling rigs to go around. Everyone connected with the business was making money.

Market forces acted without concern for these individuals. The higher prices created such a surge in drilling that new reserves were developed in the United States and elsewhere. In 1981, a record 3,970 rigs were active in the United States alone, 1,000 more than the total in 1980. For the first time since 1970, U.S. domestic oil reserves actually increased.

On the other hand, the sharply higher prices accelerated the trend by the industry and consumers to reduce energy consumption. At the same time, the world slid into an economic recession that further reduced demand for oil. Rapidly, the world energy situation moved from shortage to surplus. Despite OPEC's best efforts, prices declined. Drilling rigs suddenly became superfluous, and high-priced oil could find no market. Of even greater significance for the future, exploration in high-cost areas declined.

The extraordinary profits of 1979 and 1980 rapidly became a memory. Though the major oil companies continued to turn a much reduced profit, hundreds of independents, who had borrowed heavily when profits were high, found themselves going broke. As they did, they often took their inexperienced financiers with them. Many overextended financial institutions, such as the Penn Square Bank of Oklahoma City, were toppled by the sudden change in fortune of the oil industry.

In truth, the petroleum industry is diverse and competitive. It has thousands of principals and literally millions of owners. Its profits pay dividends to the millions of shareholders who risk their capital, chief among whom are the industry's own employees. Even these dividends are funds diverted from the huge demands for capital investment that more borrowing must make up. The borrowed funds, too, though funneled through large financial institutions, ultimately come from the savings of individuals.

Who then owns the petroleum industry? Its customers and employees do, in a very real sense. And who pays for it? The very same.

269

Millions of shareholders like the ones pictured here, are the real owners of the oil industry. They risk their savings to provide the capital that fuels the industry.

CHAPTER 26 –
COORDINATION AND SUPPLY

Once crude oil leaves the producing field, it moves "downstream" through the transportation system to refineries and marketing. Ultimately, it finds its way to the end user: a motorist buying a tankful of gasoline, a homeowner stocking up on heating oil, or an airline fueling its fleet.

Each day about 14 million barrels of crude and 16 million barrels of product are in transit in the United States, en route from wellhead to consumer. To coordinate this massive logistical exercise requires thousands of people in hundreds of organizations. It demands much planning, monitoring, accounting, and salesmanship, as well as a wealth of skills, experience, and technologies. The process sometimes requires the instincts of the gambler and the horse trader. And if the telephone had never been invented, it would all be a great deal harder, if not impossible.

Large, integrated oil companies, working individually and through networks of brokers throughout the world, do most of this coordination and supply. The work

is done by specific company departments which, while they go by many names, are the supply operation.

Supply is a seldom-seen function, not only because much of it is out of sight — in buried pipelines and seagoing vessels — but also because it does not have the glamour of exploration and production. The daily orchestration of the worldwide flow of crude oil and products, however, is just as critical to economic health as is completion of a discovery well.

The supply man's primary job is to select the most efficient channel by which to move petroleum from oil field to refinery to consumer. The choices are almost infinite. A chart of the oil industry's production and distribution system seems an incomprehensible maze to the layman. Yet to the supply expert, the chart is a clear road map and a second language.

If all crude oils and all refineries were alike, supply specialists would not be needed. But crudes differ. Moreover, on some occasions, the refiner finds no local market for his products. To move them, he has to trans-

Crude oil in these pipes is moving from Prudhoe Bay to Valdez, then via tanker to a refinery, and finally to the consumer as gasoline, heating oil, or electricity.

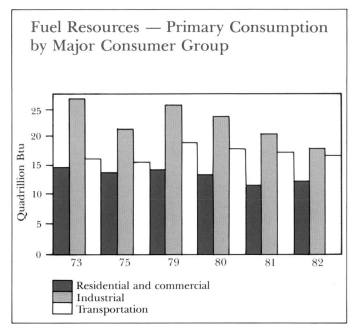

Fuel Resources — Primary Consumption by Major Consumer Group

Quadrillion Btu

73 75 79 80 81 82

■ Residential and commercial
▨ Industrial
☐ Transportation

U.S. energy consumption rose in the 1970s and peaked in 1979. Then, in response to rising prices and recession, it declined, with industrial usage falling the most.

271

port them physically, at considerable cost, to a distant market.

Unlike other areas of this capital-intensive business, supply and coordination manage few physical assets. These functions rely on swift, accurate communication, intimate knowledge of the marketplace, and, like most modern businesses, on computers.

Oil wells produce constantly. The primary supply function is keeping the crude and the product moving smoothly and continuously through the downstream phase.

Dislocations and interruptions occur, but, in general, the transfer of crude from well to consumer flows like some great Amazon of oil. Most crude oil or oil product has a predetermined destination. About 60 percent is already committed when it is produced or manufactured. The producers allocate given quantities to their own refineries or sell other amounts in the open market to other companies or to independent purchasers. The other 40 percent or so "off-the-top" crude is available for open purchase by buyers who want to route it to any number of different refineries. A notable exception is the crude from Alaska's Prudhoe Bay. Because its producers retain most of this petroleum for their own use, there are few off-the-top barrels for other buyers.

Many buyers are brokers. Some represent smaller individual refineries. Others represent trading companies that shop around for particular crudes for resale. Some are employees of major oil firms that buy from one another to make their own supply logistics as flexible as possible.

Crude-oil buyers tend to cluster around major domestic and foreign production areas as well as U.S. ports that receive large shipments.

The crude buyer has several criteria. His first concern is to find the right quality of material in the right quantity and in the right place to save transportation charges. He must know, for instance, whether a target refinery can process heavy or sour crudes or very sweet, light crudes. He must know the specifics of perhaps half a dozen crudes his refiners are seeking. More important, he must be aware of the relative market values of each, and have some understanding of the advantages and disadvantages of running a given crude in a refinery with a particular configuration. Finally, he must have a general notion of the refined products demanded by the marketing department in the region served by each refinery and of the suitability of a given crude for refining into the required slate of products.

The refiner has his constraints, too. A barrel of Saudi Arabian crude oil may differ markedly from a barrel produced in Indonesia or West Texas. Oil chemists have isolated many unique properties that identify particular crudes, including gravity, viscosity, waxiness, and sulfur content. Indeed, the U. S. Coast Guard identifies ships that have spilled crude oil at sea simply by analyzing the spillage and matching it against known cargoes passing through the area.

Matching a certain crude to a particular refinery is an important part of the supply function. In the early days of petroleum refining, when equipment was relatively primitive and crude sources were limited, most crudes could run through any refinery. But today's sys-

tems are quite different. Their equipment is built with particular feedstocks in mind. Complex processes are finely tuned to produce optimum product yield and quality. Such plants cannot accept just any feedstock. Running a high-sulfur crude through a refinery designed to handle only low-sulfur materials soon would corrode the entire system and severely damage an investment worth hundreds of millions of dollars.

Beyond this technical constraint, the refiner considers operating costs. He wants to move maximum volume through his equipment because higher throughput means lower per-barrel cost. His goal is production of as many incremental barrels as possible. Throughput, of course, depends on what products he makes and sells. But whatever those products are, he tries to make them as inexpensively as possible. The economic pressures to maximize throughput not only create periodic gluts and price wars, with refiners running their equipment at 110 percent of capacity and cutting prices to maintain their market share, but also make the oil industry one of the most competitive in the world.

Changing a refinery's output is not simple, although, within fairly narrow limits, refiners alter the relative yields of gasoline or fuel oils by modifying processing conditions or by varying the crude mix. The object is to balance these factors against the refinery's own sales and inventory patterns.

The refinery is critical to the supply process. The transit phase, when crude oil or product is between wellhead and refinery and the market, presents fewer problems. Refining itself is more complicated and more cost-sensitive. It is very significantly affected by shutdowns, some of which, called "turnarounds," are needed periodically for extensive maintenance. A turnaround, a complex operation in itself, begins with phased shutdowns that bring equipment off line in a rational sequence, allow it to cool down, and condition it against deterioration for an inactive period that sometimes lasts weeks or months. It is a sort of active mothballing. Turnarounds last so long and require so much coordination that refiners plan them years in advance and the supply people factor them into planning.

In marketing, the first priority is moving the prod-

Employees like this, equipped with the most modern computer facilities, have the important responsibility of scheduling crude deliveries into refineries.

uct to the marketplace as inexpensively as possible. Some products, especially those owned by major integrated companies, have their own prearranged distribution channels. But some refiners have no fixed distribution system. They sell their products to brokers or jobbers who, in turn, sell to independent marketers or station owners. These brokers and marketers are the middlemen of the downstream process.

Market demand varies somewhat with the seasons, although not as much as it once did. Gasoline consumption is usually higher in summer. Sales of home heating oil and residual fuels (the heavy black oils used in heating plants and in ships' boilers) generally rise in winter. These and other fluctuations in the marketplace affect refining output, although the majors operate their refineries at steady rates, using inventory to balance swings in demand.

The supply function works smoothly because special supply departments in major organizations carefully manage and coordinate these operations. Management is essential to assure the return of the best possible profit for the shareholders, as well as to make all participants aware of the overview.

The management and coordination function (departmental names vary from company to company) is usually based on a month-to-month plan that forecasts need from six months to a year ahead. The plan, normally updated each month, considers contractual agreements for both crude purchase and product sale, names of ships and shipping schedules, refinery shutdowns, product seasonality, and overall marketing objectives.

The ability to look far ahead is limited. Despite this, supply people try to anticipate future problems. They try to direct arrangements to avoid them and change situations to produce the most profitable operation. The key to success in the supply area is involving everyone and always being prepared for an emergency or crisis.

When routing is not fixed by contract, decision making is more flexible. A manager decides that the most profitable overall course is to buy crude B, refine it in refinery F, and make products X, Y, and Z to deliver to markets in area C.

Crude and product routing is not always limited to one company's channels. For example, the supply organization of one company sometimes helps another company dispose of unwanted crude. The second company may have produced more oil than it has the ability to refine. Or it may have produced oil that is too heavy or too sour for its own system. On any given day, a group may conclude sales and exchanges that involve twice as much crude oil as its parent company consumes, considering its total transactions. So extensive are the variety and scope of these transactions that some firms have established subsidiary trading companies to handle them.

Suppose that an integrated petroleum company like the Kennelworthy Company has major producing properties in the midcontinent area, but its only refinery is in the San Francisco Bay area. Because of strict California air-pollution standards, the refinery runs only light, low-sulfur crude. The midcontinent crude would be acceptable, but moving it to the West Coast by pipeline or tanker through the Panama Canal might be too expensive com-

273

sulfur crude. The midcontinent crude would be acceptable, but moving it to the West Coast by pipeline or tanker through the Panama Canal might be too expensive compared to available alternatives.

This brings out the horse trading of a supply organization. Kennelworthy, for instance, might sell the midcontinent production to a midcontinent refiner who needs such raw material. Or it might seek an exchange with Konstables, Inc. In other words, the companies swap one batch of oil for another of equal or similar value and quality. This often-used device can become complicated. Kennelworthy may find that Konstables, although willing to accept the midcontinent crude, has no West Coast supplies to trade. A third company, Kringle Corporation, may then be found with excess West Coast crude and a shortage in Wyoming, where Konstables can provide an equivalent supply.

One objective, of course, is to avoid situations where like crudes flow against one another and, in the process, increase transportation costs. Once sales or exchanges are worked out, the participants must find ways to move the incoming or outgoing crudes. The choices include chartering tankers and arranging for pipeline transportation. Since these deals are nearly always made by telephone or teletype, the principals often hire auditing or inspection teams to visit storage sites and go aboard tankers. There they measure volumes, test quality, and verify transactions.

Crude-oil trading is common throughout the industry. Trading specialists are highly knowledgeable and proficient people, and must of necessity have contact with a number of other companies to negotiate favorable business deals. A trader, for example, may find his company's refinery facing a turnaround, or periodic shutdown of facilities, during which the refinery must store its supply of crude. But tankage may not be enough for the amounts of crude accumulating. He must find someone who needs that particular crude in that particular month and will take it on loan, and then return the loan later when the refinery is back on stream.

Since there are several hundred refineries in the United States all operating on different schedules and having different needs at any particular moment, the knack for crude-oil trading is to know a lot of people in the industry, to know generally what their situation is and how they tie into the system so that when a trade has to be made it can be made on the basis of barrels, qualities, location, or time.

A vital information source for the trader is the *Weekly Statistical Bulletin,* published by the American Petroleum Institute. This report aggregates both crude and product levels in five U.S. petroleum districts. The report does not divulge individual company stocks, but it reflects availability by geographic area.

Ideally, integrated companies prefer refining and marketing their own raw material. When such an opportunity occurs, companies go to great lengths to capitalize on it. A prime example was the discovery of the 10-billion-barrel oil field at Prudhoe Bay. Assured of a long-term crude supply, the owner companies built an entire petroleum system around it, including refineries, tanker fleets, and major retrofitting of existing refineries. But discoveries like Prudhoe Bay, whose producing life will exceed 20 years, are rare. Eventually, even supergiants peter out, and new supplies must be found for the refineries originally built to process their crudes. The supply organization anticipates these shifts and makes the transition from one source to another as smoothly and efficiently as possible.

Purchases of foreign crude are handled in much the same way as domestic purchases, with a few twists. Buyers travel farther afield to do their buying, perhaps to Rotterdam for European supplies or Singapore for Asian crude. Occasionally, they deal with an international oil broker, an independent foreign company, or, much more likely, a state-owned oil company. Then the merchandise usually travels by sea, which is sometimes the trickiest part of importing foreign oil. With modern supertankers, that can mean absorbing two million or three million barrels of crude into the supply network in one gulp.

Major companies owning large refineries are better able to handle such inputs than are smaller operators. Sometimes the shipment is offloaded into smaller vessels and distributed to more than one company-owned refinery. But if a major can use a large supply at one location, it may send the entire tanker load to a single refinery.

U.S. Crude Oil Imports as a Percent of Total

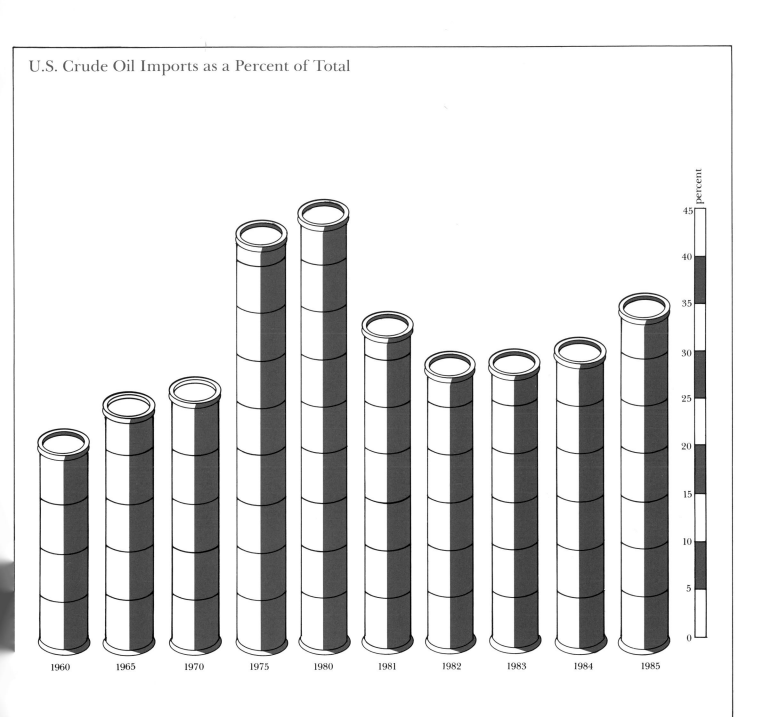

U.S. imports of crude oil reached their peak in 1980. At that time, brokers could sell all the crude they could get. Because of higher prices and conservation, the situation changed in the early 1980s, when international brokers found themselves with more crude than they could sell at the established price.

275

World Crude Oil Prices*
(Dollars per bbl.)

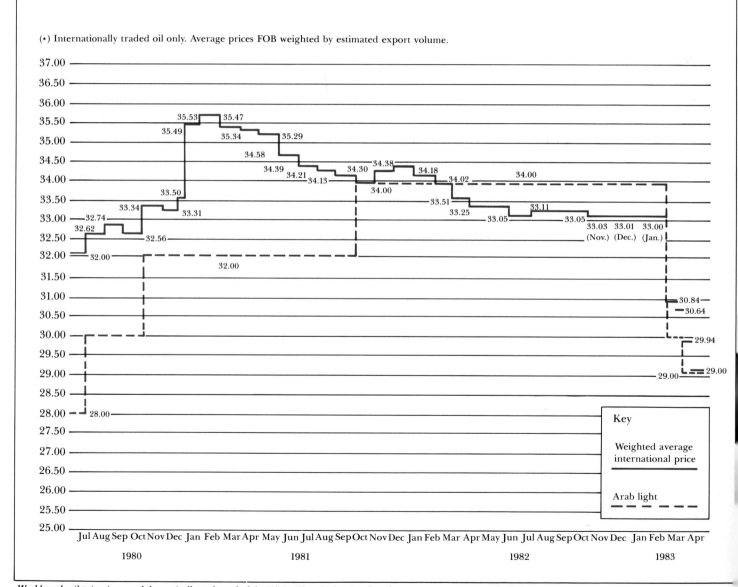

(∗) Internationally traded oil only. Average prices FOB weighted by estimated export volume.

Key

Weighted average
international price

Arab light

276

World crude oil prices increased dramatically at the end of the 1970s. These high prices forced consumers to conserve or find alternate energy sources, leaving more oil available than could be sold at those high prices. Finally, in March 1983, OPEC countries lowered prices in an attempt to sell more of the oil they produced.

patch their crude to the United States without having a buyer. Usually a sale has been arranged by the time the ship arrives. If not, the horse trading may rapidly escalate into a high-stakes poker game. In that case, a broker may post an increasingly attractive price to avoid renting storage space for his oil. In a rising market, he has few problems, but if the market should fall, he may be in serious trouble. International brokers keep a particularly wary eye on political and economic developments that affect the price of crude.

The amazing thing is that the system works as well as it does. It seems logistically impossible to move 150 million to 200 million barrels of crude and product through a complex system every day and, with few interruptions, maintain a relatively constant supply.

The system is designed not to stand still but to flow. Idle material in storage tanks, ships, or pipelines does not earn its keep. The supply man's goal is always crude in, product out, at minimum expense, and the faster the better. Because it is a flow system, the supply network has limited storage capacity, only that needed to keep things running for a reasonable period of time. How long is reasonable? About two weeks of working inventory is the most an efficient operator can afford if he wants to remain competitive.

If the total U.S. petroleum inventory is divided by demand, there would only be something like a 20- to 30-day supply. But this is misleading. About half of that supply is always in pipelines or tankers, where it must remain simply to keep the system operating. Remember that it took a month just to fill the Alaska pipeline before any oil came out the other end. The net supply is thus less than two weeks'.

The petroleum-supply function has few if any parallels in other industries. It certainly differs from the automobile industry, which generally produces much larger inventories. Despite petroleum's large volumes, producers and refiners strive to match supply to demand and hold inventories at a safe minimum. If inventory creeps upward, they lose money.

Thus the system is sensitive to fluctuations from any cause, be it a seasonal drop in demand, a change in motoring habits, price resistance, or battles in some foreign production zone. The system is also sensitive to uninformed tinkering, as when government seeks to impose allocation controls. The result of the 1979 attempt to control petroleum-products distribution was to create surpluses where demand was low and shortages where it was high. The allocation plan was based on years-old statistics, and in the meantime, shifting population and other factors had altered demand patterns.

The supply system succeeds because skilled professionals know their jobs and the system extremely well. The system has long since proved itself to be exceptionally responsive to the marketplace.

It works best when the market is allowed to follow its natural course.

CHAPTER 27 – MARKETING OIL AND GAS

Through marketing, the petroleum industry shows its public face. All the visible manifestations — the company emblems, the trademarks, the signs that distinguish the company from its competitors — are usually the responsibility of the marketing department. Product advertising, package design, and technical service also come under this department. Marketing departments have many functions. They keep in constant touch with customers to determine their needs and preferences. They monitor the performance of products in the field. They advise the research and manufacturing departments on changes needed in products specifications or new product opportunities. And they estimate future quantities needed so that production and distribution departments can adjust their operations.

Marketers help distribute the enormous amount of petroleum produced by the petroleum industry. This enormous production fuels the world's 324 million passenger cars, 86 million commercial vehicles, 20 million agricultural tractors, 200,000 locomotives, 270 million tons of shipping (including about 180 million tons to move petroleum) and the 5 billion commercial airline miles flown annually.

Still more petroleum products heat the world's homes, offices, and factories; cook its food; produce electrical energy; lubricate and fuel its production plants as well as its military machines; clean its clothes; light many of the 4 quadrillion cigarettes made each year; and, still, in the age of electricity, provide the only source of light after dark for millions of people.

The numbers are enormous, hard to comprehend. Yet as petroleum moves from the depths of the earth through refineries to the electrical generating plant, the service station, the heating-oil distributor, the camper's lantern, and the sailor's kerosene stove, the ultimate beneficiary is the customer, a person who is clothed, fed, warmed, transported, and given light and energy for the things he or she wants to do.

Marketing departments fulfill especially important functions in the nongovernment- and even the

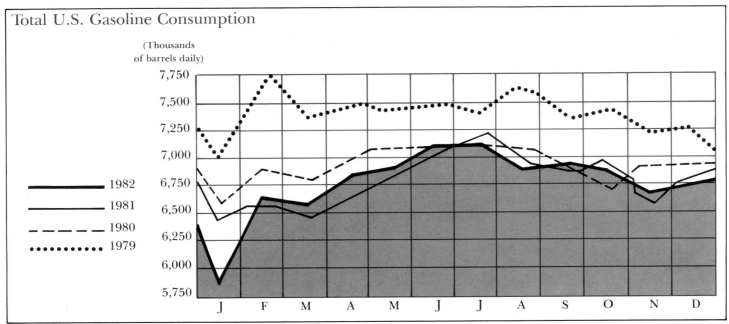

Total U.S. Gasoline Consumption

(Thousands
of barrels daily)

——	1982
——	1981
– – –	1980
••••••	1979

7,750
7,500
7,250
7,000
6,750
6,500
6,250
6,000
5,750

J F M A M J J A S O N D

Every year since 1979, when gas prices rose at an accelerated rate, gasoline consumption in the United States has declined.

government-owned companies of the West. They constantly test their markets, detect changes in customer preferences, and accommodate them. Even in a sellers' market, they actively market their products and services; they persuade their customers to do business with them.

Corporation marketing departments are organized by marketing sectors, such as retail, wholesale, lubricant, and specialties. They are also divided according to end users, such as domestic heat, automotive, industrial and commercial, aviation, marine, transport, and government. In addition, they are divided into geographical marketing organizations for local, regional, national, and international sales.

Though not the largest sector at the company level, the leading edge of the business is retail. In the United States, the largest part is automotive. In the world, boiler fuel occupies the largest single market.

In the United States, passenger cars consume more petroleum products than any other sector. In automotive retailing worldwide, there is more direct contact between the petroleum companies and the public than there is in any other marketing sector. In fact, for most people, the true face of the petroleum industry is not print and broadcast advertising, most of which supports and promotes sales of automotive products, but the ubiquitous gasoline station with the familiar company emblem. Indeed, of all retail outlets in the world, probably only dispensaries of food and drink are more widespread than gasoline filling stations. They developed with the automobile.

The first drive-in gas station appeared in Saint Louis in 1905. In 1911, the first gas station specifically designed for the purpose of pumping gas opened in Detroit. It dispensed 100 gallons a day from a 1-gallon hand pump. In six months, sales increased to 2,000 gallons a day, and the company opened a second station, to become the first chain. Thus the race of gallons pumped began. Every refiner wanted to sell all of the product that he could make. This culminated in the late 1960s, when there were over 200,000 retail gasoline outlets in the United States alone. Then every major road intersection had at least one gas station on each corner, and each belonged to a different company.

Oil was cheap, and gasoline was too. The price dropped from an average of 29 cents a gallon for 55-octane in the 1920s to 20 cents a gallon for 94-octane in the 1960s. Signs promising "Five gals one dollar" were common. Sometimes the "gals" were even cheaper in the occasional no-holds-barred price wars that often broke out as one company or another tried to increase its share of local market or penetrate a new one.

Profit drove the great U.S. gasoline sales wars. Crude that stayed in the ground earned no money. Idle plant capacity cost money. Companies that moved more gasoline and increased their market share made more money. The chimera of the incremental barrel seemed the marketers' rule. The theory was simple and, on its face, correct. Every barrel of crude processed or gallon of gasoline produced in a given plant is cheaper than the last one. Once throughput is enough to cover a plant's fixed costs, the refiner then processes incremental barrels (each additional barrel) for only the cost of materials and energy. When sold at the same price as the barrels that carry all the other costs, the incremental barrels earn a much greater profit. The more a refiner produces from a given plant investment, the more money he makes. Conversely, plants operating at less than capacity reduce the profits.

These perceptions and policies drove the industry to produce enormous quantities of gasoline. Demand was there. The American love affair with the automobile was in full flower. Manufacturers built ever larger and heavier cars to satisfy that demand. The cost of fuel was the last concern. Automakers gladly sacrificed fuel economy. Gasoline mileage of American automobiles declined from better than 20 miles per gallon in the 1930s to under 15 miles per gallon in 1970.

Elsewhere in the world, in places where taxes kept the cost of gasoline high compared to the cost in the United States, where competition was reducing it, automobiles remained small and frugal in their use of gasoline. Ownership of private vehicles was not so widespread as in the United States.

280

On August 14, 1914, Atlantic opened its first drive-through station on Baum Blvd. in Pittsburgh, Pa. By 1921, it pumped a record 6,223 gallons in one day.

Built in the 1920s at Mont Ave. and Levering Mill Road in Cynwyd, Pa., this Atlantic station was noted for its Classical Revival architecture.

The race for gallonage had some unexpected results. Industry doomsayers constantly claimed that the profit on the incremental barrel was illusory, and it was. There was so much gasoline that all of it was being sold at barely profitable prices. A refinery running at 110 percent of capacity (achieved by cutting back on normal maintenance downtime) barely broke even. Gas wars and other competitive factors caused losses in a true free-market fashion. Companies could not compete for very long on quality. All cars had to run on all gasolines, even on mixtures of them. Any slight advantage one marketer gained for his product through a better octane, a detergent additive, or better cold-starting characteristics, his competitors soon matched. What was true for gasoline was also true for lubricants and other automotive petroleum products. They were interchangeable.

Nor could companies compete on price indefinitely. Except for the occasional and sometimes ruinous price wars in which one company subsidized the below-cost sales of its local distributors, prices equalized around a point that both the very efficient and the not so efficient could support. In any free market with unlimited supply, open pricing finds its level at a point that allows the efficient to make a profit, the less efficient to make less profit, and the inefficient to go out of business. No one can sell even the incremental barrel below cost for very long.

Gasoline marketers competed effectively on service, image, and availability. A complete range of branded products, heavy advertising support, clean restrooms, friendly and competent attendants, premium stamps, giveaways and a gas station in every likely location were the standard tools of the gasoline marketer.

Selling more gasoline, however, is not the only concern of the marketer. Gasoline is a seasonal commodity. Demand is greater in summer than in winter. Gasoline does not store well. And demand does not always equal supply in a given locality. It never pays a refiner to run his plant at varying capacities to follow local fluctuations.

Refineries are efficient over only a fairly narrow range of production levels and product mix. Outside that range, the unit must shut down. That means that the manufacturing arm drives the marketing arm, unlike the situation in most other businesses. The refinery must make a certain minimum amount of products, and marketing must then move those products. To do that, a secondary gasoline market grew up to help even out local and seasonal fluctuations in demand and to allow refiners to continue operating at optimum capacity year round.

Refiners buy from or exchange with one another to even out supplies. A marketer situated far from a company refinery buys from or exchanges with a local refiner to supply his outlets during the busy season. It costs more, but it is better than paying freight. Conversely, the refiner supplies competitors in his locale in similar circumstances. The product is made to the specifications of the buyer.

At another level, major refiners, and even minor ones, use jobbers to help move the product. The jobber is an intermediate independent wholesaler — a middleman — who buys from a refiner, usually on contract, then moves the product through his own string of branded or independent service stations. Sometimes he has his own customers to whom he wholesales that product. In some

281

This early Standard Oil of New Jersey service station began operating May 13, 1923 at Third Ave. and 17th Street in Huntington, W. Va.

instances, the jobber buys and blends the products of several refiners. In others, he deals with one refiner exclusively. He is a ballast in the marketing system that absorbs and disperses oversupply and often finds supplies to fill shortages. Within limits, the jobber helps the refiner run his plant efficiently by cushioning short-term market fluctuations.

This free-market system, which prevailed into the early 1970s, assured ample supplies of gasoline at a fair price. By 1973, more than 225,000 outlets were deriving most of their income from retail automotive petroleum products sales, mainly gasoline. Many thousands more also sold gasoline, although it was not their main product. Total retail outlets numbered more than 340,000 at the high point, according to government statistics.

The system also gave the motorist ample, convenient service and top-quality products of all kinds, from tires to antifreeze, motor oil, and lubricants. The neighborhood service station, owned or leased and operated by a local businessman, with its loyal local clientele, came to have its place in the family phone book along with the grocer, the doctor, and the dentist. He was another local retail supplier who valued his customers and whose customers trusted him.

This all changed beginning in 1970 when U.S. domestic oil production peaked. In addition, static growth in world markets during 1971 and 1972 and the prospect of heating-oil shortages in the United States in early 1973 influenced marketers to alter plans to increase sales by price wars and cut-throat competition. The oil embargo of 1973 made the public aware for the first time that oil was a depleting resource and that alternate forms of energy would have to be found. This changed the status of the neighborhood gas station.

Some believed that price increases were inevitable anyway, since petroleum was becoming a steadily cheaper commodity as measured against the over-all pattern of long-term world inflation. Pursuit of the incremental barrel in a continuing state of gasoline surplus and impending fuel-oil shortage was not in anyone's best interests.

The oil embargo, nationalization of crude supplies, and price increases by producing countries vastly

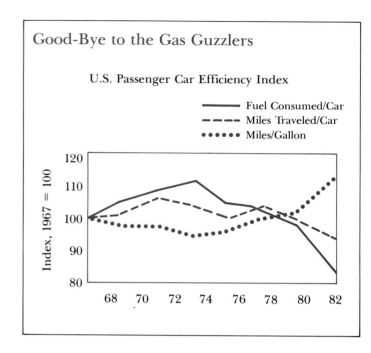

Good-Bye to the Gas Guzzlers

U.S. Passenger Car Efficiency Index

—— Fuel Consumed/Car
- - - Miles Traveled/Car
•••••• Miles/Gallon

changed marketing. On the supply side, government intervention and control at many levels stifled the free market. However, although demand continued to increase throughout the 1970s despite the still-well-remembered gasoline lines of 1974, growth rates slowed as prices accelerated. Products again went into surplus, but this time the theory of the incremental barrel did not work. The high and rising cost of crude, maintained in the face of surplus, set such high floor prices on products that oversupply had negligible effect on prices so far as the consumer was concerned. Five cents a gallon on a 30 cent price has much more market influence than 5 cents on a price exceeding a dollar.

In 1982, however, a second relatively prolonged surplus combined with soft economies and the producing countries' now chronic need for money did break the trend of ever-increasing prices. For a while, consumers dreamed of a return of retail prices of less than a dollar a gallon, while wholesalers hopefully foretold the end of OPEC's control of prices. Nonetheless, in consumer

282

January, 1974, saw drivers lining up to buy gasoline, which was in short supply following the first Arab oil embargo of October, 1973. These shortages had little real impact, however, until prices skyrocketed in 1979. After that, consumers began to be more conservation-minded.

terms, in the mid-1980s summer product prices tended to remain far above the levels they had been only a few years earlier.

Other factors, which were always present, have become increasingly important to marketers. These are the activities of the oil traders and the spot market. Oil traders, who operate for themselves or as special trading departments of major companies, buy and sell quantities of crude and products in the open or spot market. The price they charge is determined by what the buyer is willing to pay and the seller willing to sell for "on the spot."

Traders normally perform a minor function — their dealings are but 8 percent of the total — but an important function nonetheless. They help bring buyers and sellers together, ensure that refiners and marketers whose temporary demands exceed their assured supply can meet their commitments to their customers, and help those in oversupply to move products. The market is not orderly. Prices fluctuate wildly according to changes in weather and world politics. Prices reflect what people think will happen as much as what actually happens. The spot market tells the world what the real prices are and in this way influences the posted prices.

Spot prices turn on a single cargo, on one operator's desperate need either to buy or to sell. The official price, on the other hand, is the price of crude, specifically the price of Arabian Light of Saudi Arabia, which serves as a benchmark. The U.S. Department of Energy calculates and publishes a weighted average international price based on many oils from many different countries. In the shortage of 1979, spot prices reached $50 a barrel, while official prices hovered around $20. In the surplus of 1980 and again in 1981-82, spot prices fell far below the official prices as the latter approached $40. In spring of 1983, the official price was around $29, and the spot-market price was around $28, reflecting the reality of that period. The official price and the spot price constantly reflect one another, despite occasional trade well off the market.

Traders keep the oil moving. If a refinery expects a tanker load of crude and something happens to the tanker, the refinery cannot close down. The company must find another cargo of a suitable quality of crude or product. The refinery will not run on just anything, nor

284

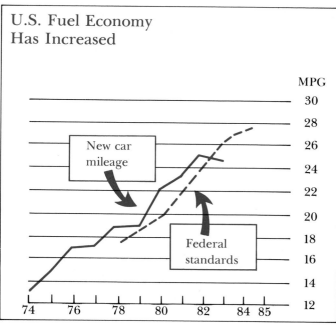

In 1974, cars averaged only 13 miles per gallon of gasoline. By 1983, fuel-efficiency had increased to over 24 miles per gallon.

will customers accept substitutes. The trader goes to the spot market to see who has a cargo to sell at what price. The seller, another trader, may have acquired a cargo on a long-term contract from a supplier, but the cargo is in New York, where the refinery is. The second trader may have a customer in South Africa, close to the point where the first trader may also have a cargo. Both save on transportation costs and time by exchanging the cargoes. Or the second trader may have a cargo he does not need, which he then sells to the first trader at whatever price the market will bear. Either way, the refinery gets the crude it needs, and consumers get their gasoline and heating oil. The trader, of course, tries to buy at $20 and sell at $50. Many have bought at $50 and have been forced to sell at $20 in a falling market.

Though oil is the world's largest cash commodity, there has not been a futures market in oil as there is in most other commodities. Pricing reflects the spot market, private contracts, and agreements and, of course, OPEC's administered price. On March 30, 1983, a new and perhaps important factor entered the pricing equation. This was the establishment of a futures market in New York and Chicago. There sellers and buyers agree on what the price of oil will be several months into the future, regardless of other factors or posted prices. This development means that a refiner can buy oil now for future delivery knowing exactly what his cost will be. The seller can sell today for that future delivery knowing what he will earn. His profit, or loss, will be the difference between the price he contracts for today and what he could have sold the oil for at the time of delivery.

This futures market will not replace any of the other pricing mechanisms, any more than futures markets in other commodities replace spot markets and private agreements. But the entrance of large numbers of investors and direct contract buyers into the petroleum futures market promises to have a powerful influence on petroleum prices. The first contracts, 4,000 of them in Chicago and nearly 2,000 in New York, each for 1,000 barrels, went for about $29.50 per barrel for delivery in June 1983.

By the end of 1984, trading in the futures market had become a major force in the world petroleum markets with over 20 million bpd being traded. This figure is estimated to exceed 100 million bpd by 1986 at which time futures trading will be more than double actual daily production. This tremendous speculative surge in crude oil and product futures is a totally new phenomenon for the industry and could strongly influence spot markets.

The changes in marketing since 1973, especially in automotive marketing, have been substantial. Gallonage is not the only goal, despite the multiplication of self-serve stations (even among the majors) after the price increases of 1979 in the United States. Another important factor is efficient use of resources. With the glut of 1983, companies have added such functions as convenience stores to their stations to increase sales volume. They have converted some stations to other uses and have closed tens of thousands of them. Only 120,000 stations remained in 1983. Price differentials between branded and unbranded stations have narrowed as majors cut frills and costs.

Companies that once trumpeted the universal avail-

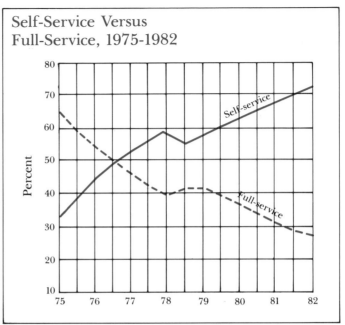

Companies have switched from full-service to self-service stations in order to keep prices competitive and to retain their share of a smaller market.

285

Major oil and gas retailers have added other functions to their service stations, such as convenience stores. ARCO's MP&G Tune Up centers are an example of this trend.

ability of their brands have withdrawn from high-cost or marginal areas, leaving them to local suppliers. Government regulations that allocate supplies and control prices crimp the free market and distort supply patterns, probably retarding the territorial contraction of individual company marketing operations.

In the United States, gasoline and automotive products are still the industry's biggest sellers. Motor gasoline represents about 50 percent of U.S. refinery output, nearly twice the quantity of distillate fuel oil, or heating oil, and six times the quantity of jet fuel. But gasoline occupies only one-third of the world's market. Heating oil is the largest-volume product in the world market.

Historically, more than half of all petroleum products sold by refiners to industrial users, resellers, and wholesalers has been at wholesale. Today, the figure is 83 percent. Typically, the wholesale customer is a jobber or a distributor, a large industrial enterprise or utility, a shipping or airline company, a railroad or trucking line, an office building, or a blending and packaging plant. These customers buy not by the gallon but by the tanker load or cargo or take delivery via their own pipelines or trainloads of tank cars.

There are many ways to organize wholesale marketing; the usual way is by product. Typical sectors in a wholesale marketing organization include gasoline and distillate sales, industrial-product sales, and lubricants and specialties sales. Since these areas overlap one another, most companies further divide into regional and specific product and industry market areas.

The largest direct users, such as airlines, utilities, chemical plants, refineries, and major jobbers and distributors, generally take direct delivery from the oil company. A tanker ties up alongside a power plant or refinery and unloads into storage tanks large enough to hold an entire cargo, hundreds of thousands of barrels. Similarly, a major jobber who handles the products of several companies or a distributor who handles the products of one company takes delivery in large amounts.

Heating oil usually reaches the retail customer through jobbers and distributors, though some refiners have their own retail subsidiaries. Jobbers get the oil wherever they can, often from distributors, and sell it

286

Business at U.S. Service Stations

Year	No. of service stations	Total sales (add $000,000,000)	Avg. per station (add $000)
1983	139,250(a)	$114.5	$822.2
1982	144,690(a)	106.5	736.1
1981	151,250(a)	101.7	672.4
1980	158,540(a)	96.3	608.0
1979	164,790(a)	70.9	430.0
1978	172,300(a)	60.9	353.0
1977	176,465(b)	56.5	320.0
1976	186,400(a)	48.0	257.0
1975	189,480(a)	43.8	232.0
1972	226,459(b)	33.7	148.8

(a) Franchising in the Economy 1981-1983, U.S. Department of Commerce; data are as of the end of the year.
(b) Bureau of Census definition of gasoline service stations: 50 percent or more of dollar volume comes from the sale of petroleum products.

While the number of service stations has been reduced by almost half since 1972, volume of sales at the surviving stations has/more than quadrupled.

under their own names. The householder seldom knows and probably cares less where the oil originated. Jobbers are more at the mercy of market forces, however, and form a spot market of sorts. Distributors generally identify with a major from which they obtain supplies and other services on contract. Although they, too, are independent businessmen, their terminals and trucks display the company emblems of their suppliers.

Generally, these people have contracts with householders and apartment owners to deliver a given amount of heating oil over a given period, plus service contracts to maintain the heating plant. As in gasoline sales, competition can be fierce. The customer has little to influence choice except service. Unlike the gasoline service-station operator, the distributor has to compete for the business with other fuels, particularly natural gas. He is not any happier in knowing that the gas his former customers are buying from the local gas and electric company is coming from the same oil company that sells him heating oil.

In the industrial products sector, individual units are

Typical Marketing Information

Industry Scoreboard 11/21

	Latest week 11-11-83	Change from Week before*		Year ago*	Percent change year ago
Demand 1,000 b/d: (4-week avg.)					
Motor gasoline	6,636	+	77	+ 161	+ 2.5
Distillate	2,653	−	69	− 14	− 0.5
Jet fuel	1,060	+	169	+ 55	+ 5.5
Residual	1,251	+	18	− 280	− 18.3
Other products	3,848	+	9	+ 601	+ 18.5
Total demand	15,448	+	204	+ 523	+ 3.5
Supply, 1,000 b/d: (4-week avg.)					
Crude production	8,679	−	8	− 4	−
Crude imports	3,508	+	107	− 43	− 1.2
Products imports	1,810	−	15	+ 338	+ 23.0
Crude runs to stills	11,674	−	3	− 71	− 0.6
Stocks, 1,000 bbl:					
Crude	360,637	+	5,598	+ 21,414	+ 6.3
Motor gasoline	227,540	+	2,759	− 2,974	− 1.3
Distillate	164,174	+	2,624	− 12,387	− 7.0
Jet fuel	42,662	−	1,013	+ 1,516	+ 3.7
Residual	47,881	+	352	− 13,608	− 22.1
Drilling: (4-week avg.)					
Rotary rigs	2,464	+	43	+ 32	+ 1.3

*Based on revised figures.
OGJ Newsletter, Nov. 21, 1983

287

The Oil and Gas Journal Newsletter, *published weekly, provides marketing planners with valuable information about the availability of crude. This knowledge can affect pricing decisions.*

Capital Spending by Marketing Industry

$ Millions

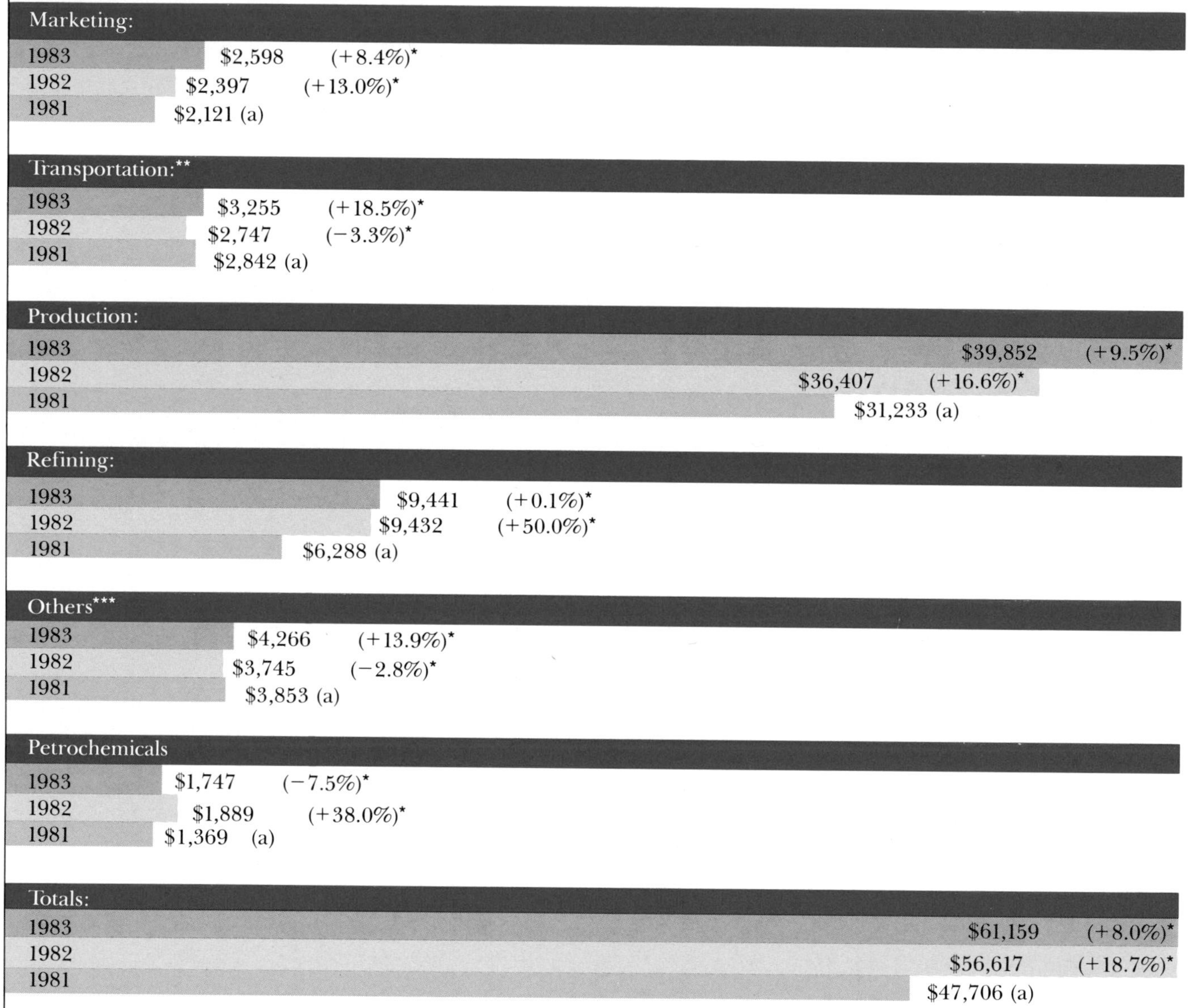

Marketing:

1983	$2,598	(+8.4%)*
1982	$2,397	(+13.0%)*
1981	$2,121 (a)	

Transportation:[**]

1983	$3,255	(+18.5%)*
1982	$2,747	(−3.3%)*
1981	$2,842 (a)	

Production:

1983	$39,852	(+9.5%)*
1982	$36,407	(+16.6%)*
1981	$31,233 (a)	

Refining:

1983	$9,441	(+0.1%)*
1982	$9,432	(+50.0%)*
1981	$6,288 (a)	

Others[***]

1983	$4,266	(+13.9%)*
1982	$3,745	(−2.8%)*
1981	$3,853 (a)	

Petrochemicals

1983	$1,747	(−7.5%)*
1982	$1,889	(+38.0%)*
1981	$1,369 (a)	

Totals:

1983	$61,159	(+8.0%)*
1982	$56,617	(+18.7%)*
1981	$47,706 (a)	

*Figures in parentheses indicate percentage of change from previous year's total.
**Includes only oil pipelines, not gas.
***Includes coal mining.
(a) Comparisons with prior years cannot be made inasmuch as the projection bases from 1981 on were changed materially and cannot be related.

designed either to sell specific products or to sell to specific industries. Unlike gasolines and distillates, which are sold to large numbers of customers in given geographic areas, industrial products go to geographically dispersed, large-volume accounts that require considerable technical expertise to service.

Typical industrial sales include asphalt for paving and roofing, residual fuel oil for utilities and marine transportation, coke for the metal and rubber industries, aviation products for airlines and governments, and natural-gas liquids for chemical, agricultural, and heating markets. If the sales volume warrants, some of these functions are combined. In many companies, natural-gas-liquids sales are set apart, either as a separate sector under wholesale marketing or as a separate subsidiary.

In major companies, some of these functions are international. For aviation and marine sales some companies maintain marketing operations in many countries to service airlines and ships around the world. Depots stock many grades of hundreds of items — fuels, lubricants, hydraulic fluids, greases, deicing liquids, and other petroleum supplies — made, stored, and dispensed to the customers' exact specifications.

The lubricants and specialty-products area, which includes waxes, additives, and agricultural oils, reflects the smallest volume of sales, but probably the most diverse marketing sector. Unlike large-volume fuels and other products, which work within broad specification tolerances, lubricants and specialties are often formulated precisely. Selling to and advising customers on the proper use of these hundreds of products takes special know-how and requires far more attention per volume of product sold than that required for any other petroleum product.

A company can conveniently sell lubricants and specialty products through local sales outlets in industrialized areas. But since they cross all lines, the marine and aviation departments, as well as the automotive department, also sell them. In addition, sometimes they are wholesaled to resellers, who sell to smaller accounts.

Distribution is integral to marketing. The product must get from the refinery to the customer reliably. Transportation alone is not distribution. In the days before the modern network of pipelines, railroad tank cars carried most refinery product to small bulk-storage plants across the country. The bulk-plant operator then delivered products by tank truck, or tank wagon, as it is still called, to service stations, industrial plants, and farms. Pipelines and highways changed this. Large bulk facilities at pipeline terminals now serve cities. Better highways and larger vehicles allow longer road deliveries. Small bulk plants became obsolete, and increasingly stringent environmental regulations hurried them to oblivion. Redesigning a large plant to comply with rules is cheaper per barrel than is reworking many small ones.

Each major company has its own terminals in important markets. Common-carrier terminals exist as well to service all companies. They are located principally at water ports. In regions where volume is too small to justify a terminal, some companies exchange product with others through another company's terminal. In such instances, a company takes the exchange base product, gasoline for example, and blends it with its own additives to its specifications.

While the shape of the marketing man's world has changed, the needs of the industry have not. More than ever, the industry needs funds to find and develop future supplies that are as free as possible from external control. Those funds can come only from sales. The marketing department's job is to analyze and forecast trends in markets to assure that the products needed to generate those funds will be available and can be sold.

Further, industry changes in recent years have not lessened competition or changed the fundamentals of the free marketplace. Each company still must satisfy its stockholders and vindicate its management by increased efficiency, profitability, and market share. The marketing department no longer can produce funds and profits through ever-increasing sales at ever-lower prices. The myth of the incremental barrel is dead.

The marketing manager's future challenge is to provide the money the industry needs for exploration and development as well as profits. He must do this from a shrinking resource base at surely increasing prices in an ever-more-competitive environment and under the continuing threat of increased government regulation.

289

CHAPTER 28 –
WHO WORKS FOR OIL

The function of the oil-and-gas industry is to find, produce, process, and market oil and gas and their products. To do so, the industry needs tools and services of every kind and at every level. The oil companies, of course, could invest their own money to provide these tools and services for their own use, but that ties up capital. Oilmen can better employ the money and manpower in their own businesses than duplicate what others can supply at a lower cost.

Drake knew that. So did every oilman after him. Though Drake was his own drilling contractor, he did not try to make his own tools and machinery. He bought them from suppliers who were already in the business of providing goods and services to drillers of water wells. If Drake or his early successors needed a special tool, piece of equipment, or service, astute local businessmen were quick to provide it. The odd piece that could not be had off the shelf was sketched or described to the local blacksmith, who made it for him. If the smith was smart, he refined the prototype and sold it to other drillers.

Sometimes the blacksmith and the driller were one and the same. There is the famous story of one smith who, with his brawny sons and his own tools, drilled a well on his own land and lived happily ever after off the proceeds. His investment in tools, equipment, and services from outside vendors was virtually nothing. He was a rarity then and would be more of a rarity now, when even the largest petroleum and gas companies buy or lease most of their tools, equipment, and services.

Today, the oil-field service industry is immense, and its boundaries have never been easy to define. The old-fashioned millwright or blacksmith who built equipment and made tools for the oil patch did the same for railroads, canal companies, shipping lines, and other industries. The extent to which his customers were petroleum producers probably had more to do with the happenstance of his being where the oilmen struck oil than with any conscious intent by him to participate in the "oil supply and service industry."

When Sir Henry Bessemer invented and patented

the Bessemer furnace for making steel, he hardly considered himself an oil-industry supplier. There was no oil industry in 1856. Yet his process and his product, and the companies that manufacture it, are as essential to the petroleum industry as are the companies that use steel to make drill bits, drill pipe, derricks, barges, platforms, and pipelines.

While it is not an oil-field service industry per se, the steel industry is by far oil's largest supplier. Of the $8 billion the oil industry spent in the United States in 1978, $1.6 billion, or about 20 percent, went to the steel industry for pipe, casing, and tubing. During the peak drilling year, 1981, the nearly 4,000 active rigs in the United States alone required more than 3.7 million tons of steel pipe and tubing, much more than the industry produced. Increased imports of foreign-made steel, whose share of the U.S. tubular-goods market increased from less than 20 percent to more than 30 percent, filled the gap.

The reemergence of China as an economic influence

Training is an integral part of every company's personnel policies. At this Los Angeles center, a student works with an audio-visual instruction system.

Members of ARCO's staff went to Lame Deer, Mont., the headquarters of the northern Cheyennes, where they provided seismic training, pictured here, to members of the tribe.

testifies to the interdependent nature of this international business. China's drive to develop its own energy sources could change the worldwide market for steel goods because Japanese steelmakers will try to redirect their own manufacturing capacity to serve that expanding economy.

When the petroleum industry is slack, the malaise extends all the way back to the mines from which the ore comes to make the steel. When the petroleum industry prospers, the mines and the companies that make machinery for the mines also prosper. Jobs in those sectors reflect employment in the oil-and-gas industry, which is in turn directly proportional to the number of active wells in the United States.

As drilling activity expanded in the 1970s — in 1971, there were only 976 active rigs; in 1977, more than 2,000; by 1981, nearly 4,000 — something of a labor shortage developed, particularly among drilling contractors. There was pressure to run rigs on two shifts, or even around the clock, to increase their productivity. But the skilled workers were not available then. Contractors were able to find workers to handle two shifts, but they could not always find experienced hands. Some contractors undertook training programs so that they could field capable crews but then found that the oil company that owned the well had not kept pace because the oil company's engineer was woefully short of experience. Neither the contractor nor the oil company wanted to run the rig with inexperienced or untrained crews, who could jeopardize themselves and the considerable investment in the well.

Thus training is an essential part of any company's personnel policies. Perhaps it is even more so for the petroleum and petroleum service industries. The value of the product has increased so much in recent years, and the technology has become so complex and costly, that there is less room than ever before for relatively unskilled personnel at any level. At the same time, the industries are mature and their employee profiles reflect that. Many skilled and experienced workers at every level are close to retirement. Their places must be taken by new employees, who learn the trade not in the field, as in the old days, but in school.

Studies by Sun Oil presented in a paper at an American Petroleum Institute meeting in 1979 showed that that company expected to lose as much as 42 percent of its personnel within the next ten years. In this study, which is not untypical for the industry as a whole, between 54 and 59 percent of the senior operating employees will leave by 1986. They must be replaced by workers mostly from the company's lower-level field staffs, which also will suffer an attrition rate of 23 percent. The practical effect is that where formerly the industry developed its employees over a 10- to 20-year span now the same group has a mere 2 to 3 years of experience before it has to assume the same responsibilities. In that time, the value of a 500-barrel tank of oil for which those employees are responsible has soared from $1,000 to $10,000 and more. The value of the equipment that they operate has similarly inflated.

Job opportunities in the petroleum and related industries continue to increase. According to the Contractors' Association, in 1981 a good worker could move

The oil industry provides many jobs in the U.S. economy. An example is the construction of this chemical plant in Texas.

from floorman to derrick worker in six to eight months and then, in a similar span of time, become an operator. Such rapid advancement is a far cry from the old days, particularly considering the value of the rig and the well. A company today may place a $10 million rig and a well that costs several million dollars under the supervision of an operator who less than two years earlier was a green roughneck.

Similarly, a brand-new graduate engineer employed by an oil company or one of the service companies may find himself at a very youthful age wholly responsible for the construction of a huge project or the operation of a plant on the off shifts. Experience like this soon matures a young person and tests his training and his management and personnel-selection abilities. It also assures that competence and merit are the only criteria for advancement.

The oil-and-gas industry is easily described as the exploring, drilling, producing, processing, transporting, and marketing of crude petroleum, gas, and their products. Total direct employment in the oil and gas industry in 1981 was 1,835,000, according to the U.S. Bureau of Labor Statistics. Those figures include everyone from chairmen of the board, explorationists, and financial advisors to service-station attendants. The service industry is not so clearly defined.

Thousands of companies comprise the oil-field services industry. These supply every conceivable kind of equipment and service to the oil-and-gas industry. Communications, food and catering, entertainment for offshore and remote drilling and production crews, housing, transportation, medical services — there is no end to the list.

For every million dollars' worth of petroleum products produced in the United States there are nearly 30 jobs throughout the economy. About 11 of those jobs are in the petroleum industry itself. The other 19 include the drilling contractor's crew, the coal miners that produce the raw material for the coke that goes into the blast furnaces from which eventually comes drill pipe, the secretary in a public-relations firm that has one oil-related account among its many clients, and free-lance writers preparing articles for business publications.

Drilling activities expanded in the 1970s, providing many job opportunities for young men and women.

Building a pipeline is a costly and complex operation, requiring hundreds of workers and the cooperation of many companies, from steel manufacturers to construction companies.

293

In addition to employment in the oil-and-gas business itself, there are approximately five million jobs directly or indirectly dependent on the industry. Most of these jobs support families. Thus, conservatively estimated, more than ten million people in the United States alone depend in some manner on the oil-and-gas industry for their livelihoods. Those companies whose fortunes are tied most closely to the oil-and-gas industry are those whose business depends entirely or almost entirely on the petroleum industry. They prosper or suffer greatly with the ups and downs of the oil business. If the oil companies do not or cannot spend money for exploration, development, and maintenance, then these companies and their employees lack work and jobs. When the oil industry sneezes, the closely allied oil service companies catch pneumonia.

This ripple effect, both the good and the bad, of course, is greatest for companies most closely tied to the oil industry. The effect becomes less marked in the companies on the outer edges, which have more business with other sectors besides oil. But the ripple never becomes imperceptible.

The drilling and oil-well-servicing contractors, whose business is clearly 100 percent oil-and-gas-related, make up the most numerous sector, since it requires a relatively small investment to buy a drilling rig and hire oneself out as a contractor. The Association of Oil Well Servicing Contractors, for example, has 425 member companies that, collectively, operate about 5,000 rigs. This represents about 60 percent of the total number of rigs in existence, a number that fluctuates with economic conditions.

This segment of the industry is highly individualistic, perhaps the last refuge of the legendary independent oilman. While mergers and buyouts are common in the rest of the corporate world, including major sectors of the oil service industry, the contract drilling industry remains one of mostly small operators. About 80 percent of the Contractors Association members operate ten units or fewer.

Acquisitions and mergers do happen. The greater financial power of a large operation gives it a competitive advantage and allows it to operate in areas that it could

294

Drilling offers an opportunity for the independent oilman, the last of the rugged individualists. A small operator can buy a workover rig for $100,000 or less and strike it rich if he's lucky.

not otherwise enter. Deep-hole rigs are expensive and beyond the resources of small operators, as is virtually any kind of offshore work. But as the association itself says: "There are a lot of rugged individualists in our business. You couldn't buy them for a half interest in China." And as fast as a company disappears, through merger or acquisition, a new one appears, perhaps founded by the principals of the defunct one or by their well-educated, entrepreneurial, business-minded sons.

Small operators dominate the highly competitive land-drilling business. Since a small workover rig can be bought for $100,000 or less, contract drilling is an ideal field for an entrepreneur with a little money and a lot of knowledge, even though success or failure is tied directly to the state of the oil business in general. Upscale, rigs are costlier but still within the reach of a small operator. A shallow drilling rig costs about $2 million, while a top-of-the-line deep-drilling rig costs around $10 million.

The high costs of drilling offshore, as a practical matter, limit entry to the larger, financially stronger firms.

Even a shallow-water jack-up rig can cost $28 million, and the prices go up from there to over $100 million for a very large, "blue water," all-weather, semi-submersible rig. Such prices, combined with the much higher costs of offshore operation, exclude the small, low-capital operator.

While the oil-field service industry is as diverse as American industry itself, there are certain tools, equipment, and services that are clearly oil-field-oriented and whose suppliers thus rank high in the service industry. The companies that supply these tools and services tend to be the larger ones in the industry. Their sales are in the hundreds of millions, even billions, of dollars annually. Most of them are conglomerates, the results of many mergers and acquisitions over the years. The business interests of some extend far beyond oil and gas, though their names remain most closely associated with that industry. Some, like Pullman and Combustion Engineering, have names that at first one does not associate with the petroleum industry.

295

Workers on the inside of the rig prepare to hoist the crown block into place.

The tool-bit industry is the cutting edge of the oil-and-gas industry. A few major companies such as Hughes Tool, Reed, and Smith Tool dominate this sector. These companies and many others also supply most of the other special downhole tools, such as packers, fishing tools, and shoes.

Companies like NL Industries and Baroid are best known for drilling muds. Others, like Mission Manufacturing and Cameron Iron Works, are well known for blowout preventers, well-completion equipment and wellhead assemblies. Still others, like Dowell, are specialists in cementing, stimulating, and remedial services. Schlumberger is a famous old name in well-logging services, but today this $2 billion company provides a broad range of products and services.

Other sectors, with their specialized suppliers, include rig outfitting; offshore rig construction ; contract drilling; marine construction of all types, including pipeline laying, platform construction, and undersea production systems; downstream services, such as process plants and equipment; and support services, such as geophysical services, service boats, and helicopters.

Many of the companies in these areas provide services and equipment across several segments. One thing leads to another. A firm that obtains a contract to build a platform, for example, may then be asked to undertake additional work in related areas. That work eventually expands to another division.

In 1981, of about 45 major companies in the oil-field service industry, the smallest had sales of over $67 million. Seventeen had sales of well over $1 billion dollars each. Halliburton, which includes the construction firm Brown and Root, had sales of over $8.5 billion. J. Ray McDermott's sales exceeded $3.1 billion. Dresser Industries topped $4.6 billion. Combustion Engineering, which has business in many areas besides oil, had sales of $3.8 billion.

During the same year, 1981, the domestic oil-and-gas industry spent $83 billion in capital expenditures. Of that amount, $58 billion, or 70 percent, was spent on exploration and production. The industry spent over $36 billion just to drill and service some 80,000 wells. That money went to buy drill stem and bits, lease equipment — the

pipe, pumps, and tanks that take the production from the well — seismographic and geological equipment, offshore rigs, and drilling vessels, and much, much more.

Worldwide figures are about double the domestic figures. Much of that business is American, too. Many U.S. companies do as much oil-field service work overseas as they do at home; some do more of their work abroad — this despite the unsettled political and economic conditions in many oil-producing nations. Indeed, though the business is worldwide, the state of the art is that as practiced by U.S. companies. The oil industry in the United States escaped much competitive pressure from foreign companies, despite increasing involvement by foreign government-owned oil groups. Where technology is a large component, specifically in equipment and tools and in deep-water offshore work, the U.S. firms remain on top. Companies in Italy and France, however, are becoming increasingly competitive in offshore work. American publications also carry advertisements for oil equipment made by foreign firms.

Where the labor component is high, as in construction work, American firms are beginning to experience much more competition from construction and engineering companies based in Korea, Japan, West Germany, Italy, and even India. The higher wages of American personnel, U.S. income tax laws, and national subsidies of foreign firms by their governments, which often own the oil companies that hire the contractors, are increasingly hampering U.S. firms abroad.

At the same time, the worldwide demand for energy enlarges the market for American oil-field equipment. The number of active drilling rigs, on a swooping roller coaster in the early 1980s, will increase toward the end of the decade. The number may approach 6,000, with over half in the United States. The expansion in domestic activity that began in 1979, peaked in 1981, and stalled in 1982 will continue, though in more measured style. It reflects the nation's determination to reduce its dependence on foreign oil and shows the immediate effects of incentive pricing and changes in regulations that make it attractive to drill at home for oil and gas.

Oil-field machinery follows the rigs. Since U.S. technology is still dominant, the United States remains

Capital and Exploration Expenditures

(U.S. $ Million)	1973	1974	1975	1976	1977	1978	1979	1980	1981	1982
Amerada Hess	242	423	283	291	422	353	659	985	1,261	875
Atlantic Richfield	500	1,751	1,751	1,826	1,681	1,358	1,823	3,370	3,559	3,953
British Petroleum	930	946	1,751	2,167	1,765	2,409	2,766	2,416	8,215	6,402
Exxon	2,235	4,520	3,654	5,100	4,600	5,300	7,400	8,000	11,095	11,412
Getty	437	520	513	618	684	865	1,239	1,345	1,975	1,860
Gulf	979	1,564	1,546	1,742	3,013	2,129	2,513	2,968	4,325	3,740
Kerr-McGee	113	228	235	333	269	270	389	575	606	674
Mobil	1,334	1,652	1,469	2,485	1,786	2,175	3,812	4,649	4,469	3,821
Occidental	109	267	321	370	390	510	566	1,080	1,342	1,159
Pennzoil	251	236	217	247	264	344	369	524	604	606
Phillips	329	587	679	716	1,078	940	1,454	1,666	2,664	2,132
Shell Group	1,357	2,071	3,113	4,188	5,227	5,252	9,034	7,959	9,320	9,231
Standard of California	1,217	2,271	2,360	1,705	1,429	1,692	2,258	3,599	5,568	4,666
Standard of Indiana	1,100	1,876	2,114	1,774	1,874	2,240	3,028	4,326	5,211	4,438
Sun	284	537	657	616	667	717	852	3,800	1,519	1,800
Tenneco	402	663	550	620	714	1,008	1,477	1,825	2,359	2,103
Texaco	1,334	1,674	1,864	1,583	1,564	1,581	1,642	3,077	3,166	3,768
Union	391	686	680	798	813	782	1,088	1,700	1,961	1,917

the major world supplier of equipment. In 1982, U.S. exports of oil-field equipment exceeded $7.5 billion, three times the $2.5 billion total of only three years before. Moreover, though the 1982 amount was a precipitous drop from the record $9.7 billion in 1981 exports, the industry expects that such exports will continue to increase rapidly to nearly $14 billion per year by 1991.

Further down the road, the maturing of new technologies in the oil-and-gas industries will create even greater demands for equipment and services. The 930 billion barrels of oil locked up in Canadian and U.S. tar sands and the 600 billion barrels in oil shales will require a huge investment if they are to be exploited. A survey in the early 1980s indicated that expenditures for tools and equipment alone would exceed $8 billion over ten years in this one segment.

The business of oil-field support will continue to grow in the decades ahead, adjusting to the inevitable changes in technology and in the oil-and-gas business itself. The real business, after all, is energy. For a long time to come, most energy will come from fossil sources of one type or another, to be consumed in liquid or gaseous form. Those companies that have developed the technology and the people who can deal with petroleum and natural gas will be the ones best suited to handle future liquid and gaseous fuels derived from other sources.

The opportunities, personal and commercial, that are expanding in gas and oil today will not contract in the foreseeable future.

CHAPTER 29 –
GOVERNMENT AND THE INDUSTRY

The petroleum industry was not very old when legislators and others began to hem it in with regulations and laws. Naturally, an industry born in the nineteenth century, an era of laissez faire that rewarded high risk but had few formal controls, did not take easily to outside constraints. The constraints came nonetheless, sometimes in the public interest, sometimes at the behest of a special-interest group, and sometimes because the industry asked for them.

The industry's involvement with the legislative process preceded the first well. In April 1859, the Pennsylvania legislature passed a law permitting companies to organize for the manufacturing of oil from coal; this may well be the first legislation concerning the refining industry. The gas industry came next, when Massachusetts established the office of Inspector of Gas Meters and Il-luminating Gas.

In 1863, after Oil Creek caught fire a few times, the Pennsylvania legislature passed the first antipollution laws. They prohibited the running of oil and tar into waterways in oil-producing areas. Even in remote Baku there was a rough form of regulation. If a gusher buried its neighbor's property in sand and oil, the owner of the offending well had to pay to clean it up. The cost could and did bankrupt the owners of prolific wells.

From those small beginnings an enormous regulatory structure has developed, governing the industry to the smallest detail. Of course, much of today's regulation and legislation applies to other industries as well. Antitrust and fair-business-practices laws; legislation and regulations governing employment, environment, and safety; securities-and-exchange regulations; general tax laws and regulations; laws governing natural monopolies — these cover all industries. The cost of all this regulation, which authorities estimate to be as much as $100 billion in the total economy, is a major financial burden for even the largest companies.

Much early petroleum-industry regulation was self-imposed. Unable to agree among themselves on such matters as conservation and pollution control, producers

In the early days of oil, fire was one of the great hazards of the business. Lacking facilities for the containment of fire, companies suffered enormous losses in property, stock, buildings, and sometimes even lives. Fires like this one at Oil Creek prompted the Pennsylvania legislature to pass in 1863 the first laws regulating the oil industry.

turned to government to impose regulations the industry as a whole needed for its greater health. Everyone knew what the "level playing field" of the times should be; what was needed was an umpire to see that all players observed it.

Then, as now, some of the legislation and regulation was instigated by people who had no interest in level playing fields. When oilmen wished to build pipelines to transport oil more cheaply, teamsters who feared for their jobs worked against and defeated legislation that would have permitted the pipelines to be built.

Times have not changed in that regard. Recently, a court decision allowed a pipeline company to convert a natural-gas line to petroleum and to sell the line to another operator. Shipping interests vigorously objected. The judge acknowledged that the conversion would have a damaging effect on the maritime industry but concluded that this would be more than offset by benefits to the public.

In a true free market, of course, such an issue would not arise. The pipeline operator and the ship operators would compete solely on cost, service, and convenience — without government intervention. The survival of the less efficient, more costly method means higher prices and, frequently, lower availability, especially when it occurs as the result of regulation. Worse, government protection of the less efficient dampens incentives to reduce costs and increase efficiency throughout the industry. The results again are higher prices and reduced supply.

In the early days, oil-and-gas regulation was largely a state responsibility. The federal government confined itself to overseeing tax matters, to arbitrating among the states, and to ruling on actions that came under the general powers of one or another of the existing regulatory agencies. Pipelines came under federal scrutiny fairly early, in 1906, when the Hepburn Act provided for their regulation as common carriers. The federal government, under its antitrust powers, intervened massively in the industry when it finally dissolved the Standard Oil Trust in 1911. The Federal Trade Commission approved a marketing compact designed to end chaotic conditions in gasoline marketing in the 1920s. The Supreme Court confirmed the powers of the Texas Railroad Commission to regulate the oil industry in that state. But as early as 1916, federal control of the industry had been proposed not by radicals but by prominent leaders within the industry itself. And when federal regulation did arrive, it came by invitation.

By 1929, many states had conservation laws that regulated drilling and production rates. The federal government then began requiring unitized production on its own lands. Still, there was no real federal regulation of the industry. The industry rejected a plea by President Herbert Hoover to join in an interstate compact for conservation purposes.

Then Dad Joiner discovered the East Texas field, and the world was awash in oil. Prices plummeted. The Texas Railroad Commission's attempts to regulate production in the name of conservation were fruitless, despite the calling out of the National Guard. "Hot oil" went to market illegally across state lines, keeping production levels at reservoir-ruining rates and, incidentally, depressing prices. Untrammeled competition was wrecking the industry's structure.

In time, of course, the "invisible hand" would have settled matters, as inefficient, undercapitalized companies went broke, demand rose to meet supply, and supply shrank with the ruination of reservoirs. Whatever happened to the companies was their problem. But destruction of the reserves was a national concern. The government had to step in.

The industry acted. First, it formed the interstate oil compact that it had spurned a few years earlier. In addition, it prevailed upon Congress to pass the Connally Oil Act, better known as the Connally Hot Oil Act. This law made it a crime to transport oil produced illegally under the laws of a state in interstate or international commerce. In effect, it made producers absolute captives of state regulation, though Supreme Court decisions had earlier validated the enforcement of such regulations under state police powers.

The Interstate Oil Compact, a conservation organization, was a cooperative move by the states themselves. Intended to forestall federal regulation in this area, it was not the regulatory foot-in-the-door that the Connally Act was.

Government regulatory fervor began in the 1930s and accelerated after World War II. Often the intention of the regulation was to restrain competition, maintain prices, restrict supplies, or conserve resources. It was meant to protect or give advantage to a sector of industry that was in favor at the time or to establish supportive floors and ensure "orderly" markets. On this level playing field, the players were the companies and the industries. Consumers and employees were merely spectators. They elected the referees but had little influence on either the rules or the conduct of the game.

The practical effect of state proration and the Interstate Oil Compact, armed with the teeth of the Connally Hot Oil Act, was to replace market forces with government price and production decisions. Oil prices, which had fallen from a range of 75 cents to 95 cents a barrel to 10 cents and less, rose to $1.00 to $1.50, a price higher than the price a free market would have supported.

At the same time, huge new low-cost fields were appearing in the Middle East, even though the United States had adequate domestic capacity. Because the price of crude in the United States was relatively high, cheap oil from abroad was attractive. This put pressure on domestic prices. In 1959, after voluntary import restraint was unsuccessful, the industry turned again to the government for help and supported a Mandatory Oil Import Program. This program was designed to assure "the viability of domestic petroleum and production efforts to . . . meet the national defense requirements of the United States."

In fact, the program further distorted the free market by shielding the domestic industry from competition and encouraging the exploitation and depletion of domestic reserves. The net result was the opposite of that intended. As is inevitable with regulation, the program opened a new market, a market in import rights, or "tickets."

All companies were allowed to import a certain amount of cheaper foreign crude, but not all had a need for it. Those without need could trade their rights. As long as foreign oil was cheaper than domestic oil, these tickets had real value, about $1.25 a barrel.

The federal government has always paid special

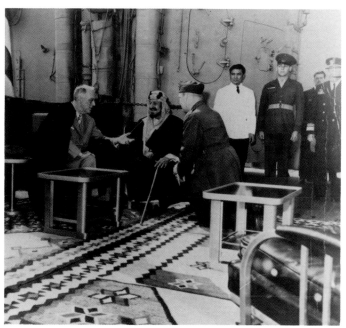

President Roosevelt met Saudi King Abdul Aziz on February 12, 1945, signalling a new phase in the U.S. government's interest in petroleum.

301

When, in the early 1930s, unrestrained production threatened to ruin the East Texas oil field, Gov. Sterling ordered operations to cease and sent in troops to enforce his order.

attention to natural gas because it travels through common-carrier pipelines that cross state lines. Gas fell under the purview of the oldest federal regulatory agency, the Interstate Commerce Commission, established in 1887 mainly to regulate the railroads as common carriers. The powers of the ICC were greatly strengthened by the Hepburn Act of 1906, and as early as 1912 it declared pipelines to be common carriers.

When Kansas shipped gas to Missouri and increased the price, Missouri protested. But the Supreme Court ruled that Missouri could not prevent the Kansas company from raising the price to Missouri customers because the gas had traveled in interstate commerce. That decision gave rise to the Natural Gas Act of 1924. Then, in the regulatory climate of the 1930s, Congress passed the Natural Gas Act in 1938. It marked the formal entrance of the federal government into regulatory control of the natural-gas industry.

In 1954, following yet another Supreme Court decision upholding the validity of the 1938 law, natural-gas production came under the aegis of the Federal Power Commission (FPC). The FPC set the price of gas sold in interstate commerce, but not that of gas sold intrastate. In the increasingly consumer-oriented regulatory climate of the 1960s, interstate prices were held at ever more unrealistic levels to encourage use of natural gas in place of coal or oil and to ensure low costs to consumers. The inevitable happened. Producers reduced their exploration and development and sold more of their product in the intrastate market. Eventually, gas shortages appeared nationwide.

The late 1960s and early 1970s found two trends converging. The regulatory zeal of the government was becoming a juggernaut just as production of domestic petroleum peaked in 1970. The Trans Alaska Pipeline was just getting under way when the National Environmental Protection Act became law in 1969. The project was brought to a halt as a result of suits against it by environmentalists, as well as efforts by the pipeline company to comply with the 1969 law. Its completion was delayed for years.

Not coincidentally, foreign producers became aggressive, and in 1970, for a variety of reasons, the prices of imported oil began rising. In August 1971, in an effort to restrain inflation, the government imposed wage and price controls on all industries. As a result, the price of domestic oil was controlled while imported oil prices were not. This anomalous situation naturally increased strains on the market. Shortages of oil and gasoline existed even before the oil embargo of 1973.

Attempts to relieve the strains with still more regulations only increased the distortion. For example, a ruling of the U.S. Cost of Living Council in effect established two types of oil companies — controlled and uncontrolled. The rule limited price increases to 1.5 percent for firms which had annual sales of $250 million or more, thus allowing uncontrolled companies to buy at controlled prices and sell at uncontrolled prices. In that situation, the bigger controlled companies did not care to sell to the smaller uncontrolled companies. That, in turn, led to yet another rule. In addition, multinationals had little incentive to pursue domestic business when there were greater profits and less paperwork abroad.

Pipe sections, intended for TAPS, are covered with snow while awaiting the go-ahead on construction. Approval was held up almost five years, until the environmental impact of the line could be determined.

302

In 1972, for the first time since it instituted prorationing, the Texas Railroad Commission raised allowable production to 100 percent. The domestic industry was running flat out. The restrictions of the Mandatory Import Control Program were loosened, and oil imports rapidly increased to almost 50 percent of U.S. consumption. Foreign oil was now costlier than domestic. Import tickets were worthless.

As fix followed fix, matters went from complex to impossibly convoluted. A formal two-tiered price structure for crude appeared in 1973. Its goal was to encourage domestic oil production while holding down large price increases. Special price controls were instituted on producers, refiners, resellers, and retailers.

The embargo of 1973 brought further changes. The actuality of shortages spurred the government to give the go-ahead to the Trans Alaska Pipeline project. It did this in a special act that also freed stripper wells from controls. That in itself introduced more distortions. The law defined a stripper well as one that produced 10 barrels or

Shortages which caused 1974 gas lines like this were due to the 1973 oil embargo as well as government regulations which complicated production and distribution.

303

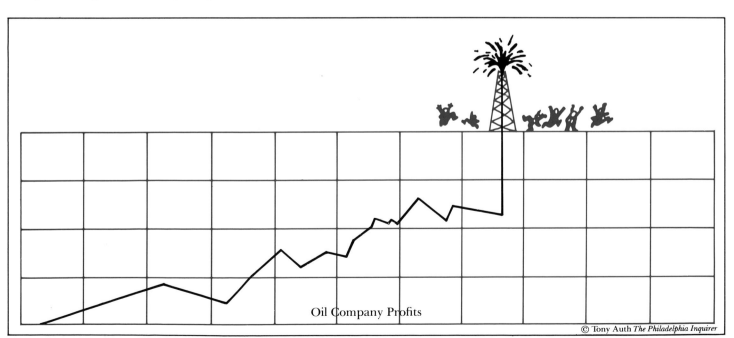

Oil Company Profits

© Tony Auth *The Philadelphia Inquirer*

Throughout the 1970s, oil companies were seen as villains because of shortages, rapidly rising prices, and the well-publicized high company profits. Editorial cartoons like this one encouraged the public, unaware of the government's role in these problems, to demand even more regulation of the industry.

less a day. Some owners found it expedient to cut back a slow well to less than 10 bpd and qualify it as a stripper. By this means, they reduced output but increased income. While such tactics were wholly contrary to the intent of the exemption, they were the expectable results of tampering with free-market pricing.

Attempts by both the oil and the gas industries to convince government and the public that the only sure route to energy sufficiency was less regulation generally failed. Shortages, rapidly rising prices, and large published company profits convinced both the public and the government that the industry was a villain. In the period after the 1973 oil embargo and before the 1975 Energy Act, almost 800 bills in Congress concerned themselves with the energy crisis in one way or another. Few favored a free market. The Energy Policy and Conservation Act of 1975 itself contained 99 pages of authorizations, mandatory pricing provisions, regulations, reporting requirements, and calls for planning. This act made it clear that government was now a full partner in the industry.

Later legislation established the Department of Energy (DOE) as the industry's own voice in the cabinet. The new department assumed the energy responsibilities of the Interstate Commerce Commission, the Federal Power Commission, the Atomic Energy Agency, the Department of Commerce, the Department of Transportation, the Department of the Navy, the Energy Research and Development Administration, and the Federal Energy Administration. The last two were newborn children of the energy crisis. This unwieldy melding of many existing bureaucracies at first proved so large, so ineffective, and so unmanageable that there was serious talk of eliminating the DOE and redistributing its functions. However, as the nation and the government realized that energy is a vital national interest that requires a unified approach, and as the department itself settled down, the talk became muted. Energy at the cabinet policy level appears to be here to stay.

While the DOE's mandated functions are many, regulation remains one of them. Its Federal Energy Regulating Commission (FERC) is the linear descendant of the FPC, with much broader powers and wider sweep.

The first reaction of government to shortage and

rising prices was, indeed, to impose controls. These were riddled with exceptions and special cases. As time has passed, however, and as both the public and Congress have been able to gauge the practical effects of controls and regulation, attitudes have shifted. The industry's pleas to be left alone have begun to be heard, though not without some skepticism.

The Natural Gas Policy Act of 1978, while extending government regulation to intrastate gas sales, also began steps toward decontrol of gas pricing. Results were immediate. Exploration and drilling for gas soared, leading to talk of possible gas surpluses rather than shortage. The industry's approach was vindicated. The 1978 law, however, made no changes in crude-oil pricing or regulation. It provided no incentives. In 1980, under discretionary authority granted in the 1975 control law, President Jimmy Carter began a phased decontrol program that was intended to bring domestic prices up to world prices and end price control. But what the president gave he also largely took away in the "windfall profits" tax (in

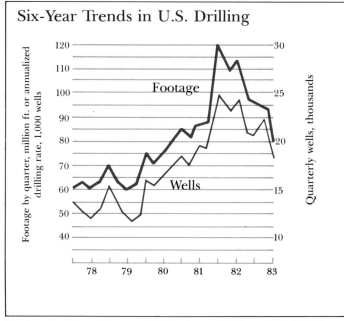

Six-Year Trends in U.S. Drilling

The promise of deregulation spurred exploration and drilling activities in 1981. Because demand decreased and prices fell, this activity soon tapered off.

304

How 1983 Drilling Compares With Previous Years

1980

Type I	Footage	47,716,597
	Wells	21,144
Type II	Footage	181,889,751
	Wells	37,753
Type III	Footage	70,557,901
	Wells	6,370
Total	Footage	300,164,249
	Wells	65,267

1981

Type I	Footage	60,808,621
	Wells	27,411
Type II	Footage	217,476,718
	Wells	44,085
Type III	Footage	89,255,955
	Wells	8,094
Total	Footage	367,541,294
	Wells	79,590

1982

Type I	Footage	64,468,409
	Wells	29,602
Type II	Footage	232,452,309
	Wells	47,473
Type III	Footage	110,944,048
	Wells	9,915
Total	Footage	407,864,766
	Wells	85,990

1983

Type I	Footage	45,170,629
	Wells	22,358
Type II	Footage	142,313,632
	Wells	31,041
Type III	Footage	
	Wells	
Total	Footage	
	Wells	

1983 drilling statistics show a trend toward less drilling and shallower wells, reflecting weaker demand because of conservation and recession.

reality, a large-scale excise tax) contained in the Energy Security Act of 1980.

Decontrol was not total; it applied only to gas from certain formations and depths. Multiple prices for the same commodity remained in place. Shortage and higher prices brought more gas on to the market. Reserves were refilled, eliminating any real shortage in supply. However, a web of regulations and marketing practices enmeshed suppliers, pipelines, and consumers and caused enormous dislocations. Despite surplus gas, prices accelerated. Cheap gas stayed in the ground, while consumers were forced to take the more expensive product.

Eventual deregulation is intended to stabilize gas pricing at the wellhead and allow the fuel to find its level among competing forms of energy. The road to that point will not be smooth because gas, unlike oil, is inextricably entwined with the pipelines that transport it. Pipelines, practically speaking, are natural monopolies. As such, they and the product they transport will never be free of some form of regulation.

But even as President Reagan began to accelerate decontrol of crude oil in 1981, much regulation continued and with it all the strange situations that would never arise in a free market. For example, a 1973 rule froze oil supplier-purchaser relations regarding sales, purchases, and exchanges of crude oil. The rule generally required a supplier to keep selling to the same customers he was selling to at the time of the embargo or, in the case of a new well, to the buyer of the well's first production. The rule was slightly amended from time to time, but remained in effect until 1981. In how many other industries are suppliers required by law to keep their customers?

In another strange case, the Standard Oil Company of Ohio (Sohio) had to sell its gasoline for 10 cents a gallon under the prevailing price in Ohio. The reasons are complex. They involve incentive pricing of Alaskan oil, prices of imports, price regulation, and carry-forwards earned by not charging the maximum permissible price. All the circumstances have little to do with market forces but

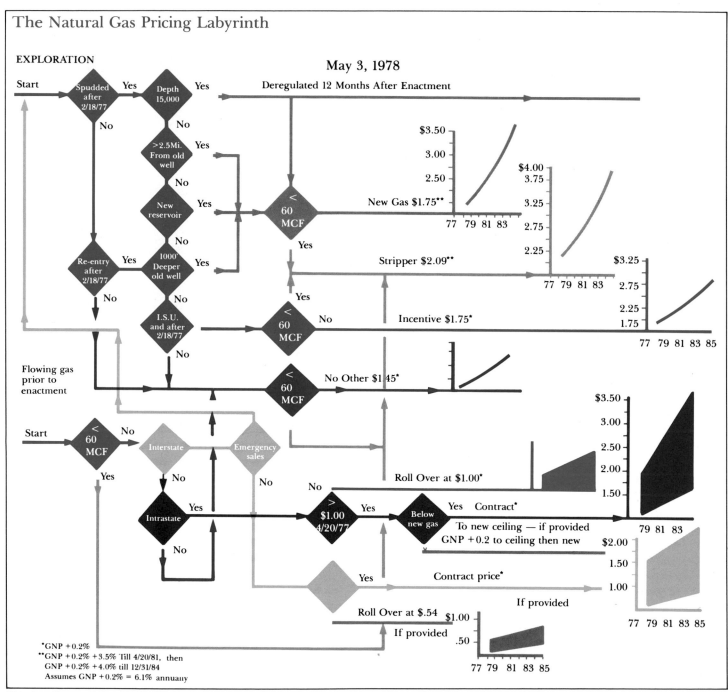

The Natural Gas Pricing Labyrinth

EXPLORATION

May 3, 1978

306

*GNP +0.2%
**GNP +0.2% +3.5% Till 4/20/81, then
GNP +0.2% +4.0% till 12/31/84
Assumes GNP +0.2% = 6.1% annually

This chart illustrates in graphic detail the maze of complex regulations governing producers under the Natural Gas Policy Act of 1978. Drawing the chart was difficult, but operating under the provisions of the law is even more difficult.

everything to do with regulation. Naturally, Sohio's competitors were not too happy that their customers were flocking to Sohio service stations. The government then ordered Sohio to raise its prices — and to pay the government the extra 10 cents a gallon.

The Sohio example is a case study in the unforeseen, cumulative results of sometimes conflicting rules and regulations applied over a number of years. Apparently, there is no such thing as a simple regulation or interpretation. Under the Natural Gas Policy Act of 1978, "agricultural use" of natural gas is exempt from the incremental pricing provisions of the act. This means that for such uses the price remains controlled. Naturally, every gas user would like to have his usage considered agricultural so that he may have access to continuing low-cost energy. Thus the Federal Energy Regulating Commission must spend time determining what is and is not agricultural. The results sometimes resemble medieval discussions on theology.

Industrial plants pay the higher prices. But if a plant uses the gas as boiler fuel for an ultimate purpose defined as "agricultural," it is exempt. The processing of wood pulp into cellulosic man-made fibers is exempt. Rayon is such a fiber. Processing of man-made fibers, including rayon, in textile mills is not exempt, though processing of cotton in the same mills is. Manufacture of food-grade wax into containers is exempt, while the same process for coating and lining containers is not.

That government is now a full partner in the energy industry is particularly evident in the creation of the $15 billion U.S. Synthetic Fuels Corporation, which represents the government's first full-scale venture into research, development, and production of energy. In future years, government probably will become less deeply involved. But as the antagonisms of the 1960s and 1970s ease, and as this unwieldy private- and public-sector collaboration gradually swings around on the new course toward energy sufficiency, all industry personnel hope that relations among the sectors will become mutually supportive and respectful.

Some regulation and legislation are necessary. Energy security is too important to be left entirely to the free market. Private priorities are not necessarily concurrent with national ones. Nor is there always time for the invisible hand to work its leisurely will. If even the gentlemanly game of tennis must have umpires, referees, and linesmen, then how much more so must the game of capitalism have them, where the stakes are so high?

Yet regulation should be a matter of common sense and, in rare and special circumstances, of arbitration. Regulation should not be a mindless and literal interpretation of narrow rules written for another time and applied to broad sectors in a fast-changing age.

307

CHAPTER 30 –
TAXES

In response to the normal growth of population and the level of services demanded by the people, tax loads have grown enormously since World War II. But they have also grown as government at all levels has taken on more and more of the functions once discharged by the people themselves, by the private sector, or by lower levels of government.

State and local taxes, once mere pinpricks in the public hide, now flay the taxpayer. As federal and local jurisdictions sought new revenues, business was an early target. And the extractive industries in particular were especially attractive targets since they could not easily move their operations to escape burdensome taxes.

The petroleum industry has long been the most inviting of industrial targets. Basically, it is taxed in the same manner as other industries, but more heavily. For example, it pays a severance tax unique to extractive industries and is the only industry on which the government has imposed a windfall-profits tax. Special tax-computation rules apply to it as well.

As one of the world's largest and most international industries, the petroleum industry pays taxes to the federal government; to states, counties, school districts, and other local jurisdictions; and to foreign governments. It pays income taxes, payroll taxes, state and local franchise taxes, property taxes, and production taxes. The government also imposes costs on it, such as royalties on foreign production and contributions to "spill funds" for each barrel of oil transported by water and by pipelines.

Where a tax may not exist today, it is a fair bet that some taxing authority will create one tomorrow. Consider the 5-cents-per-gallon federal tax on gasoline, imposed on consumers in the early 1980s just as competition was driving down the price of gasoline. The proceeds of this tax are slated for road maintenance and construction. Petroleum, in short, may be the answer to a tax man's dream.

In recent years, not only has the industry's total tax burden increased tremendously but many of its tax incentives, such as percentage depletion, have been eliminated or sharply curtailed. Hence, as the industry continues to shoulder the growing cost and risk of finding, producing, refining, and marketing urgently needed oil and gas, a major portion of its income is going to government in the form of taxes. For example, the total tax revenue paid to the U.S. government in 1979 was more than seven times the amount of dividends paid to industry stockholders and about two and a half times the amount the industry reinvested in its own operations. U. S. taxes are actually penalizing the industry at a time when the cost of doing business is climbing to unprecedented levels.

What is the industry's burden in actual tax dollars? In December 1980, the Chase Manhattan Bank reported that the industry paid $42 billion in federal and state income taxes in 1979, compared with $25 billion to $30 billion annually from 1975 to 1978. Other taxes were about $15.5 billion, giving the industry a total 1979 U.S. tax burden of nearly $58 billion, compared to about $7.3 billion in 1968. In addition, consumers paid

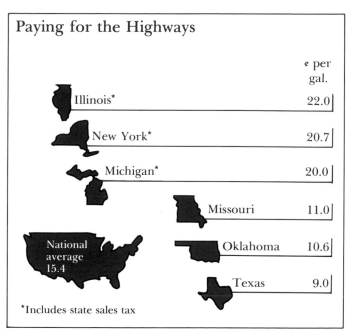

Paying for the Highways

	¢ per gal.
Illinois*	22.0
New York*	20.7
Michigan*	20.0
Missouri	11.0
National average 15.4	
Oklahoma	10.6
Texas	9.0

*Includes state sales tax

Illinois, New York, and Michigan impose the highest taxes on gasoline sales; Missouri, Oklahoma, and Texas charge the least.

309

approximately $48.8 billion in sales and excise taxes on gasoline and other refined products in 1979. This tax burden is still growing. The windfall profits tax was enacted in 1980. During the remainder of that year, the 20 leading U.S. oil companies alone paid $4.5 billion in taxes. In 1981, this tax generated $15.3 billion in revenue; and in 1982, the top 20 paid $9.6 billion. In 1982, the top 25 companies in the industry paid over $64.4 billion in federal and state taxes.

Taxes paid to states and local jurisdictions are also increasing rapidly. In 1975, state taxes of various kinds exceeded $1.3 billion. These included $370 million in income taxes and various other levies, such as severance taxes, a special levy on the extractive industry for the privilege of severing a state's oil or gas resources from the earth. That year, the industry also paid $943 million in property taxes to local governments. Recently, a growing number of states have begun enacting special taxes on oil companies. New York has levied first a 2 percent and then an additional 0.75 percent tax on oil companies' gross

Total Taxes Paid by 25 Leading U.S. Oil Companies
($ Millions)

Year	Total Taxes
1973	19,325.5
1974	30,829.1
1975	31,788.1
1976	29,846.3
1977	34,150.6
1978	37,539.1
1979	46,481.5
1980	67,009.2
1981	76,661.0
1982	64,430.2
First half 1983 (app.)	25,000.0

The petroleum industry looks like the answer to a taxman's dream, especially because the amount it has paid in the last ten years has increased by more than 300 percent.

receipts. Other states have considered a variety of similarly discriminatory laws.

Foreign tax obligations are substantial. According to the American Petroleum Institute, the petroleum industry paid $21.75 billion in income taxes to foreign governments in 1979. Adding significantly to overall foreign costs are government equity participation, royalties, severance taxes, bonus payments, and the risks of investment or reinvestment, which are initially borne by petroleum companies.

At one time, the U.S. petroleum industry, and the nation, benefited from the depletion allowance, a tax provision that freed the capital needed for additional exploration and drilling. Now, most recognize that the petroleum business involves extraordinary risks, perhaps greater risks than those of most other businesses. Even today, discovery methods are not sophisticated enough to assure a better than 5 to 10 percent chance that an exploratory well will prove commercial. In recognition of these risks, Congress created an incentive for oil-and-gas exploration in the form of discovery value depletion in 1925. The incentive was based on the sound principle of depreciation, which holds that assets wear out or become obsolete and sooner or later must be replaced.

Under this law, a taxpayer could recover either the fair market value or the cost of the property, plus certain capital improvements, through annual deductions against income. The deductions could be made over the lifetime of the property, but were limited to 50 percent of taxable income in any single year.

Despite its good intentions, the law proved difficult to administer. It had overlooked, for example, the problem of determining the fair market value of a newly discovered reservoir. Thus, in 1926, Congress replaced discovery value depletion with percentage depletion, which allowed a producer to deduct from income an amount equal to a percentage of his property's gross income. Unlike discovery value depletion, which only the oil discoverer could claim, percentage depletion went with the property and applied to anyone having an economic interest in that property. This proved a powerful incentive to investment, as Congress intended.

In the decades that followed, however, public outcry swelled against the highly visible oil industry and its depletion allowance. The public labeled it an unjustified profit denied other enterprises. Even other extractive industries escaped the attack.

Thus, in 1969, when domestic oil production was peaking and beginning its inexorable decline, Congress reduced the percentage depletion allowance for oil and gas. In 1975, political pressures forced Congress virtually to eliminate percentage depletion for oil and gas though not for many other minerals.

Similar problems have arisen around intangible drilling and development costs (IDC). Generally, a company or individual capitalizes costs incurred in acquiring and preparing for production any type of property. In the building of a factory, all costs up to the point of production, including labor and materials, are capital costs recoverable through a number of years of depreciation or amortization.

Since 1917, however, Congress has allowed the oil industry to deduct intangible drilling costs such as wages, fuel, repairs and supplies in the year incurred, again recognizing the risky nature of exploration and development. This is not an extra deduction in the sense that a taxpayer could take both IDC and depreciation on the same asset. The law makes it possible to write off intangible costs sooner than otherwise would be possible. Accelerated deductions are not unique to IDC. They have been applied to other types of assets, such as railroad rolling stock and pollution-control facilities. In 1980, Congress indicated a move in the direction of faster write-offs for industry in general. Tangible drilling assets, such as drill rigs and tanks, are capitalized and depreciated as in any other business.

Tax authorities also have questioned the validity of IDC exemptions for offshore equipment prefabricated onshore and for preplatform wells used in locating the best sites for permanent offshore installations. At issue is an understanding that costs associated with offshore wells are really similar to those of onshore wells but that offshore gear differs in the way it is constructed. In general, the courts have supported IDC deductions for offshore operations and have reaffirmed congressional intent to encourage offshore production.

310

State governments are proving that they, too, can find ways of exerting tax pressures on the petroleum industry. Of particular concern is the "unitary taxation" principle, which some contend enables a state to pursue income earned outside its boundaries. Under the concept, companies operating in a "unitary" state must report their total combined income, wherever earned, to that state. The state then uses a complex formula, usually involving sales, payroll, and property owned, to determine how much income is attributable to the local operation and how much elsewhere. The actual earnings of the local unit are irrelevant. Naturally, this formula works out to the benefit of the taxing authority. Unitary taxation has been enormously profitable to states and immensely costly not only to the oil industry but to industry in general.

Some states are more aggressive than others; some have extended the reach of unitary taxation beyond U.S. borders to a company's worldwide operations, although opposition to such practices has been brewing in Congress since 1966. In 1983, the U.S. Supreme Court upheld the right of California to apply a combination-unitary principle to a taxpayer who was, in fact, headquartered in another state.

At face value, the principle is not inequitable if fairly applied. But the rules vary from state to state. Unfortunately, attempts to unify the rules have been unsuccessful, leaving industry open to substantial risks of double taxation.

Another example of state efforts to maximize taxes on oil companies occurred in Alaska in 1978. The state passed a law that taxed oil and gas income in a way that was unlike the manner in which income of any other enterprise was taxed. It prescribed tax accounting methods that have no parallel in the United States. The law included a special application of the unitary concept. Alaska realized that the provision could result in so heavy a burden that it could jeopardize future development of the state's oil and gas reserves. It could also be found unconstitutional. So the state amended the law, effective in 1982, so that it more closely resembles unitary taxation in other states.

Today, the American petroleum industry — small

State Tax Collections — 1982

State	Gasoline tax rates per gal. (a) 12/31/82	Total state motor fuel tax revenues (Add 000)		Percent change
		1982(b)	1981	
Alabama	11.0ᶜ	$244,003	$247,127	−1.3%
Alaska	8.0	16,258	15,019	8.2
Arizona	8.0	139,035	127,339	9.2
Arkansas	9.5	129,770	131,001	−0.9
California	7.0	816,212	812,071	0.5
Colorado	9.0	139,914	123,618	13.2
Connecticut	11.0	148,161	159,221	−7.0
Delaware	11.0	36,110	30,767	17.4
Dist. of Columbia	13.0	24,933	20,294	22.9
Florida	8.0	445,607	419,750	6.2
Georgia	7.5	247,904	248,501	−0.4
Hawaii	8.5	25,513	27,085	−5.8
Idaho	11.5	56,879	54,764	3.9
Illinois	7.5	365,846	367,484	−0.4
Indiana	10.5	354,470	292,590	21.1
Iowa	13.0	210,254	157,818	33.2
Kansas	8.0	149,890	115,286	30.0
Kentucky	10.1	193,029	194,026	−0.5
Louisiana	8.0	197,899	191,665	3.3
Maine	9.0	49,544	48,131	2.9
Maryland	9.0	191,858	187,574	2.3
Massachusetts	11.2	258,071	266,097	−3.0
Michigan	11.0	448,091	444,483	0.9
Minnesota	13.0	286,066	248,791	15.0
Mississippi	9.0	135,150	118,397	14.1
Missouri	7.0	194,375	194,218	0.1
Montana	9.0	51,973	48,175	7.9
Nebraska	13.9	123,320	125,531	−1.8
Nevada	10.5	64,429	44,890	43.5
New Hampshire	14.0	55,838	52,698	6.0
New Jersey	8.0	279,659	279,621	0.0
New Mexico	9.0	83,886	76,581	9.5
New York	8.0	464,004	466,425	−5.2
North Carolina	12.0	386,912	329,343	17.5
North Dakota	8.0	34,074	34,490	−1.2
Ohio	12.3	587,873	435,660	34.9
Oklahoma	6.5	140,131	132,256	6.0
Oregon	8.0	120,656	87,973	37.2
Pennsylvania	11.0	562,435	638,889	−12.0
Rhode Island	12.0	44,000	43,194	1.9
South Carolina	13.0	227,086	205,245	10.6
South Dakota	13.0	55,571	53,606	3.7
Tennessee	9.0	288,751	249,200	14.7
Texas	5.0ᶜ	489,313	487,022	4.7
Utah	11.0	92,135	75,959	21.3
Vermont	11.0	27,280	22,662	20.4
Virginia	11.0	321,734	319,206	0.8
Washington	13.5	255,295	256,822	−0.6
West Virginia	10.5	97,440	96,456	1.0
Wisconsin	13.0	333,242	240,743	38.4
Wyoming	8.0	36,901	38,039	−3.0
Total	9.5ᶜ	$10,725,729	$10,083,810	6.4%

(a) Tax rate for highway use: does not include federal or local taxes;
(b) Preliminary.

companies as well as large ones — has extensive operations abroad. There it is subject to all the laws, rules, regulations, and whims of foreign governments, including heavy taxes.

Before 1950, oil companies developed the reserves of the Middle East under concessions that yielded bonus payments and royalties to the various governments. But over time, the producing countries sought new ways of increasing their shares. In 1950, they initiated a tax arrangement based on a 50-50 profit-sharing method introduced by Venezuela in 1948. Under this formula, which became a standard for producing countries, a government received 50 percent of the net profits from a producing operation. A country's royalties were offset against its income taxes so as to limit a government to no more than 50 percent of net income.

By 1957, just before the establishment of the Organization of Petroleum Exporting Countries (OPEC), producing countries obtained participation interests in production operations that were similar to partnerships. But the oil companies still assumed most of the risks and costs, along with substantial increases in income tax. Although terms differed from country to country, the result was a larger government take, as well as greater government control.

Producing countries eventually sought to control pricing, too, which led to the formation of OPEC in 1960. As OPEC gained strength and increasingly directed the terms and prices under which companies could lift oil, company payments to governments became less like royalties and more like income taxes. These taxes were based on the posted prices that the producing countries themselves set through OPEC. Since 1974, foreign producers have continued to capitalize on their position, increasing income taxes to rates as high as 85 percent and unilaterally and dramatically increasing the posted prices on which taxes are based.

Even as the foreign tax load multiplied, U.S. tax authorities began to scrutinize foreign operations more closely. A basic premise of U.S. tax law is that income of all U.S. nationals, corporate or individual, is subject to U.S. tax regardless of where earned. Foreign nations also tax the same income, although taxpayers can deduct from their U.S. taxes a credit for taxes paid on the foreign earnings.

Before 1976, a company computed its total foreign tax credit on a country-by-country basis. Losses from oil-and-gas operations in one country did not reduce the credits that could be taken for income in another country. Credits from oil-and-gas operations sheltered income from other operations in that country.

In 1975 and 1976, Congress changed the law so that foreign tax credits were no longer computed on a country-by-country basis. Losses in unprofitable countries are not netted against gains in profitable countries, thus reducing the net amount of foreign-source income that yields usable credits. If a taxpayer has a net loss from all foreign operations in a given year, he must "recapture" that loss in future years by using it to reduce foreign-source income that otherwise would be sheltered by credits. These rules are particularly hard on the oil-and-gas industry, which often incurs heavy losses when it is entering a new area. In addition, foreign tax credits must now be applied separately to oil-related and nonoil-related income, meaning that credits from the former no longer can be used to shelter income from the latter.

Overseas taxpayers, and especially oil companies, make many kinds of payments to foreign jurisdictions, including license fees, property taxes, and, particularly, royalties. These are considered not taxes on income but business expenses and, as such, are deductible from income before taxes. For example, in 1981 a royalty deduction reduced taxes by 46 cents for every royalty dollar paid. The same payment made as an income tax is deductible dollar for dollar from U.S. taxes.

The issue of whether a payment is a royalty or a tax has been highly controversial. Foreign governments frequently designate these payments as taxes. The U.S. Internal Revenue Service does not always agree and has been known to reverse a particular position after extensive study. When the verdict is "royalty" instead of "tax," the net effect deprives oil companies of funds needed for further exploration and production.

Some argue that foreign income taxes only should be a deduction against income instead of a dollar-for-dollar credit, just like any other cost of doing business.

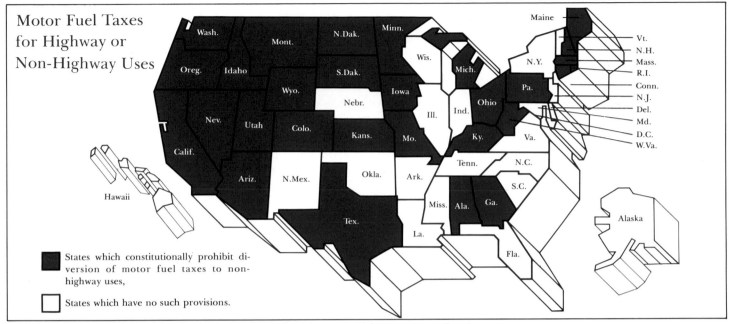

Motor Fuel Taxes for Highway or Non-Highway Uses

Legend:
- ■ States which constitutionally prohibit diversion of motor fuel taxes to non-highway uses,
- □ States which have no such provisions.

Twenty-seven states have provisions which prohibit diversion of motor fuel taxes to non-highway uses; twenty-three do not.

313

The argument fails, however, because taxes are not like any other business expense. They are the *only* significant business expense which can be imposed twice — by two different countries — without corresponding benefit to the taxpayer.

Many major oil-and-gas companies traditionally conduct their foreign enterprises through foreign-incorporated subsidiaries. Many other American businesses work through wholly or partly owned foreign corporations. Such decisions are based on legal considerations or various other economic, social, or political factors frequently having nothing to do with U.S. income tax laws. Yet the establishment of foreign corporations, especially in countries with zero or low tax rates, is often viewed as a device for avoiding U.S. taxes.

From time to time, various groups and even administrations seek to end all tax deferrals on "controlled foreign corporation" (CFC) income. The Carter administration sought such a complete ban, but it did not become law.

Federal Collections (In thousands)

Year	Gasoline (a)	Diesel and special fuels (a)	Lube oil (b)	Total
1982 (c)	$3,854,363	$589,933	$77,822	$3,854,363
1981	3,801,795	570,477	74,695	4,446,967
1980	4,019,481	518,352	78,835	4,616,668
1979	4,282,919	508,595	82,335	4,873,849
1978	4,381,950	488,064	79,047	4,949,061
1977	4,229,721	452,631	78,830	4,761,182
1976	4,140,924	408,560	70,135	4,619,619
1975	3,884,483	362,675	66,124	4,313,282
1974	3,854,505	373,302	78,155	4,305,962
1973	3,958,682	358,120	86,266	4,403,068
1972	3,736,545	321,551	85,841	4,143,937
1971	3,513,927	275,159	67,130	3,856,216
1970	3,408,505	264,815	65,862	3,739,182
1969	3,172,243	240,650	86,338	3,499,231

(a) 4¢ gal.; (b) 6¢ gal.; (c) Figures for 1982 are fiscal year gross collections that went to the Highway Trust Fund, since net IRS calendar year information was not available. All other figures represent calendar year net collections after refunds and include the taxes the federal government pays on its own purchases.

U.S. Federal Tax Burden of 20 Leading U.S. Oil Companies
($ Millions)

		1982	1981	1980	1979
A	U.S. net income before provision for U.S. income taxes	$21.8	$27.9	$26.4	$20.9
B	Provision for U.S. income taxes	$ 8.4	$12.1	$12.5	$ 9.1
C [=A−B]	U.S. net income after provision for U.S. income taxes	$13.4	$15.8	$13.9	$11.8
D	Windfall profits tax	$ 9.6	$15.3	$ 4.5	—
E [=A+D]	U.S. net income before provision for U.S. income taxes and windfall profits tax	$31.4	$43.2	$30.9	—
F	Current U.S. federal income tax	$ 4.0	$ 6.3	$ 8.3	$ 5.5
G [=(D+F)−E]	Current federal income and windfall profits taxes as a percent of U.S. net income before provision for U.S. income and windfall profits taxes	43.3%	50.0%	41.4%	
H [=D×0.46]	Estimated additional current federal income tax assuming no windfall profits tax	$ 4.4	$ 7.0	$ 2.1	—
I [=(F+H)−E]	Estimated current federal income tax as a percent of U.S. net income before provision for U.S. income taxes assuming no windfall profits tax	26.8%	30.8%	33.7%	—
J [=F−A]	Current federal income tax as a percent of U.S. net income before provision for U.S. income taxes	18.3%	22.6%	31.4%	26.3%

Oil companies represented in the study:

Amerada Hess Corporation
American Petrofina, Inc.
Ashland Oil, Inc.
Atlantic Richfield Company
Exxon Corporation

Getty Oil Company
Gulf Oil Company
Kerr-McGee Corporation
Mobil Corporation
Murphy Oil Corporation

Pennzoil Company
Phillips Petroleum Company
Shell Oil Company
Standard Oil Company of California
Standard Oil Company (Indiana)

Standard Oil Company (Ohio)
Sun Company, Inc.
Superior Oil Company
Texaco Inc.
Union Oil Company of California

314

Pressure continues toward ending tax deferrals on all CFC income. In many instances, such a move could be softened by the taxpayer's ability to shelter CFC income with foreign tax credits. Yet tax timing differences and other structural differences in foreign tax laws could limit the availability of credits to shelter fully such income from U.S. tax. Thus, abolishing tax deferrals entirely would weaken the overall ability of U.S. companies to compete abroad with businesses of other Western nations, most of which do not levy taxes on the unrepatriated income of their CFCs.

Obviously, governments tend to treat the petroleum industry differently from the way they treat other industries. Today, the avowed intent is to take away as much as possible of the industry's allegedly unjustified gains and divert that money to better, socially useful purposes.

One recent example is the "windfall profits" tax. The objective was to tap excess industry profits anticipated after the removal of petroleum price controls in 1979. Profits did rise, but in the process several points were overlooked. In the first place, government-imposed price controls had thwarted free-market forces, discouraged exploration and production, and greatly distorted many U.S. industries.

Controls had also held prices at unrealistically low levels at a time when they would have risen gradually and naturally. Rising U.S. prices would have generated additional supplies and perhaps would have exerted a moderating effect on world prices. As a result, there would have been no "windfall profits" to be taxed. All available funds would long since have been drilled back into the ground to find more oil.

Further, without such a tax, government, in the normal course of events, would get 50 to 60 cents of every dollar in existing taxes and royalties, leaving to the industry 40 to 50 cents for exploration and production. Under the windfall-profits-tax law, which will be phased out over ten years or after it has collected $227 billion, *whichever is later,* the industry will be left with only 22 cents of each dollar.

But the bulge in industry profits that excited such public resentment has subsided. A change of administrations in 1981 brought new appreciation for the necessity of increasing exploration activity to head off serious energy shortages in the future as well as to improve American self-sufficiency in petroleum production. Many feel that the bulge actually represented inventory profits, a temporary situation soon overtaken by the rising prices of new inventory. Finally, there was renewed faith in an old axiom: tax policy works better when it encourages private investment rather than bureaucratic growth.

The windfall profits tax has not been repealed, but it has been modified. The goal remains $227 billion or ten years, whichever comes later, but the rates on newly discovered oil will drop from 30 percent to 15 percent by 1986. That will save the industry $11.6 billion in taxes. The money will go to discover more new oil. In addition, independent stripper wells became exempt in 1983.

While the industry remains a tempting tax target, particularly to state and foreign governments, tax credits and subsidies that stimulate more investment in energy-productive activities are becoming increasingly common.

In the long run, however, the industry needs to keep a wary eye on the course of petroleum taxation, not only because tax laws are always changing and growing more complex but also because they foster self-defeating attitudes. The petroleum industry is so visible to the public and so convenient to the architects of tax law that a great many unwieldy, hastily considered, unnecessarily punitive, and easily misunderstood measures have already found their way into state, national, and international tax codes. One result is a frame of mind that considers one of the nation's most productive and indispensable industries fair game for any conceivable new form of taxation.

As oil and gas become increasingly difficult, risky, and expensive to find, the public must be more aware of the consequences of such thinking. To the extent that heavier taxes on an industry eventually translate into higher prices in the marketplace, the trend to excessive taxation only leads to higher indirect taxes on consumers, as well as disincentives to a vital industry. The Economic Recovery Tax Act of 1981, many provisions of which were designed to aid business in general and not a particular industry, is a positive step forward.

CHAPTER 31 –
THE FUTURE OF OIL

No forecast is ever exact, no matter how detailed, up-to-date, and accurate the basic data. Petroleum and energy forecasts are even less exact than most others. Accurate data are simply not available.

Unfortunately, oilmen cannot see what is under the ground or sea or ice. Although much of the world's economic and political future depends on the world's ultimately recoverable reserves of fossil fuels and minerals, oilmen cannot observe them directly. They know how much oil, gas, and coal has been produced. They have a good idea of how much is left in deposits already tapped and can estimate fairly closely how much of that can be recovered as a function of price and available technology. But beyond these knowables, oilmen begin to tread very uncertain ground.

At the World Energy Conference held in 1978, 29 "experts" were asked to estimate the size of the world's remaining oil-and-gas resources. The experts believed that plenty of oil and gas was still to be found, but they came up with 29 different estimates of how much. The mean estimate for oil alone was 2.1 trillion barrels, of which 1.0 trillion barrels have already been discovered. The optimists believed that the total reserves discovered and to be discovered are around 3 trillion barrels. The pessimists estimated 1.2 trillion barrels, which would mean that there are only 200 billion barrels yet to be found. The majority range was between 1.6 trillion and 2.1 trillion barrels, a 500-billion-barrel difference of opinion.

This is no mere argument among academics that time will resolve. At current world-consumption rates of about 60 million barrels per day, the high estimate gives enough oil and gas for over a century; the low estimate, enough for only half that. Consumption will not remain static, of course, though projections are just as hazy as those for reserves. Exxon has estimated that oil demand will grow only 0.4 percent per year for the next 20 years. A United Nations forecast expects oil-and-gas production to peak in the year 2000 and to contribute only 23 percent of world energy by 2020. Oil-industry estimates show oil alone accounting for 23 percent of world energy in 2000, compared to 47 percent in 1980. Gas will provide 19 percent, about the same as now.

This caliber of guesswork is the basis for plans for the expenditure of billions of dollars for exploration and development of petroleum as well as the development of alternate energy sources. How much faith is placed in any of these numbers has tremendous implications for future planning, policy, and investment. The choice of numbers may mean the difference between an orderly, long-term transition to other sources of energy or the institution of horrendously costly, politically disruptive, and possibly futile crash programs to prepare for the day when the oil runs out.

If anything, recent experience has made all prophets much more cautious. No one expected the tremendous change in energy markets that took place during the late 1970s and early 1980s. The prophets and people who made large investments based on the rosy predictions of ever-expanding markets at ever-higher

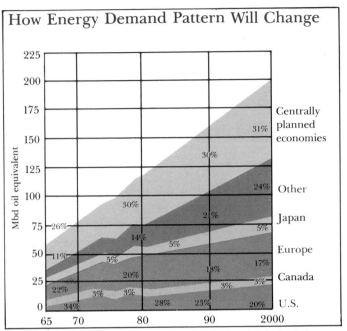

Energy demand is expected to increase by the year 2000, mainly in developing countries and centrally planned economies.

317

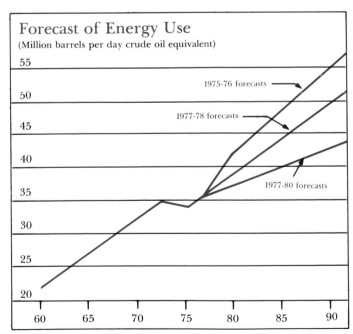

This chart, based on published predictions from government, business, and industry, shows how the trend toward energy conservation has caused forecasts of energy usage to drop.

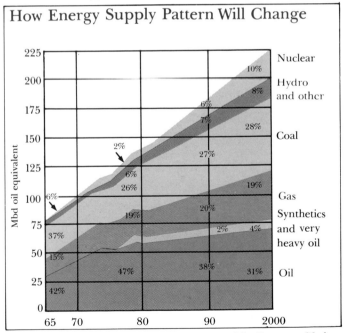

Supply patterns will change as well. Oil, which accounted for 42 percent of fuel use in 1965, will decline to 31 percent of the total by the year 2000, while gas, nuclear, and hydropower will bridge the gap.

prices were badly burned. Thus, current projections for growth tend to be conservative. Their projections for growth now reflect not sustained booms but inexorable growth in world populations and accompanying industrial development that energy must fuel.

Estimates for alternative fuels are even hazier. Gas, coal, shale, and tar sands have not been sought nearly so intensely. Their known reserves are larger, however, and their eventual exhaustion further in the future, making their uncertainties not quite so critical.

Learned people have been predicting the exhaustion of the world's oil since 1863, about the time Drake was complaining about overproduction breaking the market price. Forecasters have repeated these predictions regularly ever since. Somehow, the day of reckoning has always seemed to be about 20 years in the future. Somehow, a new series of major discoveries has always put off that day.

The forecasters have been correct in that the world's supply of oil is finite; only their timing has been off. The problem is not so much that less oil is found or that significant new discoveries may not be made as that world consumption patterns have changed drastically in the last 20 years. Forecasts based on data known to be accurate and reasonably projectable — population growth and economic expansion — also show very wide ranges for the next 20 to 40 years. But they agree on one point: energy consumption will continue to increase faster than population growth, just as it has since World War II. Economists and demographers estimate that world population will about double by 2020. There are low and high estimates here, too, but the statistics start from a known base and are built on a fairly reliable model. The variations are nowhere near those of energy resource forecasts.

Either world economic activity will grow at a faster rate than population, or else standards of living will decline. With these figures, planners construct models that reflect energy demand based on low, moderate, and high economic growth — with or without significant conservation efforts — and then apportion the projected demand to projected available supplies.

By the most optimistic projections, the supply of

318

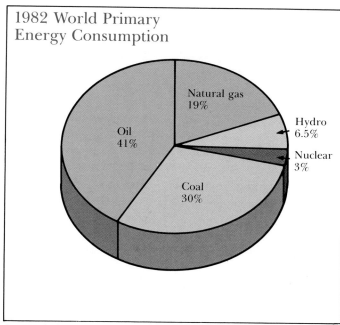

In 1982, hydropower and nuclear power had a small but increasing share of world energy consumption. Oil usage had declined about 10 percent since 1978.

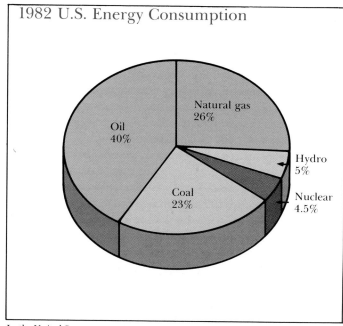

In the United States, energy consumption patterns paralleled worldwide usage. Oil usage had declined about 7 percent since 1978, while coal, hydropower, and nuclear power took up the slack.

energy from all sources is limited. In even the moderate scenario, demand could greatly outstrip supply. Again, the forecasts vary greatly. The International Energy Agency (IEA) of the Organization for Economic Cooperation and Development (OECD) estimates that demand in the non-Communist world in 2000 will range between 62 billion and 90 billion barrels of oil equivalent per year (170 million to 247 million barrels per day). Another estimate for the same year shows a range between 156 million and 209 million barrels per day. An estimate for the entire world made by the World Energy Conference shows a range between 378 million and 600 million barrels of oil equivalent per day for the year 2020. The higher estimate comes to 219 billion barrels per year.

If oil were the only energy source, such demand would use up even the most optimistic reserves in 15 years. Today, oil accounts for 41 percent of all worldwide energy use; gas, 19 percent; coal, 30 percent; hydropower, 6.5 percent; and nuclear power, 3 percent. In the United States the figures are 40 percent oil, 26 percent gas, 23 percent coal, 5 percent hydro, and 4.5 percent nuclear power.

If the world has already used about half of all the oil energy believed ultimately available and about a third of the gas, and the usage rate continues to increase, oil and gas cannot be expected to retain their pre-eminent energy positions until 2020. Most estimates agree that global oil production will peak by about 2000 and decline thereafter. In the United States, oil production peaked at 9.6 million barrels per day in 1970. No one expects domestic production to reach that level again.

Certainly, oil and gas will cease to be major sources of energy within the lifetimes of many now living. New deposits will be found, some of them quite large, at ever-increasing costs. Production from older fields will continue, through expensive enhanced-recovery methods. As demand grows and supplies become scarcer, prices over the long run can only continue to increase, despite sometimes dramatic fluctuations. The higher prices will promote more efficient use and intensify the search for alternate fuels, a development that is already well under way and has already produced some remarkable effects. For example, domestic oil imports dropped from over 7

million barrels per day in 1977 to just below 4 million barrels in 1982. Growth in world oil usage also slackened, accommodating itself to reduced Middle East production caused by the Iranian revolution and the Iran-Iraq war.

This very flexibility in the face of what should have been shortage could itself prove a problem. Too often, it removes a sense of urgency and sharpens the ax that producing countries hold over consuming countries.

OPEC is a major influence on world supplies and world market prices and will continue to be for the foreseeable future. By increasing or withholding supplies, OPEC countries can control the price of oil as long as they act in concert. And even though worldwide shortage is inevitable before the end of the century, producible supply will outstrip demand for many years to come.

Naturally, non-OPEC consuming countries try to increase their own production and thereby reduce OPEC's share of the market. But these same countries will consume all they produce and more. As surplus and discretionary producers are driven primarily by their own

319

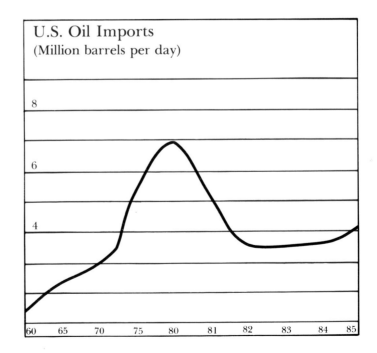

U.S. Oil Imports
(Million barrels per day)

financial and political needs, OPEC countries will control the incremental barrel and thus the price. However, and fortunately for the rest of the world, OPEC spending inevitably rose to meet income, and the countries cannot easily cut back. The pressure to maintain income in the face of declining oil consumption, however temporary that might be, has forced individual OPEC members to reduce prices in order to move more oil. The countries have less discretion than either they or the rest of the world supposed.

OPEC could hold the price at a level that makes development of alternative sources not yet economic. For instance, oil shale, tar sands, and heavy-oil deposits have long been regarded as backup sources as the cost of conventional oil goes up and up. Known reserves of these unconventional sources far exceed those of conventional oil, and the technology to exploit them is not new, but it has always been costly.

In the 1960s, when conventional oil sold at $3 a barrel, theorists estimated that oil from tar sands and oil shale would be competitive at $8 a barrel. In 1978, with world oil at $22, the cost of oil from unconventional sources was estimated at $30. It is the carrot at the end of the stick, always just out of reach. OPEC has the power to keep moving it beyond reach.

In 1982, the estimated cost to establish a complete synthetic-fuel facility to produce only 50,000 barrels per day was more than $3 billion. The range of cost estimates per barrel for fuel from various sources had again climbed. For shale oil, it was $30 to $45; for medium-Btu coal gas, $35 to $42 per barrel of oil equivalent; for coal liquids, $50 to $75; and for alcohol, $52 to $60. These prices cannot compete in a free market with even the most expensive of current oil-and-gas production. Only when Saudi Arabia is producing at 100 percent capacity and cannot cut back to control the price will these synfuels be competitive.

At that crossover time, probably in the 1990s, the combined productive capacity of all conventional sources, including OPEC, will no longer equal demand at a price OPEC can influence. Demand will pull prices up, and no producer will have the power to depress them. The cost of producing unconventional sources will con-

verge with the inexorably increasing price of conventional oil. Technologies that are now uneconomic or unprofitable, however technically feasible they may be, will become attractive.

Late in this century, other sources of energy, conventional and unconventional, will begin to enter the market in amounts large enough to compete with oil and gas. Some of these sources — sunlight, waterpower, ocean thermal energy, and other, more exotic ones — are virtually infinite. They require huge capital investment to realize, however, and for many uses are neither convenient nor practical. One hardly sees an African or South American farmer going to the corner energy store to pick up a can of solar energy as he does a gallon of kerosene or a stack of wood.

Coal, in all its forms, and even uranium will deplete. But the reserves are so great that users who switch to such sources will not need to worry about running out. Factors besides supply govern their economic choices.

While oil was in its ascendancy, essentially since World War II, it had no viable competition as commercial energy. Oil vanquished coal in the United States, shrinking its energy share from 75 percent in 1900 to a postwar low of 13 percent. It held off gas, until environmental considerations gave the latter an edge in some markets. It ignored hydropower and left nuclear power an also-ran. No other energy source could compete with oil's price, convenience, and flexibility.

But as the crossover point approaches, these other sources will reemerge, and new ones will mature. Oil will retreat to those markets for which there is no adequate substitute — transportation and chemicals. To paraphrase an old expression, oil may become too costly for any purpose *but* burning.

Large markets which oil and gas captured from coal will switch back — and, in fact, are already doing so. Many new utility plants burn coal where formerly they would have burned oil or gas. Many older plants have switched from gas to coal, encouraged by the Powerplant and Industrial Fuel Act of 1978, which was designed to force utilities out of oil and gas and into coal or other fuels. That act has since been repealed, as gas supplies became more plentiful. Pulp and paper companies in the south-

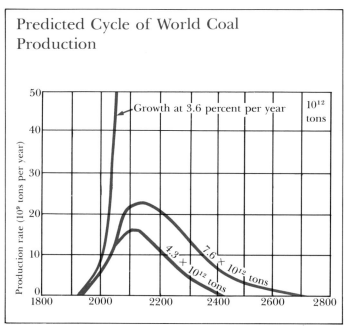

Predicted Cycle of World Coal Production

At the present rate of growth of coal production, coal will be depleted as an energy resource by about the year 2050.

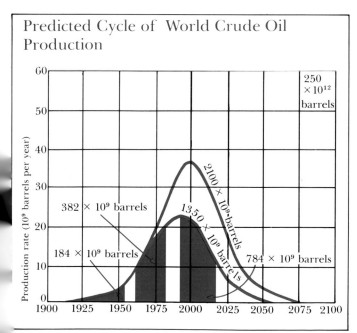

Predicted Cycle of World Crude Oil Production

If used conservatively at current lower rates, crude oil may last through the year 2100.

east United States are building new and expensive multi-fuel plants that can run on coal, gas, or oil. The trend has been particularly strong in Europe, which had not switched from coal to the same extent that the United States had.

Even in transportation and chemicals, oil and gas will see increasing competition from synthetic fuels made from coal, as well as oil produced from unconventional sources. These in turn will face competition from new and renewable nonfossil energy as it becomes cheaper to generate power from a windmill than a steam plant, whatever the fuel.

Because the world's energy appetite will remain insatiable, diminished only relatively by more efficient use among once-wasteful users, the end of the twentieth century will see many more sources of energy. The relative importance of oil and gas will be lessened, as will their influence on the cost of energy from whatever source. Oil will become the price follower instead of the price setter. Coal or synthetic oil will take over that role in the general market.

Today, because they have so much oil and can produce it so cheaply, some oil producers like Saudi Arabia can make prices fall as well as rise. Future energy producers, however, are not likely to have that power. Producing oil and gas is not like producing coal and shale. Though oil and gas are increasingly costly to find and develop, production costs are low. The wellhead cost of the incremental barrel from a large onshore field is negligible and increases only slightly during primary production, even after the well goes on artificial lift. Costs begin to rise when the operator resorts to enhanced recovery. Production then becomes vulnerable to market forces. Thus, every oil producer or alternative-energy producer is at the mercy of the lowest-cost, largest-volume producer, such as OPEC.

Other energy sources can have no such downside market impact because they are all inherently costly. Oil-and-gas production is primarily a recovery operation, meaning that the economic return on a successful well often bears no relation to the money spent to drill it. The risk lies in the number of wells that must be drilled before a successful field is found. Every other source of energy,

321

however, provides a return in direct proportion to the investment and requires a large and continuing investment for every equivalent barrel of oil produced. The cost of such investment sets an absolute floor under the future price of energy. Today oilmen cannot predict what that cost will be, but at first it will bear a close relationship to the lowest price of the incremental barrel of oil in world markets. Thereafter, the price of that barrel will track the costs of other energy.

To produce commercial energy from any source other than oil or gas requires a fundamentally different approach. The sun is free; every day it bestows 20,000 times as much energy as is used. But that energy can be captured and used only in proportion to the amount of land or other investment made. The investment, whether for a rooftop water heater, a solar cooker, photovoltaic cells to generate electricity, or a high-temperature furnace, is directly proportional to the energy captured. Before that solar energy can be put to any but the most primitive use, something must be manufactured.

Recovery of energy from coal, shale, and tar sands first requires a mining operation, an activity far different from well drilling. Though experiments with in-place conversion of these resources to energy are being carried out, most schemes require that the material must first be dug out of the earth and moved somewhere else for further processing. The capacity of a coal, shale, or tar-sand mine is directly proportional to the investment and requires constant investment in men and machinery to mine the deposit. The first shovel costs about the same as the last. This sets a bottom limit on the cost of producing a ton of coal or a barrel of shale oil. Large operations, the only kind feasible, allow some economies of scale. But the cost differences between a large and a small mining operation working on the same deposit of coal or tar sand, while enough to put the less efficient at an economic disadvantage, are not significant in the overall market.

Coal is used as it comes from the mine to cut transportation cost. Utilities have built power plants at the mouths of the mines. Pipelines carry liquid slurries long distances. New furnaces burn oil-like mixtures of coal and water. There has been speculation about using such a mixture for ship fuel to replace bunker oil.

Coal is going to regain some of the markets it lost to oil and carve out some new ones. Worldwide, coal production will increase from its current level of about 3 billion tons to around 5.8 billion tons in the year 2000, and 8.8 billion tons in 2020. Other estimates project even larger inceases.

Converting to coal will not be cheap. In the United States alone, projections show that an expenditure of $120 billion will be needed to increase coal production by 15 billion tons in 2000. That expansion will take another 600,000 workers in mining and transportation. Similar investments are needed just to establish large-scale mining facilities for oil shale and tar sands.

Processing is also expensive. If coal is converted to liquid fuel or gas, or if the oil is extracted from shale and sand, additional plants must be built to process the raw materials. These plants will be more complicated and far more costly than the refineries they will precede in the processing stream. They will produce synthetic crude, which then must be processed further to yield the desired products — gasoline, heating oil, and chemical feedstock.

Today a 100,000-barrel-per-day refinery costs about $1.3 billion to build from scratch, with known and tested technology and equipment. Current estimates peg synthetic-fuel plants of similar capacity at $3 billion. If, over the next 20 years, all imported oil were to be replaced along with the shortfall in the inevitably declining domestic production, the United States would require the equivalent of 120 plants to produce synthetic crude. That would cost $360 billion at today's prices, in addition to the $120 billion for expanded coal production and a like amount for mining shale and sand.

Thus, in the United States alone, an expenditure of some $1,000 billion — a highly speculative and undoubtedly underestimated figure — would be needed over the next 20 years merely to assure that needs for commercial energy can be met from domestic fossil sources. That is $60 billion per year, beginning right now. Moreover, those figures do not include additional expenditures to further develop nuclear and nonfossil energy sources, which could be equally high. Such numbers are staggering. They begin to give a perspective on the real dimensions of the energy picture in the not-very-distant future.

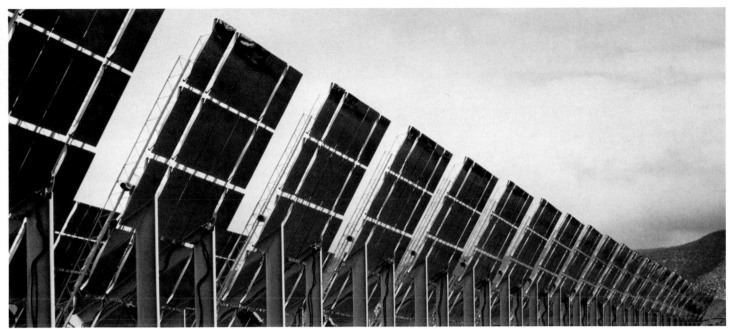

Heliostats like these at Sandia, N.M., are used to concentrate the sun's heat on a central receiving station from which steam can be generated to produce electricity. A single heliostat can deliver the heat equivalent of a barrel of oil per week.

Because shale oil most closely resembles conventional crude oil, it was considered a very desirable alternative, especially when the price of oil was high enough to justify the cost of extracting 34 gallons of shale oil per ton of ore. Projects like Union Oil's Parachute Creek Shale Oil Program in western Colorado have been preparing for such production.

Beyond the economic factors, planners must consider the environmental and physical problems associated with unconventional energy sources. Processing oil shale creates disposal problems — what to do with the mountains of spent sand and shale created in the processing of millions of tons per day year after year except to shovel it back into the pit?

Coal mining, too, presents serious environmental problems when practiced on a large scale, not to mention the atmospheric pollution associated with burning it, especially coal with high sulfur content.

These problems have solutions, but at a price.

Against this backdrop, it becomes clearer why the world has shown such reluctance to embrace the new technologies as long as oil and gas are available, regardless of price. The petroleum industry is spending enormous sums to press the search for more oil and gas, tapping ever-smaller reservoirs, leaning heavily on expensive enhanced-recovery methods, improving processing technology to upgrade a greater percentage of lower-grade crudes, while also pursuing alternative technologies and energy sources. The industry's total annual investment in developing new oil production is expected to reach $70 billion by 2000 and $110 billion by 2020, not considering inflation. Further development of oil and gas has an assured return. While billions of dollars are required, the industry is well aware of the costs and the risks and knows that, as costs increase, so will the selling price.

Eventually, alternative-energy pursuits will justify their investments. But until that time, ten years or more in the future, no industry with responsibility to its owners and stockholders, either privately owned or nationalized (in which the taxpayers are, in effect, the stockholders), can justify diverting from the search for and improvement of conventional energy sources the huge sums needed to commercialize unconventional energy. So great are current capital demands on the petroleum industry that companies must go to capital markets to fund their programs, despite the profits and cash flows of recent years.

Of course, for many years the industry has worked on synthetic fuels from unconventional sources.

Research and development necessary to prove processes and yield cost figures are essential to the survival and growth of any industry. But investing several hundred million dollars in pilot-plant facilities that produce a few thousand barrels a day is a long way from building multibillion-dollar production plants that yield hundreds of thousands of barrels per day. Until the product can be sold at an assured profit, no industry can afford to make investments of such scale, particularly when there is no hope of return for 10 or 20 years.

This is an area where government, as a matter of national policy, must lead the way. The National Synthetic Fuels Corporation, a U.S. government corporation, is the vehicle. The corporation was funded at the seemingly huge level of $15 billion. However, as the foregoing figures indicate, that level is modest for the job to be done. It is a level of modesty, nevertheless, that is well beyond the reach of any private corporation or industry.

The Synfuels Corporation was designed to take the lead in developing viable processes for producing energy from unconventional sources, assuming that the price of petroleum made such a move feasible. Only government can assume the financial burden and manufacture enough product at a loss to prove the processes, products, and side effects under real-world conditions.

The petroleum industry has changed in the last 10 years, and it will continue to change in the next 20 to 40 years. Since 1973, the influence of major oil companies has declined as their share of OPEC exports has dropped from 90 percent to 45 percent. Oil-producing countries now market much oil directly, outside the major companies. Although this practice already has many implications and will have more, the industry, through the large international oil companies, remains essential in the management of production operations, logistics, capital, and technology and the exploration and development of new fields. As one study says, "Their ability to coordinate and operate very large projects of growing cost and complexity and to move immense volumes of crude from the field to the market is of vital importance." And, despite its declining influence in world-government circles and the increasing politicization of oil and gas, the industry's importance in these areas will continue to be substantial.

In the 1970s, Synthane coal gasification plants like this one at Bruceton, Pa. hoped to produce a substitute for natural gas. But these projects lost favor in the 1980s when oil prices fell and natural gas prices rose forcing users to turn to cheaper alternatives.

Oil companies are large organizations responsible to hundreds of thousands of employees and shareholders whose welfare the industry must safeguard. The industry, despite its need to borrow money so that it can continue to serve expanding world energy requirements, is the repository of huge sums of money, not all of which it can efficiently invest in energy projects. Its fiduciary responsibility to both employees and shareholders demands that the industry prudently invest these sums in areas that will grow, provide future employment and return on investment, and address world needs.

As a dealer in oil, the petroleum industry will be in business for generations to come. The liquid fuels produced from unconventional sources will continue to be refined, distributed, and marketed in the same manner as oil-and-gas products are now. Natural crude will continue to be an important commodity well into the next century, though sharing the throne. But the industry will also expand across the energy spectrum. It has already begun using its resources to fund development and improved manufacture of other energy sources, such as solar energy, that promise an earlier return for smaller but more numerous ventures than those available through the synfuels or nuclear approach.

No other industry has the resources, the technology, or the experience to manage the floods of capital and the unimaginable scale of construction that will be needed to produce and distribute energy in the near and distant future.

325

CHAPTER 32 –
ENERGY FOR TOMORROW

All energy is equal. It can be measured and described in identical terms regardless of its source. Energy sources, however, are not equal, nor are the ways in which they are used. Some energy sources are more convenient, cheaper, and cleaner than others. Some can be used more efficiently, though efficiency alone is far from the only criterion. Some require huge capital investment before they can be used at all, while others need no more investment than a lighted match.

There is no real hierarchy of energies or energy sources. Human beings have always used what was most convenient, rather than most efficient or least costly, to do whatever work or function needed doing. Indeed, for most of history, people have used energy inefficiently, and still do in most of the world.

For the past two centuries, most useful energy has come from materials that first make heat. The heat then does the work. Until fairly recently, animals, water wheels, windmills, sails, oars, and human beings pumped the water, ground the grain, plowed the fields, trans-ported goods and people, and hammered out the swords and ploughshares. People used heat mostly to warm themselves, cook food, smelt metals, and burn their enemies. For fuel, they used anything at hand, usually wood, but also charcoal, coal, peat, oil, gas, and pitch if available. For much of the world, this is still so.

Such use of energy is not efficient, nor does it need to be. In preindustrial societies with small populations, solar bounty grew trees for firewood faster than people could burn them. Slavery and serfdom provided great pools of low-cost labor. Stately windmills ground the grain and pumped water at a leisurely pace.

The steam engine changed all this. This heat engine differed fundamentally from any other type of prime mover used until then. Steam heat allowed people to have energy wherever it was needed. Traction engines soon followed stationary pumping engines. Steam engines allowed mills to be located where there was no water power. Coal was plentiful and cheap, as was wood.

The steam engine is a relatively efficient energy converter. But it is a two-stage process in which fuel burns to boil water. The resultant steam, not the fuel itself, does the work. In the later internal-combustion engine, the burning fuel does work directly. Gasoline and diesel engines and, in time, turbines have thus replaced steam in all but the largest applications, where capital cost and other factors actually make a steam plant more cost effective. More of its waste heat can be recaptured.

Internal-combustion engines demand liquid or gaseous fuels. Petroleum production therefore grew along with the burgeoning use of internal-combustion machines. Coal lost markets to oil because oil was clean-burning and easily transported and handled. Ships, utilities, and manufacturing plants shifted from coal to oil and gas even for their steam-generating plants.

Efficiency was not the only consideration. Centrally generated in huge, efficient plants, electrical power can be transported conveniently to any point of use, provides enormous flexibility, and can do just about anything steam once did and a great deal more. When fuel was cheap, utility-supplied electrical energy cost users less than they could generate it for themselves and less than another power source.

Thus, by 1970, the world was electrified and, in the United States as elsewhere, largely dependent on oil and gas not only for transportation and heating but also for electric power. Energy was so cheap that people employed it without regard to efficiency. Nonpetroleum energy was common only where some other source was plentiful and cheap compared to oil and gas. And those were quite conventional: coal, water, and nuclear power. People thought little then of other nonpetroleum energy sources.

Two major categories of nonpetroleum energy sources are nonpetroleum fossil energy and nonfossil energy. The first category includes coal in all its varieties — anthracite, bituminous, lignite or brown coal, and peat — and oil shale. Tar sands and very heavy oil deposits sometimes fall in this category also, since they too must be mined before further processing.

Nonfossil energy covers a very wide range: sunlight, wind, water, ocean currents, tides and thermal gradients, alcohol, wood, biomass, chemical, electrochemical,

geothermal, geopressure, nuclear and — far, far in the future — nuclear fusion.

Governments and energy companies around the world are pushing development of all these sources. Some, indeed, are already widely used in special circumstances where location or other factors justify the costs or inefficiencies.

In fossil nonpetroleum energy, coal is by far the leader. And coal as coal will substantially increase its share of the energy market in the years ahead. With 200 years of known reserves in this country alone, coal again will be one of the most important energy sources. Coal-fired locomotives are not likely to appear again, but electric locomotives whose power comes from coal-fired central plants are on the horizon. Coal-fired ships are a strong probability, though in the future they will burn a pumpable mixture of coal and oil or even coal and water.

Anticipated coal development goes in two directions. One seeks to make coal itself a more efficient, cleaner-burning fuel. Another approach is magneto-

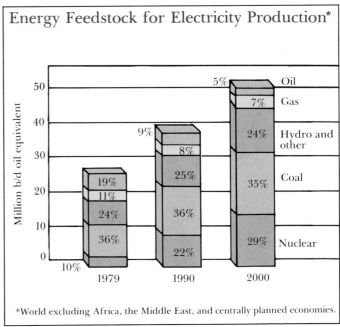

In 1979, oil accounted for 19 percent of electrical generation, while nuclear's share was 10 percent. Forecasters in the 1980s predicted that by 2000 oil's share would be 5 percent and nuclear's would increase to 29.

hydrodynamics (MHD), which burns powdered coal.

In the first option, which tries to make coal cleaner-burning and more efficient, much of the work is mechanical and physical rather than chemical. The goals are finding cheap ways to remove impurities before burning the coal and better furnaces that will burn the fuel more efficiently to recover more of the heat. One such is the fluidized-bed furnace.

The fluidized bed is an old principle in petroleum-refining technology. Simply, it is a bed of very fine particles suspended in a gas moving up through the mass. In catalytic crackers, the bed is the catalyst, and the fluid is the vaporized petroleum stream being cracked. In a fluid-bed furnace, the bed is pulverized coal, and the gas is air.

As the air moves through the coal, the tiny particles burn much hotter, more completely, and more efficiently than the same coal would burn in a standard grate. The energy recoverable from the coal is thus much higher, and the output of noxious gases and ash is much lower. Unfortunately, coal is a complex material prone to clinkering — making hot, sticky masses that agglomerate and defeat the fluidizaton. Coal is also abrasive. These characteristics, together with great heat, create difficult technical and materials problems.

Therefore, another approach, magnetohydrodynamics (MHD), is employed as well. For many years American and Russian scientists have been working in cooperative competition on developing this method. By this process finely powdered coal is burned at such a high temperature that the resulting gases become a plasma, a cloud in which the atoms are torn apart and free electrons swarm around. The cloud has an electrical charge and can be the prime element in an electrical generator. As a result, an MHD plant needs only the coal-handling system, the MHD generator, and equipment to remove the power. Efficiency of thermal conversion — conversion of the heat energy in the coal into electrical energy — runs better than 90 percent, while that for the very best of modern electrical plants approaches only 60 percent.

The United States has such reserves of coal that it will not have to turn to poorer grades in any great quantity for generations. For example, there are large reserves of peat, a vegetable material with a variable but respectable heating value, found in boggy land. It is little used for fuel in the United States. Elsewhere, though, peat is a major fuel, particularly in Ireland and Russia. Ireland generates 30 percent of its electrical power from huge generating plants set up next to peat bogs, which are also the major source of home heating fuel.

Yet, there are always exceptions. The oil country of Texas has one of the world's larger deposits of lignite. A large aluminum smelter, a very energy-intensive operation, is located in the heart of that deposit. The aluminum company decided that it was cheaper to bring the raw materials to the energy than the energy to the raw materials. Several utilities in the area also generate power from the lignite, and have for years.

Today, public attention focuses on processes that will turn coal and other nonpetroleum fossil sources into liquids and gaseous materials that can replace oil and gas, not only for energy but also for chemical raw materials. Petrochemicals trace their origins to the complex chemistry of coal. The first refineries were converted coal-oil plants. Germany made gasoline from coal during World War II. The old German technology underpins the new technology.

The technologies for gas and for coal-based hydrocarbons of high quality date from the 1920s and 1930s. They are well known and well tried. Methanol can be made from coal and then used as a fuel or raw material for gasoline and chemicals. Now South Africa makes synthetic fuels successfully from coal at prices competitive with world oil prices.

But despite so long a history, current success, and expanded efforts, coal-conversion processes are energy-consuming operations, an extra step on the way to making hydrocarbons. To do anything with coal other than to burn it in its natural state is costly. Certainly coal liquids will become an important factor in the world energy picture before the turn of the century, if only as a lurking threat that keeps the lid on OPEC oil prices.

Oil shale and tar sands are another story. Neither is much used in the natural state, though rich shale burns well. They must be mined and processed to extract the hydrocarbons, which are of heavy density and must

This is the gasification section of the Sasol II synfuels plant built by Fluor Corporation in 1980. At this plant, reactors break coal into its gaseous form before it is cleaned and turned into petroleum liquids.

undergo further processing to yield a range of otherwise conventional liquid-petroleum products.

The processes appear simple: dig up the sand or the rock, put it into a retort, heat it to extract the fluids flow, remove the fluid, and refine it. And indeed, that is about all there is to it — oil shales have been mined and processed around the world since the early 1800s. The problem again is cost. Even though these processes are simpler than those required to make oil and gas from coal, they represent more steps, consume more energy in their recovery, and require more capital investment than does a barrel of crude oil. Inflation continually keeps those costs rising just beyond break-even or profit.

In 1983, the concept of retrieving petroleum from tar sands took a step backward. As oil prices turned downward, Exxon shut down its Montana project, which was on the brink of commercialization, even though it had spent millions of dollars building this facility.

Nonfossil energy sources encompass a far wider range than that of fossil sources. Government and industry are leaving almost literally no stone unturned to find sources of energy that can substitute for oil and gas at reasonable cost in specific or general situations. Most nonfossil energy sources are not very efficient. The energy often exists at a relatively low level or is so costly and difficult to extract that much is not usable. Thus, for any but such low-level uses as home and water heating, the generation of high-level energy suitable for industry and commerce from nonfossil sources requires huge capital investment.

One of the oldest nonfossil energy sources is water power, traditionally obtained by damming water so that it falls from a height to a lower level. The energy available is a function of the distance the water falls—the head—and the quantity of water. A waterwheel or turbine recovers kinetic energy from the water. The greater the head, the faster the wheel will turn and the smaller it needs to be for a given output. Geared to shafts or generators, the turning wheel produces useful mechanical work or electricity.

Small hydraulic plants powered the Industrial Revolution. Immense plants on large rivers throughout the world produce great quantities of electrical power. In the United States, hydropower accounts for about 4 percent of total electrical generation. This figure will not increase significantly because dams and hydroelectric plants cost too much. The energy provided, however, is constantly renewable. And today in some parts of the country, enterprising people are rehabilitating nineteenth-century water wheels for small-scale generation.

Power from the sea has also intrigued human beings. In some parts of the world, large tides produce daily flows of enormous quantities of water that have great potential energy. Various private corporations and governments have spent many millions of dollars studying these sources and attempting to commercialize them. So far, only two projects exist — one each in France and Russia. The needed capital investment places most such projects beyond economic feasibility and undoubtedly will continue to do so.

Again, the idea is simple enough: build dams that impound the water at high tide, flowing it in through turbines, and then release it at low tide through the same turbines. The quantity of energy in a suitable estuary is very great, some 40 megawatts in Nova Scotia's Bay of Fundy, but it is at relatively low potential. The difference in height between high and low tides at their extremes, some 40 feet in the Bay of Fundy, does not produce enough kinetic energy per unit of water to be economically feasible. Nonetheless, people have been trying to harness the Fundy tides since 1915. A new $25 million pilot project raised fresh hopes of success in the 1980s.

Many other projects also seek to tap the energy potential in water. Some of the more imaginative plan to exploit the temperature difference between surface and deep ocean waters, as in the Ocean Thermal Energy Conversion (OTEC) project. In this concept, a chemical, ammonia, for example, that is liquid at deep-ocean temperatures but gaseous at the surface is evaporated in heat exchangers by warm surface water and then condensed by cold water drawn from several thousand feet down. As the ammonia evaporates and recondenses, it does work that drives an electrical generator. The ocean temperature difference provides the heat to drive the system, just as a fire provides the heat to make the steam that drives a conventional power plant.

The energy potential in a steam plant is very great

Shasta Dam, in California, is an example of a large supplier of hydroelectric power. Besides controlling floods and storing winter run-off, the dam has a powerplant with five main generating units for a total capacity of 422,310 kW.

Hydro-electric power is also supplied by smaller dams such as the Nimbus Dam on California's American River. This dam, 76 feet in height, has a two-unit powerplant that generates 13,500 kW.

The type of low head hydroelectric dam pictured here could supply more and more of the nation's electric power in the future. It can be constructed on smaller streams at relatively low cost.

because the difference between the temperatures of the steam, which is at thousands of pounds of pressure and degrees of temperature, and the condensed water is great. The energy potential of the Ocean Thermal Energy Conversion project that uses ammonia is low because the difference in temperature between the liquid and gaseous ammonia is less than 40 degrees. Thus, the capital investment needed to realize any significant quantity of power from such a system is large.

Other systems that use the difference in salinity, and thus density, of water in ponds or the wave energy of the ocean suffer similarly from low-energy potential. Nonetheless, the energy in such sources is gigantic in quantity, free, and virtually infinite. If construction costs could be stabilized and the plants made durable enough, such units could prove economically feasible in suitable locations. But they will never be universal energy sources any more than hydro or geothermal power are. All are site-limited.

Nuclear power offers the virtues of site flexibility and, for some types of nuclear-plant technologies, virtually unlimited fuel supplies. The concept was greeted with great enthusiasm and optimism shortly after World War II when it first went to drive naval vessels, then into electrical generating plants. Hundreds of nuclear power plants exist around the world. Unfortunately, many are now shut down, and few are currently under construction. The concept has fallen under a cloud.

A nuclear plant is a heat plant. A cooling medium such as water, helium, or liquid sodium captures the heat produced by the fission of heavy radioactive metals such as uranium, thorium, and plutonium. That medium then transfers its heat to another medium, usually water, which turns to steam and drives turbines to make electricity in generating plants or to drive ships.

The nuclear plant is little different from a conventional steam plant, except for the heat source. The most common type of plant, the pressurized-water reactor, actually is a step back in technology at the generating end. The steam it produces is of relatively low pressure and temperature. The quantity of waste heat in relation to energy output is much higher than that of a modern, oil-fired steam plant. The overall thermal efficiency of a

This drawing shows an ocean thermal conversion plant ship which could make up to 586,000 tons of ammonia a year using solar energy in tropical seas.

nuclear plant is very low, only a small percent of the energy in the fuel.

Nuclear plants were attractive because the energy potential in a pound of uranium is enormous. A plant can run for months or years on a single loading of fuel, an obvious advantage for ships, particularly for submarines, since the "boiler" needs no oxygen. And some types of reactors, the "breeders," actually produce more fuel than they consume. However, only a small percent of the available energy can be used in the fuel before it becomes too contaminated with fission products. At that point, the fuel must be removed and the reactor refueled.

At one time nuclear plants were actually cheaper than fossil-fueled plants. A great future was predicted, and many orders were placed. But increasing public apprehension over safety, ever more rigorous regulation that stretched out construction times, inflation, soaring costs for nuclear fuel, and the costs, difficulty, and danger in handling nuclear wastes and recycling spent fuel have driven the costs of nuclear power to noncompetitive

Increasing public apprehension over the safety of nuclear plants, due particularly to the 1979 accident at Three Mile Island, Pa., put a brake on nuclear plant construction in the United States.

levels. Faith in industry assurances of perfect safety was damaged by an incident in the Three Mile Island plant in Pennsylvania, where human and mechanical failure caused a reactor to overheat and release radioactive materials to the environment.

Some experts predict that despite all the problems nuclear energy will be the prime central energy source of the future. Some estimates place the nuclear share as high as 20 percent by the year 2000, compared to its present 4 percent. But if that is to happen, the industry must generate much new, cost-effective technology, solve the nuclear-waste problem, and provide the public with believable safety guarantees.

Today coal-fired power plants are more-cost effective than are nuclear plants. And the gap is widening. Given the enormous reserves of coal and the high level of conventional steam plant technology, utility executives see little reason to invest in nuclear plants that not only cost more but are proving to be public-relations headaches.

Still, nuclear power will continue to be an important and growing tile in the energy mosaic because nuclear-generated electrical energy costs about the same, no matter where the plant is. Thus, such plants are attractive where transportation costs make fossil plants more expensive.

The major nuclear problem is radioactive waste. Mankind has never before faced quite such a puzzle. These wastes accumulate at every stage of nuclear operations. They range from high-level, short-lived, and extremely dangerous to low-level and still dangerous even after expending their half-lives of thousands of years. Unlike chemical wastes, atomic residues cannot be incinerated safely or otherwise neutralized. Only time reduces their virulence, though packaging can contain it.

But who can design a package and guarantee its integrity for a millennium or more? Who will keep watch for 50 generations? What organization in the year 3000 will declare as no longer hazardous the nuclear wastes stored in the years 1945 to 1985? And, not least, who will pay the caretakers over such an unthinkable span of time?

Another form of nuclear energy, nuclear fusion, could relieve the world of all energy problems for all time. The principle is simple. When two hydrogen atoms fuse, they form a helium atom and release some energy. Radiation is low. In a fusion plant, the amount of radiation and radioactive waste should be much less than in a fission reactor. There would be no spent fuel to process or the wide array of radioactive isotopes to worry about.

Unfortunately, to date, only nature has been able to make fusion reactions continue for any length of time — in the sun and stars — while human achievement has been limited to the manufacture of hydrogen bombs. The fusion reaction requires temperatures of 100 million degrees. In very costly experimental equipment in large goverment-funded programs, scientists in the United States and abroad have achieved such temperatures for mere millionths of a second and have managed to fuse under controlled conditions only a very few hydrogen atoms.

Though research has now proceeded for more than 30 years and billions have been spent, controlled fusion energy is still at the stage where the scientists do not even know the questions to ask. And once controlled fusion has been reliably demonstrated in the laboratory, there are as yet only vague ideas about the engineering design of a commercial-scale fusion generating plant. The practical problems encountered in turning a laboratory process into a commercial reality could take many more years. Few doubt that one day fusion power will be in use. The lure is an inexhaustible reservoir of fuel — water. But the technology and the plant required to turn a free glass of water into a few megawatts of electricity could prove so complex and costly that fusion plants, too, will remain economically prohibitive, even when they are technically feasible.

Solar power is another nonfossil energy source much in favor today. It also seems easy enough to tap — the sun is warm and shines much of the time in much of the world. Its average flux of 40 Btu per square foot over the whole earth adds up to 20,000 times as much energy as humans use from all sources. It is an intriguing

334

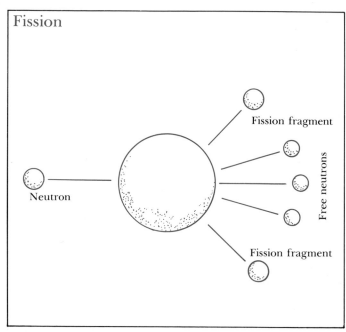

Nuclear reactors today use fission—splitting the atom—to produce energy. They are expensive to run, and the wastes they produce are difficult to dispose of.

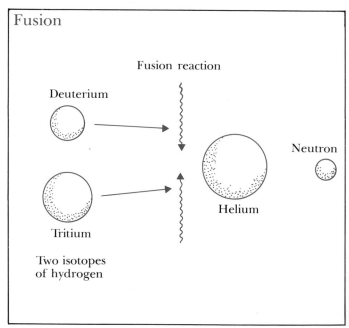

Tomorrow's nuclear reactors may use fusion—forcing atoms together—to generate power. Their raw materials are inexpensive, and the process leaves no radioactive by-products.

The world's first utility scale photovoltaic power plant, built by ARCO Solar, Inc., is located near Hesperia, Calif. Its 108 trackers produce one megawatt of electricity, enough to power 300 to 400 average southern California homes.

amount. But it is not one to conjure with.

In absolute terms, 40 Btu per square foot is not very much. If that square foot were one side of a perfectly insulated cubic foot of water and all the energy were perfectly absorbed, in 12 hours the sun would heat the water by only 7 degrees. And, of course, neither the insulation nor the absorption could be perfect, so the actual heat recovered would be much less. Air heats faster than water, which explains why greenhouses and sunny rooms get warm on winter days.

Solar energy is plentiful and free, though cyclic. Gathering and storing it is not free. And the more energy sought, in either high temperature or quantity or both, the more costly the means needed to recover it.

The cheapest solar energy is wholly passive. People build greenhouses, offices, dwellings, or solar evaporation ponds sited toward the sun. In a slightly more sophisticated system, large flat-plate collectors angled to the sun can expose a thin film of collector fluid to a large area of sunlight. The energy thus concentrates in a small volume

of fluid and heats it enough to be useful. That is the principle of domestic water heaters. But even at this level, capital investment is necessary for collector plates and plumbing, and perhaps pumps, which themselves require energy.

More effective, and costly, units can be motorized to follow the sun. Some of the energy collected goes to run the motor. Still more effective units will have lens-shaped collectors that concentrate the energy and can produce temperatures high enough for industrial purposes and for cooking.

Most solar-energy effort today aims at more efficiently using the sun's low-level heat to substitute for fossil fuels in all sorts of low-level heating situations, such as hot water for homes and building interiors. But for a long time considerable work has been directed toward converting sunlight directly to electrical energy.

For years the space program has powered its satellites with solar cells, thin devices of silicon or other photoelectric elements. These solar cells directly convert a portion of the sunlight that falls on them into electricity. That energy can drive electrical devices or charge batteries. Similar cells have been developed for terrestrial use. They are well suited to remote and unattended applications where the sun runs the operation by day and batteries take over at night. The cells are not very efficient. The theoretical limit is around 23 percent conversion, and most run far below that. Thus, a substantial installation needs a very large array of cells. In addition, they are expensive, though the price is dropping rapidly as manufacturing techniques improve and demand increases. These devices have already found a valuable niche in the energy mosiac, but it will ever be a specialized one, solar-driven-automobile demonstrations notwithstanding.

Gasoline and diesel vehicles could be replaced by entirely new vehicles. Research on electric vehicles, for instance, that run on batteries or fuel cells is plentiful. The drawbacks of these vehicles have been the enormous weight and cost of the batteries, the short distance of use, poor acceleration, and low top speeds. Intensive research is reducing weight and cost, as well as increasing energy storage capacity. Increasingly, such vehicles will have their place, but they are not likely to replace liquid-fueled ones in the foreseeable future.

Much more likely, nonpetroleum fuels will mix with and gradually replace petroleum fuels for transportation. The leading candidate is ethanol, a form of ethyl alcohol, already being sold in the United States, and particularly in Brazil, as a 10 percent mixture with gasoline called "gasohol."

Though it is made synthetically from petroleum, ethanol is primarily a fermentation product of agricultural raw materials. It can be made from virtually anything containing sugar, starch, or cellulose. In some countries where gasoline is particularly expensive and agricultural wastes are abundant ethanol can be made for less than the cost of gasoline. But elsewhere the initial enthusiasm for gasohol has waned. An industry established literally from the ground up cannot make ethanol more cheaply than gasoline.

Ethanol has drawbacks, each of which exacts a cost. The fuel has less energy content than gasoline and takes up more space. It is more chemically reactive, which means that engines and fuel systems must be modified at some cost. Its manufacture requires agricultural products and land, both of which might earn higher returns in other uses. Growing the agricultural raw material is not free; neither is the energy necessary to ferment and then distill the alcohol. Indeed, some studies indicate that the energy needed to produce alcohol exceeds the energy that is recovered from the process. The distilleries will require new investment in plant and facilities. Taken altogether, these many costs add up to a total cost of $52 to $60 a barrel of alcohol.

Methanol, a synthetic readily made from coal, oil, or gas, can also be used in gasohol or as a fuel itself. Methanol is not a significant transportation fuel now, but it could become so.

Many other nonfossil energy sources, some of which have had costly tryouts in the United States, will contribute locally or regionally significant amounts of energy in the future, but will be much more important to developing countries and those without significant fossil energy sources of any kind. All require trade-offs of one kind or another.

In the United States, millions of tons of trash and garbage are incinerated. People have long wondered why that energy could not be used. Plants have been built to burn the trash as fuel, but the experiment in general has not been successful. The materials handling costs have defeated the economics. Nonetheless, the source is there, cities must dispose of it, and if they can do so with a net energy gain, they will. Some European cities have reported good results, though the U.S. experience is generally negative.

Perhaps at the very lowest energy level is biomass. Animal and vegetable waste in simple digesters ferment and produce methane gas, which can then be captured and used as fuel. The concept appears cheap and simple. In agricultural countries and regions, biomass systems would seem to promise very significant amounts of energy produced more efficiently than current practices that burn the wastes directly, since the digester yields fuel and fertilizer. But even in countries like India and Korea, the system cannot stand on its own economic legs. When

Windmills existed long before people knew how to drill for and refine oil, and they will be around long after oil has depleted. Experiments such as this one in Oahu, Ha. use the wind turbine to generate electricity.

thoroughly analyzed, the concept actually proves more costly than existing seemingly wasteful practices.

These many alternate energy sources and others not discussed in detail—wind power, for example, or liquid and gaseous hydrogen—taken all together, will complete the energy spectrum in the coming generations. Pricing will eventually establish an equilibrium, probably in the 1990s, with the price of any particular source closely dependent on its location and application. But petroleum's role is shrinking.

Oil has fueled the world for a century. But it is running out, and it is becoming more costly. As oil supplies shrink, the real necessity to look to other energy sources increases. People now are choosing energy sources that previously were considered primitive, like windmills and solar panels. But use of these sources will occur at much more sophisticated level than in earlier times, and will require huge investments in capital and technology.

GLOSSARY

Acidizing a well — Increasing the flow of oil from a well by pumping hydrochloric acid into the well under high pressure. This reopens and enlarges the pores in the oil-bearing limestone formations.

Air drilling — A form of rotary drilling using compressed air instead of drilling mud.

Allowable — The amount of oil and/or gas a well is allowed to produce under state regulations in some states.

Annular space — The space between a well's casing and the wall of the borehole.

Annulus of a well — The space between the surface casing and the inner, producing well-bore casing.

Anticline — A geological term describing a fold in the earth's surface with strata sloping downward on both sides from a common crest. Anticlines frequently have surface manifestations like hills, knobs, and ridges. At least 80 percent of the world's oil and gas has been found in anticlines.

API gravity — The gravity (weight per unit of volume) of crude oil expressed in degrees according to an American Petroleum Institute recommended system. API gravity divides the number 141.5 by the actual specific gravity of the oil at 60 degrees Fahrenheit and subtracts 131.5 from the resulting number. The higher the API gravity, the lighter the crude. High-gravity crudes are generally considered more valuable.

Aquifer — An underground water reservoir contained between layers of rock, sand, or gravel.

Arab oil embargo of 1973-74 — During the Arab-Israeli conflict in October 1973, Arab oil producers cut off shipments to the United States and the Netherlands in retaliation for their support of Israel. At the same time, they cut down production. The shortage was felt by all oil-importing nations, with world prices moving sharply higher. Price and allocation controls suppressed some of this increase in the United States, but gasoline lines were still prevalent.

Associated gas — The gas that occurs with oil either as free gas or in solution. When occurring alone, it is referred to as unassociated gas.

Baghouse — An air-filtering device that removes particulate matter from furnace and process unit exhaust.

Barefoot completion — Completing a well without casing in firm sandstone or limestone that shows no indication of caving in or disintegration.

Barrel — Standard measurement in the petroleum industry. One barrel of oil equals 42 U.S. gallons.

Basement rock — Igneous or metamorphic rock lying below sedimentary formations in the earth's crust. Basement rock does not contain petroleum deposits.

Basin — A depression in the earth's crust in which sedimentary materials have accumulated. Such a basin may contain oil or gas fields.

BCF (billion cubic feet) — The cubic foot is a standard unit of measure for gas at atmospheric pressure.

Beam well — A well that lifts fluid by means of rods and a pump actuated by a walking beam.

Biomass — Any organic material, such as wood, plants, and organic wastes, that can be turned into fuel.

Bleeding core — A core sample of rock so highly permeable and saturated that oil drips from it.

Blowout — A sudden escape of oil or gas from a well, caused by uncontrolled high pressure. It usually occurs during drilling.

BOP (blowout preventer) — An assembly of heavy-duty valves attached to the top of a well casing to control pressure.

Bonus — Money paid to a landowner or other holder of mineral rights by the lessee for the execution of an oil and gas lease in addition to any rental or royalty obligations specified in the lease.

Bottom-hole choke — A device placed at the bottom of the tubing to restrict the flow of oil or to regulate the gas-oil ratio.

Bottom-hole pressure — The pressure of the reservoir or formation at the bottom of the hole. A decline in pressure indicates some depletion of the reservoir.

Bottom-hole pump — A compact, high-volume pump located in the bottom of a well, not operated by sucker rods or a surface power unit.

Bridle — The cable link between the "horsehead" and the pump rod on a pumping well.

BS&W (basic sediment and water) — Material pumped up with oil and gas which must be separated out.

Btu (British thermal unit) — A standard measure of heat content in a fuel. One Btu equals the amount of energy required to raise the temperature of one pound of water one degree Fahrenheit at or near 39.2 degrees Fahrenheit.

CAOF (calculated absolute open flow) — A figure representing a gas well's theoretical producing capability per day.

Cable drilling — A method of well-drilling that employs a reciprocating, rather than a rotary, motion to penetrate rock. In the nineteenth century, until Drake's time, power was supplied by men. Drake used a steam-powered cable rig. Today, cable rigs are powered by gasoline or diesel engines.

Casing — Pipe used in oil wells to reinforce the borehole. Sometimes several casings are used, one inside the other. The outer casing, called the "surface pipe," shuts out water and serves as a foundation for subsequent drilling.

Casinghead — The portion of the casing that protrudes above the surface and to which control valves and flow pipes are attached.

Casinghead gas — Natural gas produced from an oil well, as opposed to gas produced from a gas well.

Casinghead gasoline — Highly volatile, water-white liquid hydrocarbons separated from casinghead gas.

Cement squeeze — Forcing cement into the perforations, large cracks, and fissures in the wall of a borehole to seal them off.

Cement — To fix the casing in the borehole with cement. The cement is injected through the drill pipe to the bottom of the casing and up into the annular space between the casing and the borehole wall. Once the cement hardens, it is drilled out of the casing, which is then perforated to allow oil and gas to enter.

Christmas tree — An assembly of valves, gauges, and chokes mounted on a well casinghead to control production and the flow of oil to the pipelines.

Choke — An orifice installed in a pipeline at the well surface to control the rate of flow.

Circulate — To pump drilling fluid into the borehole through the drillpipe and back up the annulus.

Clean oil — Crude oil containing less than 1 percent sediment and water; pipeline oil; oil clean enough to send through a pipeline.

Coal gasification — The chemical conversion of coal to synthetic gaseous fuel.

Coal liquefaction — The chemical conversion of coal to synthetic liquid fuel.

Coal slurry pipeline — The pipeline used to carry slurry, a mixture of crushed coal and water.

Coal washing — Cleaning coal with water and additives before burning to remove impurities.

Cogeneration — The combined production of electrical or mechanical energy and usable heat energy.

Common carrier — A person or company in the business of transporting the public or goods for a fee. In the industry, a person or company engaged in the movement of petroleum products, like a public utility.

Completed well — A well made ready to produce oil or natural gas. Completion involves cleaning out the well, running steel casing and tubing into the hole, adding permanent surface control equipment, and perforating the casing so oil or gas can flow into the well and be brought to the surface.

Condensate — Liquid hydrocarbons separated from natural gas, usually by cooling.

Connate water — The water present in a petroleum reservoir in the same zone occupied by oil and gas considered by some to be the residue of the primal sea, connate water occurs as a film of water around each grain of sand in granular reservoir rock and is held in place by capillary attraction.

Confirmation well — A well drilled to "prove" the formation encountered by an exploratory well.

Conventional energy sources — Oil, gas, coal, and sometimes nuclear energy, in contrast to alternative energy sources such as solar, hydroelectric and geothermal power, synfuels, and biomass.

Core — Samples of subsurface rocks taken as a well is being drilled. The core allows geologists to examine the strata in proper sequence and thickness.

CO_2 injection — A secondary recovery technique in which carbon dioxide (CO_2) is injected into wells as part of a miscible recovery program.

Cratered — A description of a well that has blown out with such great force, accompanied by fire and explosion, that it has destroyed all surface fixtures and made a large crater where the top of the well used to be.

Crude oil — Liquid petroleum as it comes out of the ground. Crude oils range from very light (high in gasoline) to very heavy (high in residual oils). Sour crude is high in sulfur content. Sweet crude is low in sulfur and therefore often more valuable.

Crude oil equivalent — A measure of energy content that converts units of different kinds of energy into the energy equivalent of barrels of oil.

Cuttings — Chips and small rock fragments brought to the surface by the flow of drilling mud as it is circulated and examined by geologists for oil content.

Cycling of gas — Returning to a well the gas removed with the oil, thereby maintaining pressure in the reservoir and promoting more efficient recovery.

Dedicated reserves — Natural gas supply under contract to a pipeline company.

Deepwater port — An offshore marine terminal designed to accommodate large vessels such as VLCCs and tankers, connected to the shore by submerged pipelines.

Deliverability — A well's tested ability to produce.

Development well — A well drilled in an already discovered oil or gas field.

Differential-pressure sticking — A condition in which a section of drillpipe becomes stuck in deposits on the wall of the borehole.

Direct coal liquefaction — The production of liquid fuel through the interaction of coal and hydrogen at high temperature and pressure.

Directional drilling — Drilling at an angle, instead of on the perpendicular, by using a whipstock to bend the pipe until it is going in the desired direction. Directional drilling is used to develop offshore leases, where it is very costly and sometimes impossible to prepare separate sites for every well; to reach oil beneath a building or some other location which cannot be drilled directly; or to control damage or as a last resort when a well has cratered. It is much more expensive than conventional drilling procedures.

Disposal well — A well used for disposal of salt water. Usually located in a subsurface formation sealed off from other formations.

Dissolved gas — The gas contained in solution with crude oil in a reservoir.

Dissolved-gas drive — The pressure from expanding gas dissolved in crude oil that drives it up the well to the surface.

Distillate — A generic term for several petroleum fuels that are heavier than gasoline and lighter than residual fuels; for example, home heating oil, diesel oil, and jet fuels.

Distributor — A wholesaler of gasoline and other petroleum products; also known as a jobber. Distributors of natural gas are almost always regulated utility companies.

Domestic production — Oil and gas produced in the United States as opposed to imported product.

Downstream — All operations taking place after crude oil is produced, such as transportation, refining, and marketing.

Drill bit — The part of the drilling tool that cuts through rock strata.

Drillstem test — A test through the pipeline prior to completion to determine if oil or gas is present in a formation.

Drill string — Also called drill pipe or drill stem. Thirty-foot lengths of steel tubing screwed together to form a pipe connecting the drill bit to the drilling rig. The string is rotated to drill the hole and also serves as a conduit for drilling mud.

Drilling mud — A mixture of clay, water, chemical additives, and weighting materials that flushes rock cuttings from a well, lubricates and cools the drill bit, maintains the required pressure at the bottom of the well, prevents the wall of the borehole from crumbling or collapsing, and prevents other fluids from entering the well bore.

Drilling rig — The surface equipment used to drill for oil or gas, consisting chiefly of a derrick, a winch for lifting and lowering drill pipe, a rotary table to turn the drill pipe, and engines to drive the winch and rotary table.

Drilling platform — An offshore structure with legs anchored to the sea bottom that supports the drilling of up to 35 wells from one location.

Dry hole — A well that either produces no oil or gas or yields too little to make it economic to produce.

Dry natural gas — Natural gas containing few or no natural gas liquids (liquid petroleum mixed with gas).

Dual completion — Completing a well that draws from two or more separate producing formations at different depths. This is done by inserting multiple strings of tubing into the well casing and inserting packers to seal off all formations except the one to be produced by a particular string.

Duck's nest — A slang term for a standby drilling mud tank or pit used to hold extra mud or as an overflow in the event of a gas kick.

Electrical well logging — A method of oil exploration that originated with Conrad Schlumberger, who first tested it in 1927 on a 1,500-meter well in France. As used today, the process is very simple. Current passes into the ground, through the resistive medium and into the sonde. The resulting charts show the the varying resistance, the conductance, and the self-potential of the strata surrounding the well at every level, and geophysicists use them to assay whether petroleum is present in a formation.

Energy-GNP ratio — The amount of energy used to produce a dollar's worth of output, as measured by the gross national product, or GNP.

Enhanced oil recovery — Injection of water, steam, gases, or chemicals into underground reservoirs to cause oil to flow toward producing wells, permitting more recovery than would have been possible from natural pressure or pumping alone.

EIS (Environmental Impact Statement) — A statement of the anticipated effect of a particular action on the environment, required by the National Environmental Policy Act of 1969 for any significant action, including granting certain oil and gas permits.

ESECA (Energy Supply and Environmental Coordination Act) — An act of Congress passed in 1974 that prohibited certain power plants with coal-burning capabilities from burning petroleum. In addition, it required newly constructed fossil fuel boilers to be designed to burn coal.

Ethanol — The two-carbon-atom alcohol present in the greatest proportion upon the fermentation of grain and other renewable resources such as potatoes, sugar, or timber. Also called grain alcohol.

Exploratory well — A well drilled to an unexplored depth or in unproven territory, either in search of a new reservoir or to extend the known limits of a field that is already partly developed.

External casing packer — A device used on the outside of the well casing to seal off formations or protect certain zones. The packer is run on the casing and expanded against the wall of the borehole at the proper depth by hydraulic pressure or fluid pressure from the well.

Farmer's oil — An expression that refers to the landowner's share of oil from a well drilled on his property. This royalty is traditionally one-eighth of the produced oil free of any expense to the landowner.

Farm in — When one company drills wells or performs other activity on another company's lease in order to earn an interest in or acquire that lease.

Farm out — When one company allows another to "farm in" its area.

Fault — A break in the continuity of stratified rocks or even basement rocks. Forces on either side of a fault move in different directions or at different magnitudes. When the force becomes greater than the rock can resist, it breaks. Faults are significant to oilmen because they can form traps for oil when the rock fractures, they can break up oil reservoirs into noncommunicating sections, they help produce oil accumulations, and they form traps on their own.

Fault trap — A geological formation in which oil or gas in a porous section of rock is sealed off by a displaced, nonporous layer.

Fee lands — Privately owned, nonpublic lands.

Fee royalty — See Royalty.

Field — A geographical area under which one or more oil or gas reservoirs lie, all of them related to the same geological structure.

Filter cake — A plastic-like coating that builds up inside the borehole. Such buildup can cause serious drilling problems, including sticking of the drillpipe.

Fishing — Recovering the tools or pipe that have been accidentally lost down the borehole by using specially designed tools that screw into or grab the missing equipment.

Fishing tools — Special instruments equipped with the means for recovering objects lost while drilling the well.

Five-spot waterflood program — A secondary-recovery operation in which four injection wells are drilled in a square pattern with the production well in the center. Water from the injection wells moves through the formation, forcing oil toward the production well.

Five-year offshore leasing program — The first steps in the process of leasing offshore lands for oil and gas exploration. The U.S. Department of the Interior publishes and annually updates a five-year plan of timetables and areas that will be offered for lease.

Flange up — To complete the drilling of a well.

Flaring — The burning of gas vented through a pipe or stack at a refinery, or a method of disposing of gas while a well is being drilled. Flaring is regulated by state agencies. Venting (letting gas escape unburned) is generally prohibited.

Flooding — See Waterflooding.

Flow string — The casing through which oil in a well flows to the surface.

Flowing well — A well that produces through natural reservoir pressure and does not require pumping.

Flow tank — A tank in which produced oil is stored after gas or water has been removed.

Formation — A geological term that describes a succession of strata similar enough to form a distinctive geological unit useful for mapping or description.

Fossil fuels — Fuels that originate from the remains of living things, such as coal, oil, natural gas, and peat.

Fracturing — A well stimulation technique in which fluids are pumped into a formation under extremely high pressure to create or enlarge fractures for oil and gas to flow through. Proppants such as sand are injected with the liquid to hold the fractures open.

FUA (Power Plant and Industrial Fuel Use Act) — An act of Congress passed in 1978 which extended the provisions of ESECA and set specific goals for replacing oil- and gas-burning boilers with coal-fired power.

Gamma-ray logging — A technique of exploration for oil in which a well's borehole is irradiated with gamma rays. The varying emission of these rays indicates to geologists the relative density of the rock formation at different levels.

Gas cap — The gas that exists in a free state above the oil in a reservoir.

Gas-cap drive — The pressure exerted by the expansion of gas in the gas cap. Such pressure forces oil upward into the borehole of a well.

Gas condensate — Liquid hydrocarbons present in casinghead gas that condense when brought to the surface.

Gas-cut mud — Drilling mud permeated with bubbles of gas from downhole. The circulation of such mud can be severely impaired, seriously affecting drilling operations.

Gas injection — Natural gas injected under high pressure into a producing reservoir and part of a pressure-maintenance, secondary-recovery, or recycling operation.

Gas lift — A recovery method that brings oil from the bottom of a well to the surface by using compressed gas. Gas pumped to the bottom of the reservoir mixes with fluid, expands it, and lifts it to the surface.

Gas-lift gas — Natural gas stripped of liquid hydrocarbons used in a gas-lift program.

Gas-oil ratio — The number of cubic feet of natural gas produced along with a barrel of oil.

Geophones — The sound-detecting instruments used to measure sound waves created by explosions set off during seismic exploration work.

Geopressured brine — Salt water, unusually hot and in some instances saturated with methane, contained under abnormally high pressure in some sedimentary formations.

Geothermal energy — Energy produced from subterranean heat.

Glycol dehydrator — A facility for removing water particles from natural gas that have not been removed by the separator.

Gravimeter — A geophysical device that has been particularly useful in finding salt domes. Actually, it is a weight on a spring. The spring gets longer in high-gravity areas and shorter in areas of gravity-minus. Magnetism helps the oil geologist understand its measurements.

Groundwater — The water in underground rock strata that supplies wells and springs.

Gun perforation — A method of creating holes in a well casing downhole by exploding charges to propel steel projectiles through the casing wall. Such holes allow oil from the formation to enter the well.

Gusher — A well drilled into a formation in which the crude is under such high pressure that at first it spurts out of the wellhead like a geyser. Gushers are rare today owing to improved drilling technology, the use

of drilling mud to control downhole pressure, and oilmen's recognition of their wastefulness.

Hang the rods — To pull pump rods out of the well and hang them in the derrick on rod hangers.

H-coal process — A direct coal liquefaction process in which hydrogen and a catalyst are added to a coal slurry in a liquefaction vessel.

Head well — A well that produces best when being pumped or flowed intermittently.

Heavy oil — A type of crude petroleum characterized by high viscosity and a high carbon-to-hydrogen ratio. It is usually difficult and costly to produce by conventional techniques.

Heavy oil sands — Rocks containing heavy oils that barely flow under reservoir conditions.

High-pressure gas injection — The introduction of gas into a reservoir in volumes larger than what the reservoir itself produces in order to maintain reservoir pressure high enough to force the oil up the wellbore.

History of a well — A written account of a well's drilling and operation, required by law in some states.

Horizon — A specific sedimentary layer in a cross section of land, especially one in which a petroleum reservoir is found.

Horsehead — The curved guide or head piece on the well end of a pumping jack's walking beam. The guide holds the short loop of cable, called the bridle, attached to the well's pump rods.

Hot footing — Installing a heater at the bottom of an input well to increase the flow of heavy crude oil from the production wells.

Hydraulic fracturing — A method of stimulating production from a low-permeability formation by creating fractures and fissures by applying very high fluid pressure.

Hydrocarbons — Any of a large class of organic compounds that contain only carbon and hydrogen, such as methane, crude oil, and natural gas.

Hydrocracking — A refining process for converting residual materials to high-octane gasoline, reformer charge stock, jet fuel, and/or high-grade fuel oil.

Hydrometer — An instrument that measures the specific gravity of liquids.

Hydrostatic head — The height of a column of liquid. The difference in height between two points in a body of liquid.

Incremental pricing — A provision of the Natural Gas Policy Act of 1978 that requires natural gas price increases to be charged to large industrial users instead of to residential users, until the price of gas reaches the price of an alternative fuel.

Independent producer — 1. A person or corporation that produces oil for the market, who has no pipeline system or refinery. 2. An oil entrepreneur who secures financial backing and drills his own well.

Indirect coal liquefaction — A synfuel manufacturing method in which coal is first converted to synthetic gas, then catalyzed to produce hydrocarbons or methanol. Additional processing can convert methanol to gasoline.

Infill drilling — Wells drilled to fill in between established producing wells to increase production.

Intangible drilling costs — Expenditures, deductible for federal income tax purposes, incurred by an operator for labor, fuel, repairs, hauling, and supplies used in drilling and completing a well for production.

Interstate pipeline — A pipeline carrying oil or natural gas across state lines. Interstate pipelines are regulated by the federal government.

In situ — In its original place. Refers to methods of producing synfuels underground, such as underground gasification of a coal seam or heating oil shale underground to release its oil.

Isopachous map — A geological map showing the thickness and shape of underground formations. A tool used to determine underground oil and gas reserves.

Jack — An oil-pumping unit. The pumping jack's walking beam provides the up-and-down motion to the well's pump rods.

Jackknife rig — A mast-type derrick whose supporting legs are hinged at the base. The rig can be lowered or laid down intact so that it can be transported by truck from site to site.

Jack-up rig — A floating platform with legs on each corner that can be lowered to the sea bottom to raise or jack up the platform above the water.

Jetting — Injecting gas into a subsurface formation for the purpose of maintaining reservoir pressure.

Jobber — See Distributor.

Joint venture — A large-scale project in which two or more parties (usually oil companies) cooperate. One supplies funds and the other actually carries out the project. Each participant retains control over his share, including liability and the right to sell.

Junk basket — A magnet used to retrieve small tools lost in the well. A fishing instrument.

Kerogen — The hydrocarbon in oil shale. Scientists believe that kerogen was the precursor of petroleum and that petroleum development in shale was somehow prematurely arrested.

Keyseating — A condition in which the drill collar or another part of the drill string becomes wedged in a section of crooked hole.

Kick — Occurs when the pressure encountered in a formation exceeds the pressure exerted by the column of drilling mud circulating through the hole. If uncontrolled, a kick leads to a blowout.

Kill a well — To overcome downhole pressure by adding weighting elements to the drilling mud.

Landman — A self-employed individual or company employee who secures oil and gas leases, checks legal titles, and attempts to cure title defects so that drilling can begin.

Landowner royalty — The share of the gross production of the oil and gas on a property without deducting any of the cost of producing the oil or gas. The usual landowner's royalty is one-eighth of gross production.

Lay barge — A shallow-draft vessel used in constructing and laying underwater pipelines.

Lead lines — The lines through which production from individual wells is run to tanks.

Lease broker — An individual engaged in obtaining leases for speculation or resale.

Lease offering (lease sale) — An area of land offered for lease — usually by the U.S. Department of Interior — for the exploration for and production of specific natural resources such as oil and gas. Such a lease conveys no title or occupancy rights apart from the right to search for and produce petroleum or other natural resources subject to the conditions stated in the lease.

Lifting costs — The costs of producing oil from a well or lease; the operating expenses.

Lignite — A solid fuel of a grade higher than peat but lower than bituminous coal.

LNG (liquefied natural gas) — Natural gas that has been converted to a liquid through cooling to -260 degrees Fahrenheit at atmospheric pressure.

LPG (liquefied petroleum gases) — Hydrocarbon fractions lighter than gasoline, such as ethane, propane and butane, kept in a liquid state through compression and/or refrigeration, commonly referred to as "bottled gas."

Logs — Records made from data-gathering devices lowered into the wellbore. The devices transmit signals to the surface which are then recorded on film and used to make the record describing the formation's porosity, fluid saturation, and lithology. The filing of a log is required by the federal government if the drill site is on federal land.

Lost circulation — A serious condition that occurs when drilling mud pumped into the well does not return to the surface, but goes into the porous formations, crevices, or caverns instead.

Magnetometer — An instrument used for measuring the relative intensity of the earth's magnetic effect. Can be used to detect rock formations below the surface.

Marginal strike — A discovery well that produces somewhere between what is considered commercial and noncommercial amounts.

MBDE — Million barrels a day of oil equivalent.

MCF — Thousand cubic feet.

Metallurgical coal (coking coal) — Coal low in ash and sulfur and strong enough to withstand handling. Usually composed of a blend of two grades of bituminous coal. Used in the manufacture of coke and steel.

Methanation — The final step in high-Btu gas production in which hydrogen-rich gas reacts with carbon monoxide in the presence of a catalyst to form methane.

Methane — The simplest saturated hydrocarbon. A colorless flammable gas, one of the chief components of natural gas.

Methanol (methyl or wood alcohol) — A one-carbon atom alcohol made from natural gas, coal, or biomass.

Midcontinent crude — Oil produced mainly in Kansas, Oklahoma, and North Texas.

Milling — Cutting a "window" in a well's casing with a tool lowered into the hole on the drillstring.

Milling tool — A grinding or cutting tool used on the end of the drill column to pulverize a piece of downhole equipment or to cut the casing.

Mineral rights — Ownership of minerals under a tract of land, which includes the right to explore, drill, and produce such minerals, or to assign such right in the form of a lease to another party. Such ownership may or may not be severed from land surface ownership.

Miscible flooding — An enhanced recovery method in which a fluid, usually carbon dioxide, is injected into a well and dissolved in the oil, which then flows more easily to producing wells.

MMCF — Million cubic feet. The cubic foot is a standard unit of measure for quantities of gas at atmospheric pressure

Monocline — A geologic formation in which all the strata are inclined in the same direction.

Moonpool — The opening in a drill ship through which drilling operations are carried on.

Mud — See Drilling mud.

Mud hog — A pump that circulates drilling mud.

Mud log — A progressive analysis of the well-bore cuttings washed up from the borehole by the drilling mud.

Multiple completion — Completion of a well in more than one producing formation. The tubing of each production zone extends up to the Christmas tree to be piped to separate tankage.

Natural gas — A mixture of hydrocarbon compounds and small amounts of various nonhydrocarbons (such as carbon dioxide, helium, hydrogen sulfide, and nitrogen) existing in the gaseous phase or in solution with crude oil in natural underground reservoirs.

Natural gas hydrates — Ice-like mixtures of methane and water, sometimes found in permafrost or in sediments beneath the ocean floor.

NGL (natural gas liquids) — Portions of natural gas that are liquefied at the surface in lease separators, field facilities, or gas processing plants, leaving dry natural gas. They include, but are not limited to, ethane, propane, butane, natural gasoline, and condensate.

Naval petroleum reserves — Areas containing proven oil reserves that were set aside for national defense purposes by Congress in 1923 (located in Elk Hills and Buena Vista, California; Teapot Dome, Wyoming; and on the North Slope in Alaska).

Octane number — A rating used to grade the relative antiknock properties of various gasolines. A high-octane fuel has better antiknock properties than one with a low number.

Offset well — A well drilled near the discovery well. Also a well drilled to prevent oil and gas from draining from one tract of land to another where a well is being drilled or is already producing.

Offshore platform — A fixed structure from which wells are drilled offshore for the production of oil and natural gas.

Oil in place — The crude oil estimated to exist in a field or a reservoir. Oil in the formation not yet produced.

Oil pool — An underground reservoir containing oil. An oil field may contain one or more pools, each of which has its own pressure system.

Oil run — 1. The production of oil during a specified period of time. 2. A tank of oil gauged, tested, and put on a pipeline.

Oil shale — A fine-grained, sedimentary rock that contains kerogen, a partially formed oil. Kerogen can be extracted by heating the shale, but at a very high cost.

On the beam — Refers to a pumped well operated by a walking beam instead of a pumping jack.

OPEC (Organization of Petroleum Exporting Countries) — An international oil cartel originally formed in 1960 and including in 1983: Saudi Arabia, Kuwait, Iran, Iraq, Venezuela, Quatar, Libya, Indonesia, United Arab Emirates, Algeria, Nigeria, Ecuador, and Gabon. OPEC enjoyed its greatest strength in the 1970s, when it used oil as a political weapon and raised prices unilaterally, causing inflation and recession in oil-using countries. OPEC's power seemed to break when an oil glut developed in 1982 and it had to reduce prices.

Open hole — An uncased well bore.

Outcrop — A portion of bedrock or other stratum protruding through the soil level, indicating a fault or some other oil-bearing formation.

OCS (outer continental shelf) — A gently sloping underwater plain that extends seaward from the coast. Technically, it includes only "all submerged lands (1) which lie seaward and outside the area of lands beneath the navigable waters as defined in the Submerged Lands Act (67 Stat. 29) and (2) of which the subsoil and seabed appertain to the United States and are subject to its jurisdiction and control." In general usage, the term "OCS" is used by the government and the petroleum industry to refer to both the continental shelf and the continental slope to the 2,500-meter water depth.

Overthrust belt — A geological system of faults and basins in which geologic forces have thrust layers of older rock above strata of newer rock that might contain oil or natural gas. The Eastern Overthrust Belt runs from eastern Canada through Appalachia into Alabama. The Western Overthrust Belt runs from Alaska through western Canada and the Rocky Mountains into Central America.

Payoff — The time when a well's production begins to bring in revenues.

Payout — The amount of time it takes to recover the capital investment made on a well or drilling program.

Perforation — A method of making holes through the casing opposite the producing formation to allow the oil or gas to flow into the well. See also Gun perforation.

Permeability — A measure of the ease with which a fluid such as water or oil moves through a rock when the pores are connected. Geologists express permeability in a unit named the darcy, but oilmen use the millidarcy because most of the rocks they come in contact with are not very permeable.

Petrochemicals — Chemicals derived from crude oil or natural gas, including ammonia, carbon black, and other organic chemicals.

Petroleum — Strictly speaking, crude oil. Also used to refer to all hydrocarbons, including oil, natural gas, natural gas liquids, and related products.

Pinch out — The disappearance of a porous, permeable formation between two layers of impervious rock over a horizontal distance.

Pipeline gas — Gas under enough pressure to enter the high-pressure gas lines of a purchaser; gas in which enough liquid hydrocarbons have been removed so that such liquids will not condense in the transmission lines.

Plugging a well — Filling the borehole of an abandoned well with mud and cement to prevent the flow of water or oil from one strata to another or to the surface.

Porosity — A measure of the number and size of the spaces between each particle in a rock. Porosity affects the amount of liquids and gases, such as natural gas and crude oil, that a given reservoir can contain.

Possible reserves — Areas in which production of crude oil is presumed possible owing to geological inference of a strongly speculative nature.

Primary recovery — Extracting oil from a well by allowing only the natural water or gas pressure in the reservoir to force the petroleum to the surface, without pumping or other assistance.

Probable reserves — Areas which are unproven but presumed capable of production because of geological inference, for instance, proximity to proven reserves in the same reservoir.

Producing horizon — Where the well is actually produced, since it may be drilled to a much greater depth.

Producing platform — An offshore structure with a platform raised above the water to support a number of producing wells.

Production — A term commonly used to describe taking natural resources out of the ground.

Production platform — An offshore structure built for the purpose of providing a central receiving point for oil produced in an offshore area. Such a platform supports receiving tanks, treaters, separators, and pumping units.

Production test — A test made to determine the daily rate of oil, gas, and water production from a potential pay zone.

Proppants — Materials used in hydraulic fracturing for holding open the cracks made in the formation by the fracturing process. Proppants may consist of sand grains, beads, or other small pellets suspended in fracturing fluid.

Prospect — A lease or group of leases on which an operator intends to drill.

Proved behind-pipe reserves — Estimates of the amount of crude oil or natural gas recoverable by recompleting existing wells.

Proved reserves — Estimates of the amount of oil or natural gas believed to be recoverable from known reservoirs under existing economic and operating conditions.

Proved developed reserves — Estimates of what is recoverable from existing wells with existing facilities from open, producing payzones.

Proved undeveloped reserves — Estimates of what is recoverable through new wells on undrilled acreage, deepening existing wells, or secondary recovery methods.

Public lands — Any land or land interest owned by the federal government within the 50 states, not including offshore federal lands or lands held in trust for Native American groups.

Pumping well — A well that does not flow naturally and requires a pump to bring product to the surface.

Pump off — To pump a well so rapidly that the oil level falls below the pump's standing valve, rendering the well temporarily dry.

Quad — Quadrillion Btus.

Ram — A closure mechanism on a blowout-preventer stack.

Raw gas — The gas straight from the well, before liquefied hydrocarbons are extracted.

R&D — Research and development.

Reamer — A tool used to enlarge or straighten a borehole.

Reclamation — The restoration of land to its original condition by regrading contours and replanting after the land has been mined, drilled, or otherwise has undergone alteration from its original state.

Recoverable resources — An estimate of resources, including oil and/or natural gas, both proved and undiscovered, that would be economically extractable under specified price-cost relationships and technological conditions.

Reef — A buildup of limestone formed by skeletal remains of marine organisms. It often makes an excellent reservoir for petroleum.

Refining — Manufacturing petroleum products by a series of processes that separate crude oil into its major components and blend or convert these components into a wide range of finished products, such as gasoline or jet fuel.

Relief well — A well drilled in a high-pressure formation to control a blowout.

R/P ratio (reserve production ratio) — A way of showing how many years it would take to use up the nation's proved reserves of oil and natural gas at current production levels. Such estimates do not include oil or gas that may be discovered in the future or resources known to exist but now regarded as uneconomic to produce.

Reserve (pool) — A porous and permeable underground formation of producible oil and/or natural gas, confined by impermeable rock or water barriers, and characterized by a single natural pressure system.

Reservoir — A porous, permeable sedimentary rock formation containing quantities of oil and/or gas enclosed or surrounded by layers of less permeable or impervious rock. Also called a "horizon."

Reservoir pressure — The pressure at the face of the producing formation when the well is shut-in. It equals the shut in pressure at the wellhead plus the weight of the column of oil in the hole.

Residuals — Oils that are the leftovers of various refining processes and may undergo upgrading processes.

Residual fuel — Heavy oil used by utilities and industries for fuel.

Roof rock — A layer of impervious rock above a porous and permeable formation that contains oil or gas.

Rotary drilling — A method of well-drilling that employs a rotating bit and drilling mud to cut through rock formations.

Roughnecks — Members of the drilling crew.

Round trip — Pulling the drillpipe from the hole to change the bit, then running the drillpipe and new bit back in the hole.

Royalty — A payment to a landowner or mineral rights owner by a leaseholder on each unit of resource produced.

Running the tools — Putting the drillpipe, with the bit attached, into the hole in preparation for drilling.

Run ticket — A record of the oil run from a lease tank into a connecting pipeline. An invoice for oil delivered.

Salt-bed storage — Storage of petroleum products in underground formations of salt whose cavities have been mined or leached out with superheated water.

Salt dome — A subsurface mound or dome of salt.

Sample — Cuttings of a rock formation broken up by the drill bit and brought to the surface by the drilling mud. These are examined by geologists to identify the formation and type of rock being drilled.

Sample log — A record of rock cuttings made as a well is being drilled. A record is then kept that shows the characteristics of the various strata drilled through.

Saturation — 1. The extent to which the pore space in a formation contains hydrocarbons or connate water. 2. The extent to which gas is dissolved in the liquid hydrocarbons in a formation.

Schlumberger (pronounced "slumber-jay.") — The founder of electrical well logging, now the name for any electrical well log.

Scout — An individual who observes and reports on competitors' leasing and drilling activities.

Secondary recovery — The introduction of water or gas into a well to supplement the natural reservoir drive and force additional oil to the producing wells.

Sedimentary basin — A large land area composed of unmetamorphized sediments. Oil and gas commonly occur in such formations.

Sedimentary rock — Rock formed by the deposition of sediment, usually in a marine environment.

Seismic exploration — A method of prospecting for oil or gas by sending shock waves into the earth. Different rocks transmit, reflect, or refract sound waves at different speeds, so when vibrations at the surface send sound waves into into the earth in all directions, they reflect to the surface at a distance and angle from the sound source that indicates the depth of the interface. These reflections are recorded and analyzed to map underground formations.

Seismograph — A device that records natural or manmade vibrations from the earth. Geologists read what it has recorded to evaluate the oil potential of underground formations.

Separator — A pressure vessel used to separate well fluids into gases and liquids.

Service well — A well drilled in a known oil or natural gas field to inject liquids that enhance recovery or dispose of salt water.

Set casing — To cement casing in the well hole, usually in preparation for producing a commercial well.

Severance tax — A tax levied only on the petroleum industry paid to a state by producers of mineral resources (including oil and gas) when the mineral is removed from the ground.

Shale — A type of sedimentary rock composed of laminated layers of claylike, fine-grained sediments.

Shale oil — The substance produced from the treatment of kerogen, the hydrocarbon found in some shales, which is difficult and costly to extract. About 34 gallons of shale oil can be extracted from one ton of ore.

Shoestring sands — Narrow strands of saturated formation that have retained the shape of the stream bed that formed them. In the United States, such a formation is located in Kansas.

Shoot a well — A technique that stimulates production of a tight formation by setting off charges downhole that crack open the formation. The early wells were shot with nitroglycerine; then dynamite was used. The nitro man has been replaced today by acidizers and frac trucks.

Shut-down well/shut-in well — A well is shut down when initial drilling ceases for one reason or another. A well is shut in when the wellhead valves are closed, shutting off production, often while waiting for transportation or for the market to improve.

Shut-in pressure — The pressure at the wellhead when valves are closed.

Sidetracking — Drilling another well next to a nonproducing well and using the upper part of the nonproducer. This is one way to drill past obstructions in a well.

Skidding the rig — Moving a derrick from one location to another on skids and rollers.

Slant-hole drilling — See Directional drilling.

Solution gas — Natural gas dissolved and held under pressure in crude oil in a reservoir.

Sour crude — Oil that contains significant amounts of hydrogen sulfide and must be treated to remove the sulfur before it can be used.

Sour gas — Natural gas that contains chemical impurities such as hydrogen sulfide.

Spent shale — Shale that is left over after the shale oil (kerogen) has been removed.

Spudding in — Starting to drill the well.

State OCS jurisdiction — The area where coastal states own mineral rights on offshore lands. In general, it extends three nautical miles offshore, except off Texas and the Gulf Coast of Florida, where it extends three leagues (about 10 nautical miles) offshore.

Steel reef — Refers to the artificial reefs formed by the substructures of offshore drilling and production platforms which are inhabited by a rich variety of marine life.

Step-out well — A well drilled near a proven well, but located in an unproven area, that determines the boundaries of the producing foundation.

Stimulation — Techniques used on some completed wells to initiate production, such as acidizing, hydraulic fracturing, shooting, or cleaning out sand.

Storm choke — A safety valve installed in in a well's tubing below the surface to shut off well flow at a predetermined rate.

Stratigraphic test — A hole drilled to gather information about rock strata in an area.

Stratigraphic trap — A porous section of rock surrounded by nonporous layers, holding oil or gas. They are usually very difficult to locate, although oilmen believe that most of the oil yet to be discovered will be found in these traps.

Stripper — An oil well that yields 10 or fewer barrels of oil per day, or a gas well that produces an average of less than 60,000 cubic feet per day, measured over a 90-day period. Nearly three-fourths of the oil wells in the United States are strippers, and together they account for one out of every seven barrels of oil produced.

Structural trap — A reservoir created by some cataclysmic geologic event that creates a barrier and prevents further migration. The most common structural traps are anticlines, in which at least 80 percent of the world's oil and gas have been discovered.

Structure — Subsurface folds or fractures of strata that form a reservoir capable of holding oil or gas.

Submersible drilling barge — A vessel capable of drilling in deep water. The hull is flooded to sink the barge beneath the water level, and a drilling platform is jacked up above the surface.

Submersible pump — A bottom-hole pump for use in an oil well when a large volume of fluid is to be lifted.

Substructure — A platform upon which a derrick is erected.

Sweet crude — Crude oil with low sulfur content which is less corrosive, burns cleaner, and requires less processing to yield valuable products.

Syncline — A downfold in stratified rock that looks like an upright bowl. Unfavorable to the accumulation of oil and gas.

Synfuels — Fuels produced through chemical conversion of natural hydrocarbon substances such as coal and oil shale.

Synthetic crude oil (syncrude) — A crude oil derived from processing carbonaceous material such as shale oil or unrefined oil in coal conversion processes.

Synthetic gas — Gas produced from solid hydrocarbons such as coal, oil shale, or tar sands.

Tank bottoms — A mixture of oil, water, and other foreign matter that collects in the bottoms of stock tanks and large crude storage tanks and must be cleaned or pumped out on a regular basis.

Tanker — An ocean-going ship specially designed to carry crude oil and other liquid petroleum products.

Tariff — A schedule of rates or charges permitted a common carrier or utility.

Tar sands — Rocks (other than coal or oil shale) that contain highly viscous hydrocarbons that are unrecoverable by primary production methods.

TCF — Trillion cubic feet.

TD (total depth drilled) — The full depth of a well. Since the producing horizon may be uphole, wells are not always completed at TD.

Tectonic map — A geologic map showing the structure of the earth's crust.

Tender — 1. A permit issued by a regulatory body for the transportation of oil or gas. 2. A barge or small ship that serves as a supply ship and/or storage facility for an offshore rig.

Tertiary recovery — The recovery of oil that involves complex and very expensive methods such as the injection of steam, chemicals, gases, or heat, as compared to primary recovery, which involves depleting a naturally flowing reservoir, or secondary recovery, which usually involves repressuring or waterflooding.

Therm — A measure of heat content. One therm equals 100,000 Btus.

Thermal recovery — An enhanced recovery method that uses heat to thin oils that are too thick to flow to producing wells.

Tight hole — A well about which the operator keeps all information secret.

Tight sand — A formation with low permeability. Gas produced from a formation so designated by the Federal Energy Regulatory Commission qualifies for a higher market price.

Toolpusher — A field supervisor of one or more drilling rigs.

TAPS (Trans Alaska Pipeline System) — A crude oil pipeline that runs 800 miles from the North Slope of Alaska above the Arctic Circle to the ice-free port of Valdez in southern Alaska. An example of a joint-venture operation in which there are eight participating companies.

Tripping — See Roundtrip.

ULCC (Ultralarge crude carrier) — A large tanker built especially to carry 500,000 dwt and up of crude oil.

Unassociated gas — Natural gas that occurs alone, not in solution or as free gas with oil or condensate.

Undiscovered recoverable resources — Resources outside of known fields, estimated from broad geologic knowledge and theory.

Updip well — A well located high on a structure where the oil-bearing formation is found at a shallower depth.

Upstream — Activities concerned with finding petroleum and producing it, compared to downstream which are all the operations that take place after production.

Vapor pressure — The pressure exerted by a vapor held in equilibrium with its solid or liquid state.

Viscosity — The measure of a liquid's resistance to flow.

VLCC (very large crude carrier) — A tanker built to carry 200,000 to 350,000 dwt of crude oil.

Walking beam — A heavy timber or steel beam that transmits power from the drilling engine to the drilling tools. The walking beam rocks on a post, transmitting an up-and-down motion to the drilling line.

Wall sticking — A condition in which a section of the drillstring becomes stuck on deposits of filter cake on the wall of the borehole in a well.

Water-drive reservoir — A reservoir in which the pressure that forces the oil to the surface is exerted by edge or bottom water in the field.

Waterflooding — A secondary recovery method in which water is injected into a reservoir to force additional oil into the wells.

Wellhead cellar — An airtight enclosure in which work can be completed on an underground wellhead.

Well platform — An offshore structure that supports a well's surface controls and flow piping.

Well program — The procedure for drilling, casing, and completing a well.

Whipstock — A tool used at the bottom of the borehole to change the direction of the drilling bit. Essentially, it is a wedge that crowds the bit to the side of the hole, causing it to drill at an angle to the vertical. See also Directional drilling.

Wildcat well — A well drilled in an unproven or unexplored area.

Wildcatter — An operator who drills the first well in unproven territory.

Wilderness land — The land withdrawn from development by Congress in order to preserve its pristine characteristics as set forth in the National Wilderness Preservation System Act of 1964.

Windfall profits tax — A U.S. tax on crude oil production intended to tap excess industry profits anticipated after the removal of price controls in 1979. It is calculated as a specified percentage of the difference between the sales price in the field and a base price established by the government for each of various oil categories. Both the percentage and the base price vary depending on how a given field has been treated under price-control regulations. The base prices are automatically increased annually. This law will be phased out over ten years or after the government has collected $227 billion, whichever is later.

Workover — To clean out or work on a well to restore or increase production.

Workover rig — The rig used when oilmen try to restore or increase a well's production.

Zone — A specific interval of rock strata containing one or more reservoirs, used interchangeably with "formation."

Zone isolation — Sealing off a producing formation while a hole is being deepened. A special sealant is injected into the formation, where it hardens long enough for the hole to be drilled. Afterward, the substance again turns to liquid, unblocking the formation.

CHAPTER NOTES

Chapter 1 — The Beginnings

Page 2

Many sources chronicle Drake's life and accomplishments. Indeed, few historical events and the circumstances surrounding them are as well documented as Drake's career. One of the earliest extant detailed accounts is contained in Andrew Cone and Walter R. Johns, *Petrolia: A Brief History of the Pennsylvania Petroleum Regions, Its Development, Growth, Resources, etc., from 1859 to 1869,* (New York: D. Appleton and Co., 1870). The book was reprinted by Hyperion Press, Westport, Conn., in 1976. Hyperion reproduced several early and late books, domestic and foreign, covering the early years of the petroleum industry in the series The History and Politics of Oil.

Page 3

Regarding ancient and arcane uses of oil, virtually all texts tell the same stories, based on the same sources, documented or not. The most extensively documented modern texts are those by Robert James Forbes, first published by E. J. Brill in Leiden, Holland, then reprinted by Hyperion Press. Three of these volumes and their original Leiden publication dates are: *Bitumen and Petroleum in Antiquity* (1936); *Studies in Early Petroleum History* (1958); and *More Studies in Early Petroleum History 1860-1880* (1959).

Many of Forbes's numerous multilingual references show evidence of diligent scholarship. He lists, for example, nine names in Greek and Latin, with translation, for Greek Fire, with which Leo V, the Armenian, defeated the Bulgarians in the Battle of Mesembria in A.D. 814. None of the nine footnotes documenting these observations is in English; all are in either French or German. Even so, this ancient military secret was so well kept that not even Forbes can give the definitive recipe.

Dealing as he does with the early history of petroleum, Forbes concentrates on Europe and Asia. His references to the Western Hemisphere are meager.

Early references to Baku and Zoroastrianism are numerous. Forbes cites many of them. A most interesting firsthand account of the region at the height of the early oil boom, and the assertion that temples still existed in 1870, is Charles Thomas Marvin, *The Region of the Eternal Fire: An Account of a Journey to the Petroleum Region of the Caspian in 1883* (London: W. H. Allen and Co., 1884).

Plutarch (A.D. 46-125), who says of various matters that other writers err in their accounts, tells the story of Alexander and naphtha, pages 1296-98 of the Thomas North translation, (New York: Heritage Press, 1941). Interestingly, at the point at which Plutarch discusses the origins of naphtha, the original Greek is missing.

References to Herodotus are from *The Histories of Herodotus of Halicarnassus*, Halicarnassus 450 B.C., specifically, translated by Harry Carter (New York: Heritage Press, 1958). The discussion of the walls of Babylon and the transport of bitumen came from vol. 1, book 1, chap. 179, p. 73. The story of the island is in vol. 1, book 4, chap. 195, p. 295. The well is in vol. 2, book 6, chap. 119, p. 394.

On Belisarius, see John Bagnell Bury, *History of the Later Roman Empire*, vol. 2 (London: Dover Press, 1957).

Page 6

As noted, although references to Greek Fire are many, the most exhaustive are in Forbes, *More Studies in Early Petroleum History 1860-1880*, pp. 79-80.

Biblical references, taken from the King James Version, include Genesis, 6:14: "Make thee an ark . . . and . . . pitch it within and without with pitch"; Genesis 14:10: "And the Vale of Siddim was full of slimepits; and the kings of Sodom and Gomorrah fled, and fell there"; Deuteronomy 32:12: "oil out of the flinty rock"; Job 29:6: "the rock poured me out rivers of oil."

Maccabees is one of the books of the Apocrypha. These books are not part of the King James Bible and are generally regarded as historical chronicles.

References to early occurrences and uses of petroleum in the Western Hemisphere are readily available. See Forbes, Cone and Johns, and more recent references, such as Dorsey Hager, *Fundamentals of the Petroleum Industry* (New York and London: McGraw-Hill, 1939); and James A. Clark, *The Chronological History of the Petroleum and Natural Gas Industries* (Houston: Clark Book Co., 1963).

Forbes's references are mostly primary sources. Its early date also makes Cone and Johns, *Petrolia,* a primary source, though no more reliable for that, since the book does not cite sources but retells then-common tales. For that matter, so does Plutarch, who was writing centuries after the events he chronicled. Later works, such as those by Hager and Clark, rely heavily on secondary sources. Clark's first chronological event, for example, dated 6000 B.C., is: "Noah smears Ark with two coats of bitumen on outside and one coat inside before the Great Flood in the Tigris-Euphrates Valley." Clark's source is not the Bible, which says nothing about two coats or the location of the Flood, but Max W. Bell et al., *This Fascinating Oil Business* (New York: Bobbs-Merrill, 1940). The 1979 edition has the reference on p. 302 and cites both the Bible and a 4,000-year-old Babylonian tablet. From the latter, one can infer two coats outside and one coat inside.

Various sources mention George Washington's will as authority for the burning spring. In fact, Washington made several references to the spring. The first was in a letter dated March 27, 1775, to Colonel Andrew Lewis, in which Washington mentioned 250 acres with a bituminous spring and offered to share the property with Lewis. George Washington, *Writings of Washington,* John C. Fitzpatrick, ed., (Washington, D.C.: U.S. Government Printing Office, 1940), 3: 278. The quotation from the will is in vol. 37, p. 297-98. The quoted passage is from the will.

Page 7

Forbes's *Studies in Early Petroleum History* provides the most extensive documentation and comment on oil in the Middle Ages and later, in Europe and Asia.

Pages 8-9

Cone and Johns, *Petrolia,* is a good early source on the oil industry in the United States. Concerning oil, the authors state on p. 118, "By the settlers and Indians it was used as a medicine. . . . Beyond this no other use was discovered for the substance until its later development in the decades of the present century." Oil was an annoyance to salt miners, but they nonetheless made a profit from it as early as 1819. Also, on p. 37 the authors cite a report of 1826: "It will be a valuable article for lighting the street lamps in the future cities of Ohio."

See also "The Burning Rivers: The Story of Oil in America Before the Drake Well in 1859," American Petroleum Institute *Quarterly,* (July 1945): 3-6.

Forbes documents the early oil shale industry and development of refining processes. See Martine Ede, "An Account of the Making of Pitch, Tar and Oil out of a Stone in Shropshire," *Philosophical Transactions* 20 (1967): 544.

Kier's still is described in W. K. Cadman, "Kier's 5-Barrel Still: A Venerable Industrial Relic," *Western Pennsylvania Historical Magazine* (December 1959): 351. The article, which also discusses Abraham Geisner and the origin of the word "kerosene," says: "In 1854, at Pittsburgh, Pa., on 7th Avenue just east of the old Pennsylvania canal near Grant Street, Samuel L. Kier established the first successful commercial petroleum refinery in the Western Hemisphere, if not in the modern world."

One of the best more-or-less-contemporary accounts of the Venango County, Pa., oil enterprise, the Drake well, is in Cone and Johns, *Petrolia,* pp. 47-68. It is a fascinating account, including the complete text of Silliman's report and outlining with awesome prescience the whole of a future industry.

Pages 10-11

Spindletop is among the better-documented historical events. See *Spindletop: A Texas Titan* (New York: American Petroleum Institute, 1945); and James A. Clark and Michel T. Halbouty, *Spindletop* (New York: Random House, 1972).

Patillo Higgins's perseverance and determination are delightfully described in Boyce House, "He Started It All," *Lamp* (August 1939): 20-25.

Chapter 2 — The U.S. to 1931

Page 12

Silliman's famous scientific report has been reproduced in Paul H. Giddens, ed., *A Facsimile of "A Report of the Rock Oil, or Petroleum, from Venango County, Pennsylvania"* (Meadville, Pa., 1949). It makes instructive reading, especially for those who believe that nineteenth-century chemical technology must have been crude and primitive. It is a chemical classic that touched off an industry. Silliman went on to produce many more reports on other potential oil properties as far away as California. Two excellent overall sources for this chapter are Paul H. Giddens, *The Early Days of Oil* (Princeton, N.J.: Princeton University Press, 1964) and Daum Williamson, *The American Petroleum Industry,* vol. 1, *The Age of Illumination, 1859-1899* (Evanston, Ill.: Northwestern University Press, 1959).

The account of the founding of the Pennsylvania Rock Oil Company comes chiefly from Leonard M. Fanning *The Rise of American Oil* (New York: Harper, 1948). Cone and Johns, *Petrolia,* also has a detailed account which, interestingly, gives rather short shrift to Colonel Drake and lays the entire credit for the oil industry at the feet of George Bissel, a journalist and teacher. Bissel receives the credit, say Cone and Johns, because he had the vision and hired Silliman to bolster the vision with a scientific foundation. Another source is Samuel W. Tait, Jr., *The Wildcatters: An Informal History of Oil Hunting in America* (Princeton, N.J.: Princeton University Press, 1946).

Page 13

The "contemporary writer," whose extravagant prose is quoted extensively in this chapter, was Thomas A. Gale, a resident on Oil Creek and author of *The Wonder of the Nineteenth Century: Rock Oil in Pennsylvania and Elsewhere* (Erie, Pa.: Sloan and Griffith, 1860). His breathless 80-page book was the very first about the industry, written by a close and skilled observer of whom nothing else is known. Among his observations: "Many ladies, when circumstances require it, make first-class businessmen." Ethyl Corporation reprinted the book in 1952, complete with such original advertisements as "Oil Pipe Made to Order," by Barr and Johnson, Erie; "Oil Tools & Machinery — Very Best Materials," by Keystone Iron Works; "Manufacturers of Engines. Drilling Tools and Pipe for Oil Wells," by Liddell and Marsh. The facsimile includes explanatory notes added by Ethyl.

Tate, *The Wildcatters,* mentions Williams. A more extensive exploration of the thesis is Ernest C. Miller, "North America's First Oil Well, Who Drilled It?" *Western Pennsylvania Historical Magazine* 42 (December 1959): 317-33. Miller concludes that contemporary references say nothing about "drilling" but that a well was "opened," or dug, in 1858. People have been digging for oil for thousands of years. Williams himself never claimed to have drilled the first well.

Page 14

Most of the detail is from Gale, *The Wonder of the Nineteenth Century.* Later publications, including Cone and Johns, *Petrolia,* that tell some of the same stories have used this book as the primary source.

Gale, *The Wonder of the Nineteenth Century,* mentions the Drake well fire. Clark, *Chronological History,* cites *Derrick's Handbook of Petroleum* (Oil City, Pa.: Derrick Publishing Co., 1898), vol. 1, as the source for the conflagration story.

Production figures come from Clark, *Chronological History,* and Fanning, *Rise of American Oil.*

Page 15

Tait, *The Wildcatters,* describes the short life of the West Virginia field.

Page 16

Gale, *The Wonder of the Nineteenth Century,* describes the dam-busting method of shipping oil down Oil Creek. Cone and Johns, *Petrolia,* also does so, in far greater length and detail. Oilmen did not originate the practice. Lumbermen were shipping out timber by this method long before anyone thought of oil. Neither book mentions specific casualties, but Cone and Johns say that the carnage along the creek was enormous. They also give an idea of transportation conditions: "No one would like to believe in the doctrine of transmigration of souls after witnessing the hardships of the tow horses."

Pages 19-22

The prime, if not primary, sources for information on John D.

Rockefeller and the Standard Oil Company are Ralph W. Hidy and Muriel E. Hidy, *History of the Standard Oil Company of New Jersey: Pioneering in Big Business, 1882-1911* (New York: Harper and Brothers, 1955), and the companion volume, George S. Gibbs and Evelyn H. Knowlton, *The Resurgent Years, 1911-1927* (New York: Harper and Brothers, 1955). Others include Ida M. Tarbell, *The History of the Standard Oil Company* (New York: McClure's, 1904), a ground-breaking corporate exposé that industry-loyalist James C. Clark never cites in his *Chronological History.* Anthony Sampson, *The Seven Sisters* (New York: Viking, 1975), contains an excellent though brief account of the rise of Rockefeller and Standard Oil.

Page 22

As well as those already mentioned, other sources for this chapter include Henrietta M. Larson and Kenneth W. Porter, *History of Humble Oil and Refining Company: A Study in Industrial Growth* (New York: Harper and Brothers, 1959); and Richard O'Connor, *The Oil Barons* (Boston: Little, Brown, 1971.)

Pages 23-25

Tait, *The Wildcatters,* gives a good account of Dad Joiner's travails. Every source dealing with the period, including Sampson, *The Seven Sisters,* and John M. Blair, *The Control of Oil* (New York: Vintage, 1978), mentions some aspect of this incredible story. Blair quotes at length from a statement by a Texas attorney, Karl A. Crowley, to the Temporary National Economic Committee in 1939. Said Crowley, prefatory to a long account of the East Texas field, "The independent wildcatter must be given credit for practically every major discovery of oil in the United States."

An anecdotal account of Dad Joiner's struggles appears in the booklet, *The Fabulous East Texas Oil Field,* written by Fred R. Pass, editor of *The Texas Almanac.*

Page 25

The most thorough account I have encountered is Joe L. White, "Columbus Marion 'Dad' Joiner and the East Texas Oil Boom," *East Texas Historical Journal* 6 (March 1968): 19. On page 5, the same issue also carried an account of an early Texas wildcatter: C. K. Chamberlain, "Lyne Toliafero Barret: A Pioneer Texas Wildcatter." Barret sank the first well in Texas, found oil in 1866 at 106 feet, developed a primitive rotary drill, and went north for financing and equipment. Concludes the article: "History records many instances in which the pioneer reaped little benefit from his ideas, his vision, his energy, his investments. Those who came later received the profits."

Chapter 3 — U.S. Since 1931

References for Chapter 2 also contributed to this chapter.

Page 26

A frequently used source on East Texas is James A. Clark and Michel

Halbouty *The Last Boom* (New York: Random House, 1972). See also *East Texas Historical Journal* 2 (October 1964): 130; and Bobby H. Johnson, "Oil in the Pea Patch: The East Texas Oil Boom," *East Texas Historical Journal* 13 (Spring, 1975): 34-42. An excellent work covering this period is vol. 2 of Daum Williamson, *The American Petroleum Industry, The Age of Energy, 1900-1959,* (Evanston: Northwestern University Press, 1963).

Pages 28-29

Another good source on this period is James A. Clark, *Three Stars for the Colonel* (New York: Random House, 1954). Not Colonel Drake but Colonel Ernest Thompson, as head of the Texas Railroad Commission, established control of the East Texas field and later ran the Interstate Oil Compact Commission. He was responsible for halting the trade in hot oil.

Pages 30-31

Clark, *Chronological History,* was relied on heavily.

Pages 32-33

See Editors of Look, *Oil for Victory, The Story of Petroleum in War and Peace* (New York: McGraw-Hill, 1946), for a contemporary account of the importance of oil in World War II. Ruth Sheldon Knowles, *The First Pictorial History of the American Oil and Gas Industry 1859-1983* (Athens, Ohio: Ohio University Press, 1983), also provided important anecdotal information for this section.

Page 34

On organization and relationships of the majors to the independents, see Stanley H. Ruttenberg Associates, Inc., for the Maine Engineers Beneficial Association, *The American Oil Industry: A Failure of Anti-Trust Policy,* December 1973. See also Neil H. Jacoby, *Multinational Oil: A Study in Industrial Dynamics* (New York. Macmillan, 1974); and Committee on the Interior, *Energy R&D–Problems and Prospects,* 93rd Cong., 1st sess., 1973, p. 7. See also such works as Sampson, *The Seven Sisters;* Blair, *The Control of Oil;* and Ball, *This Fascinating Oil Business.*

An excellent historical, human, and technical treatment of the Trans-Alaska Pipeline System is James P. Roscow *800 Miles To Valdez* (New York: Prentice-Hall, 1977). For a technical overview, see Peter J. Brennan, "Startup and Operation," *Alyeska Reports* (January 1977): 4-9.

Chapter 4 — International to 1939

Pages 36-38

Much here written about Baku and the Caspian oil industry comes from Marvin, *The Region of Eternal Fire.* I am pleased to give the much-traveled Marvin full due for his excellent source-book, his many observations, and his readability.

Other sources include James Dodds Henry, *Baku, An Eventful History* (London: A. Constable and Co., 1905), as well as Samuels, Forbes, and Blair.

Page 38

For an interesting account of the contributions of the Nobel brothers to the development of oil refining and shipping and the impact of the Revolution of 1905 on the Russian oil industry, see Robert Tolf, *The Russian Rockefellers: The Saga of the Nobel Family and the Russian Oil Industry* (Stanford: Calif.: Hoover Institution Press, 1976).

Pages 38-40

On Shell, in addition to sources already cited, see *The Petroleum Handbook,* 5th ed. (London: Shell International Petroleum Co., 1966). Another useful source is Kendall Beaton, *Enterprise in Oil* (New York: Appleton-Century-Crofts, 1957). F. C. Gerretson's *History of the Royal Dutch,* 4 vols. (Leiden: Brill, 1953-57), is another useful source.

Pages 40-41

Our Industry Petroleum, 5th ed. (London: British Petroleum Co., 1977), provides a succinct discussion of D'Arcy's trials and tribulations growing out of the complicated maneuvers around the Iraq Petroleum Company. This and the Shell *Petroleum Handbook* are useful compendia of industry information. In both books, each chapter is a separate article written by an expert on the subject.

Pages 41-42

The classic work on Gulbenkian is Ralph Hewins, *Mr. Five Percent* (New York: Rinehart, 1958). A better source, frequently cited by Sampson, is Nubar Gulbenkian, *Pantaraxia: An Autobiography of Nubar Gulbenkian* (London: Hutchinson Publishing Co., 1965). As Calouste's son, Nubar was privy to many matters and insights that Hewins was not. His description of the Red Line meeting is particularly vivid. The Red Line Agreement, however, though it became significant after World War II, was hardly recognized or heralded by contemporaries. The *Times* (London), for example, does not mention it by this name and barely reports on the meeting. Only after oil was discovered in the area from which many enthusiastic companies were excluded by this arrangement did its significance emerge. An important reference that describes the effects of the Red Line Agreement in great detail, especially from the point of view of the excluded companies, is *Staff Report to the Federal Trade Commission,* submitted to the Subcommittee on Monopoly of the Select Committee on Small Business of the United States Senate, August 22, 1952 (Washington, D.C.: U.S. Government Printing Office, 1975).

Page 44

Much has been written on the early attempts for concessions and exploration of the Middle East. One of the first to recognize the implications of the events was Amin Rihani, who wrote a biography of Ibn Saud that described these events and predicted the unification of

Saudi Arabia. His book, *Maker of Modern Arabia,* is a fascinating source, especially because it is written by a non-Saudi who supported Ibn Saud's dream.

Wallace Stegner, *Discovery! The Search for Arabian Oil* (Beirut, Lebanon: Middle East Export Press, 1971), is a detailed, yet romantic account of the thrill and tedium of exploration in Saudi Arabia, written from the point of view of the explorers. A more contemporary account of these same events, written from a modern political and economic point of view is Robert Lacey, *The Kingdom: Arabia and the House of Sa'ud* (New York: Harcourt Brace Jovanovich, 1981).

Page 45

An interesting source for the discovery and exploration of oil in Mexico (and elsewhere) is Ruth Sheldon Knowles, *The Greatest Gamblers: The Epic of American Oil Exploration* (Norman: University of Oklahoma Press, 1978).

Pages 45-47

For an interesting discussion of exploration throughout the world, see Wallace Everett Pratt, *World Geography of Petroleum,* published for the American Geographical Society in 1950. Pratt, a geologist of renown equal to DeGolyer's, provides important insights.

Chapter 5 — International Since 1939

A continuation of Chapter 4, this chapter draws largely on the same sources, as well as my personal recollections.

Page 48

For an account of this misuse of fine ships, see Marc' Antonio Bragadin, *The Italian Navy in World War II* (Annapolis, Md.: U.S. Naval Institute, 1957). Major Italian naval units could not leave port for lack of fuel, while smaller vessels carried gasoline in cans as deck cargo.

Pages 50-53

For an account of this last-ditch fight by Gulbenkian and his allies, see Nubar Gulbenkian's *Pantaraxia.* Newspapers from that period, such as *The New York Times,* also provide useful insights.

Pages 53-54

The rise of nationalism and the increasingly difficult relations between oil companies and the oil-owning nations are well documented in many books and other publications. Indeed, the oil industry was a favorite punching bag for many periodicals in the 1950s and 1960s. The books cited in the notes to Chapter 4 provide excellent accounts that, in the main, support one another. The facts are seldom in dispute, though the conclusions the various authors draw — and the reactions their books provoke — often differ greatly.

Authors with close and romantic ties to the industry tend to gloss over the international aspects of the business, particularly the politics and the strong presence the industry had in the oil-producing countries. The stories and the documentation are there; these writers just don't bother much with them. While they say much about the reserves and early romantic history of the business in such countries as Indonesia, Libya, and Mexico, books like Ball, *This Fascinating Oil Business,* say little or nothing about the turmoil in the oil world since World War II.

The British Petroleum publication *Our Industry Petroleum* gives a good account of this major international company's dealings with foreign governments in its chapter on the history of British Petroleum. The term "Red Line Agreement" does not appear in that account, though that agreement made British Petroleum, nor does it mention the Achnacarry Agreement of 1923, in which the major companies agreed among themselves to conduct business as is. Clark also neglects to mention Achnacarry in his indispensable *Chronological History,* though he does mention the Red Line Agreement.

On the other hand, sources often considered critical of the industry mention Achnacarry and a great deal more. We do not know whether the scholars who tend to take a critical view of the industry are more diligent than those who defend it. The latter may be equally diligent, but more selective in their citations. The result, however, is that a writer who works largely from secondary sources and personal knowledge finds richer lore among the industry's critics than among its defenders.

John M. Blair was attacked from many quarters for the conclusions he drew in his *The Control of Oil,* published in 1976. None could attack his sources and documentation, however. Sampson's *The Seven Sisters* was also attacked; his scholarship could not be. Both books, along with others cited earlier and later, are thus major sources for these chapters.

Obviously I draw a more favorable picture of the industry to which I have devoted my life, but I do not deny or ignore the well-documented events that have helped shape the industry and the world in which it operates.

Pages 54-56

For a discussion of oil-pricing practices and complexity in international trade, see Robert B. Stobaugh *The Evolution of Iranian Oil Policy, 1925-1975*. Stobaugh's study was chap. 6 in George Lenczowski, ed., *Iran under the Pahlevis* (Palo Alto, Calif.: Hoover Institution Press, 1978).

Page 58

Figures on entry of new firms into the business came from *Multinational Oil.*

Pages 58-61

The world and the role of OPEC have changed somewhat since these words were written. The spot price of oil soared to $50 and in March, 1983 dropped back to the 1983 level of $29. Saudi Arabia learned that it could not completely control the OPEC countries. High prices brought more oil to market than was thought likely, and world demand proved more sensitive to prices than anyone thought possible. Despite these changes, the basic observations and conclusions of this chapter remain valid. Saudi Arabia, with its enormous production potential and relatively smaller financial needs, can influence prices strongly in any resurgent market simply by turning the tap on or off.

Chapter 6 — Where Oil Is Found

Given the enormous recent interest in petroleum, generated by the wild swings in supply and price, press coverage of where oil most likely is has been extensive. The author has relied on statements in the annual reports of other energy companies; policy papers of the United States Department of State (*Current Policy*, no. 69, June 1979, "Energy and Foreign Policy," for example); publications of the United Nations, particularly many issues of "Important for the Future," a periodical publication of the United Nations Institute for Training and Research (UN-ITAR), an organization that has made something of a name for itself in scholarly research on unconventional energy and less accessible fossil fuels. Current issues of major daily newspapers and business publications also served as sources. A *Wall Street Journal* article, for example (August 23, 1979, p. 44), discussing hidden supplies of petroleum in unaccounted-for current storage served as the model for the opening of this chapter.

Pages 62-65

The major source for this chapter, however, is E. N. Tiratsoo, *Oilfields of the World*, 2d ed. (Houston: Gulf Publishing Co., 1976). This comprehensive overview of the world's oil supplies, known and suspected, is lavishly illustrated with maps and tables. The volume also contains much historical and geological information on the world's oil regions, great and small. More contemporary sources for petroleum statistics are *International Petroleum Encyclopedia* (Tulsa, Okla.: PennWell Publishing Co., 1981, 1982, 1983), *National Petroleum Factbook, 1982* (Des Plains, Ill.: Hunter Publications, 1983), and *BP Statistical Review of World Energy, 1982* (London: British Petroleum Co., 1983).

Page 64

Events since these words were written have indeed followed the predicted path. Higher prices led to expanded exploration and production, which increased supplies and drove prices down again to a level closer to a free-market clearing price. The author believes that the existence of new proved reserves ready to come to market any time prices increase will lead to much more orderly markets in the future on the upside. However, prices are still higher than the lowest cost producer needs. Saudi Arabia, for example, can flood the market at any

time and break it, as the East Texas field and other giant U.S. fields did in the totally unregulated days.

Page 72

Shell Oil has a field in Thailand that illustrates the relationship between price and recoverability. Shell developed the field when the price of oil was well above $30 a barrel. Shell said in 1983 that if the price fell below about $25 per barrel, the company would have to shut in the field. It would not be profitable to produce that oil, even though the expenditures to find and develop the field had already been made.

Chapter 7 — The Making of Oil

Page 74

As a symbol, the dinosaur was not only misleading concerning the origins of oil, it was also misleading concerning the industry itself. The dinosaur in the popular mind conveys a sense of sluggish outdatedness. One could not pick a symbol less representative of this dynamic, fast-moving and advanced industry.

Pages 74-77

Many, many sources provide information on the physics and chemistry of the carbon atom and of petroleum. Any standard text on organic chemistry will do. Among those referred to for this text are Albert E. Dunstan, ed., *The Science of Petroleum*, 6 vols. (London and New York: Oxford University Press, 1938-55). This comprehensive treatise covers production, refining, transportation, and distribution. Vol. 5 specifically treats the chemical and physical properties of crude oils.

Arthur and Elizabeth Rose, eds., *The Condensed Chemical Dictionary*, 5th ed. (New York: Reinhold, 1956) proved a most handy guide to organic chemistry terminology, not to mention spelling.

Pages 79-83

Sources on geology are numerous also. Dorsey Hager, whose original *Fundamentals of the Petroleum Industry* inspired this book, was himself a geologist who published much on petroleum geology. The first edition of his *Practical Oil Geology* was published by McGraw-Hill (New York) in 1915. That edition had 149 pages. The sixth edition appeared in 1951 with 589 pages.

Another reference for this text is George D. Hobson, *Some Fundamentals of Petroleum Geology* (New York and London: Oxford University Press, 1954).

Pages 79-80

P. A. Dickey, *Petroleum Development Geology* (Tulsa, Okla.: PennWell Publishing Co., 1979) was particularly useful in the section on sediments. Another hepful source is *Geology Today* (Del Mar, Calif.: CRM Books, 1973).

Page 81

Doris L. Holmes *Principles of Physical Geology* (New York: John Wiley & Sons, 1978) is a fruitful reference on the formation of deltas.

Chapter 8 — Geology and Oil

Numerous references on plate tectonics, petroleum geology, and science were used for this and the preceding chapter. Among them are: Alfred G. Fischer and Sheldon Judson, eds., *Petroleum and Global Tectonics* (Princeton: Princeton University Press, 1975); proceedings of a conference that covered everything from geothermal heat flow to giant oil fields; Mikhail Yevgen, Yvich Al'tovski, Z. I. Kuznetsova, and V. M. Shvets *The Origin of Oil and Oil Deposits* (New York: Consultants Bureau, 1961), a translation from Russian particularly useful for a discussion of the Caspian Sea area, on the shores of which lies fabled Baku; and Walter Sullivan, *Continents in Motion: The New Earth Debate* (New York: McGraw Hill, 1974), an excellent and most readable overview of the theory and science of plate tectonics.

Other helpful texts include James C. Tatsch, *Petroleum Deposits: Origin, Evolution and Present Charcteristics* (Sudbury, Mass.: Tatsch Associates, 1974); Percy E. Spielmann, *The Genesis of Petroleum* (London: E. Benn, 1923), old but not so outdated, short and readable; William D. McCain, *The Properties of Petroleum Fluids* (Tulsa, Okla.: Petroleum Publishing Co., 1973), which has a chapter on the components of natural petroleum; Graham B. Moody, ed., *Petroleum Exploration Handbook* (New York: McGraw-Hill, 1961), is most useful, particularly chap. 5; vol. 1 of the previously cited *The Science of Petroleum,* which covers the origins of petroleum; and Walter A. Ver, *How Oil Is Found* (Wichita, Kans.: Wiebe, 1951).

Page 86

Scientific American has had several articles on plate tectonics. The one consulted here is J. Tuzo Wilson's "The Continental Drift" (April 1963): 86-100.

Page 88

For a good discussion of exploration in Alaska from a geological point of view, see Bryan Cooper, *Alaska: The Last Frontier* (New York: William Morrow Publishing Co., 1973).

Chapter 9 — Searching for Oil

Much of this chapter is based on information prepared by Atlantic Richfield executive Julius Babisak, a practicing petroleum geologist of long experience in the industry and, literally, in the field. Mr. Babisak's personal background and practical knowledge about such things as the Women's Pocket anticline are not readily documented. The author is most grateful to Mr. Babisak for his contributions to this and the following chapters.

Page 94

The 1860 theory of anticlines is documented by Clark, who cites Cecil Gordon Lalicker, *Principles of Petroleum Geology* (New York: Century-Crofts, 1949). The same source cites porous sandstone as a possible petroleum reservoir.

Page 96

A *Wall Street Journal* article, "Eyes in the Sky," April 5, 1979, p. 46, discusses satellite surveys thoroughly yet succinctly.

Pages 97-100

Many sources cite early development of theory and practice in gravimeters, magnetometers, and seismic operations. Among them are *Geophysical Case Histories*, published by the Society of Exploration Geophysicists in 1949; *Geophysics*, published by the Society in 1955; *Principles of Petroleum Geology*; Kenneth K. Landes, *Petroleum Geology* (New York: John Wiley & Sons, 1951); the general publications on petroleum already cited; and, of course, James Clark's *Chronological History*, which cites them all.

Pages 102-103

Petroleum Engineer International published Henri Doll's eyewitness account of the first electrical well logging. Doll was there with Conrad Schlumberger in 1927 at Pechelbronn, France. The article "Origin of Electric Logging" appeared in the August 1979 issue of the magazine, p. 88.

A valuable review of the history of well logging is Hamilton M. Johnson, "A History of Well Logging," which appeared in *Geophysics* 27 (August 1962): 507-527. This succint article even includes examples of original logs and photographs of the method being tried out in Rumania and Venezuela.

Chapter 10 — Reservoirs

Numerous more-or-less-standard works helped in the preparation of this chapter: L. C. Case, *Water Problems in Oil Production* (Tulsa, Okla.: Petroleum Publishing Co., 1970), a highly accessible operator's guide that, among other things, warns operators away from dubious quasi-scientific gadgets that purport to stop corrosion; L. P. Dake, *Fundamentals of Reservoir Engineering* (Amsterdam: Elsevier Scientific Publishing Co., 1978), a technical account based on lectures in The Hague at the Royal Dutch/Shell training center; G. D. Hobson, *Petroleum Geology* (London: Oxford University Press, 1975); the previously cited *Geophysical Case Histories*, which has much detail on specific formations; and Norman J. Clark, *Elements of Petroleum Reservoirs* (Dallas, Tex.: Society of Petroleum Engineers, 1960), which discusses the rise in stature of the petroleum geologist by 1914 and the reservoir engineer in the 1930s as well as exhaustive discussions of porosity, permeability, and conditions within reservoirs.

Particularly useful is Peter W. Birkeland and Edwin E. Larson, *Putnam's Geology* (London: Oxford University Press, 1978). W. C. Putnam wrote the standard geology text, *Geology* (London: Oxford University Press, 1964), but died while preparing this revision. His associates finished it and named it for him, recognizing the standard it had become.

Page 110

Another standard and very useful reference is Dickey's *Petroleum Development Geology*.

Chapter 11 — Landing the Lease

Land law and oil-and-gas law evolved rapidly in the United States. Though changed in detail from jurisdiction to jurisdiction and though no two oil leases are identical, the general shape of the law as set forth in this chapter has changed little since Dorsey Hager wrote *Fundamentals of the Petroleum Industry* in 1939. Accordingly, Mr. Hager's earlier work has been a major source for this chapter.

Other sources include Victor H. Kulp, ed., *Cases on Oil and Gas* (St. Paul, Minn.: West Publishing Co., 1924), and information supplied by Atlantic Richfield executives. Of more recent vintage, reflecting the increasing importance of government lands, is Stephen L. McDonald *The Leasing of Federal Lands for Fossil Fuels* (Baltimore: Johns Hopkins University Press, 1979). McDonald was a professor of economics at the University of Texas while the book was produced. He wrote under the auspices of the Center for Energy Policy Research of Resources for the Future, Inc., a Washington, D.C., think tank.

Readers may get a different view of oil-land leasing from an article by Peter J. Brennan, "Gambling in Oil and Gas Lotteries," *Fact* (August 1982): 20. The piece tells how the government lets everyone have a chance at leasing oil lands through organized and regular lotteries.

When this chapter was being written, the daily press was a prolific source, reflecting the public's heightened interest in the oil business. Many articles recounting the huge sums paid for offshore leases and the activities of landmen appeared in such papers as *The New York Times* and *The Wall Street Journal* from 1978 to 1980. Typical headlines are: "Supply-Wary Oilmen Here and Overseas Buy More Oil and Gas Properties in U.S." (*The Wall Street Journal*, Dec. 4, 1979, p. 48) and "To Prime the Pump, Decontrol," about independent oil men and their land deals *The New York Times*, Business Section, April 22, 1979, p. 1).

Chapter 12 — Drilling the Old Way

For the most part, sources for this chapter are the general books about the industry and its history used in the historical sections of this book. Each of those books has a chapter or more on drilling methods and a section on cable drilling. Since cable drills are now infrequently used,

little has changed since Dorsey Hager's day except the amount of space devoted to the method in a book such as this. Dorsey Hager devoted 41 pages to a system that even then was largely superseded but gave only 34 pages to cover rotary and other advanced methods.

An excellent overview of drilling history appeared in *Petroleum Engineer*, a periodical that published a special issue entitled *Petroleum 2000* in August 1977. In this issue, W. D. Moore III's article, "Ingenuity Sparks Drilling History," pp. 159-77, leaves few questions unanswered.

Another superlative overview, J.E. Brantly, *History of Oil Well Drilling* (Houston: Gulf Publishing Co., 1971), discusses every aspect of drilling both historically and mechanically. Brantly devotes chapters 10 to 13, more than a hundred pages, to cable drilling methods and tools.

The current technical press, particularly various issues of *Petroleum Engineer International*, were most helpful in this and the following chapter.

Pages 122-23

Technology may advance, but what once worked as primitive technology still works. In this decade, the author has seen in use in India just the type of man-powered drill mentioned here. It looked like hard work, but it was getting the job done.

Page 126

The value of using cable tool rigs in modern situations is most aptly described by Andrew J. Maslowski in "Shallow Wells Still Profitable in Northeast" *Drilling Contractor* (August 1982): 60-64.

Chapter 13 — Drilling the Modern Way

All sources cited for Chapter 12 were also useful in this chapter. In addition, the current technical literature, particularly *Petroleum Engineer International* and the *Oil and Gas Journal* proved most helpful. So did the technical literature of many manufacturers. The author also attended major industry technical meetings, such as the Offshore Technology Conference, held annually in Houston, Texas, perhaps the most important technical event in the industry. During my years in the industry I also absorbed some knowledge and folklore.

The August 1979 issue of *Petroleum Engineer International* chronicled the first occurrence of many technical developments. Among them the first self-propelled rig powered by a gasoline engine, the first roller cone bit, and the birth of the bit weight indicator.

Page 131

Fundamentals of Petroleum, 2d. ed. (Austin: Petroleum Extension Service, The University of Texas at Austin, 1981), was a valuable resource for this chapter.

Pages 135-37

Again, a major reference here is Brantly, *History of Oil Well Drilling*.

Chapters 16 through 25 describe every aspect of rotary drilling, from hoists, machines, and feed controls to bits, fluids, and fishing.

Pages 140-43

A valuable resource discussing directional drilling is L. W. LeRoy, D. O. LeRoy, and J. W. Raese, *Subsurface Geology* (Golden, Colo.: Colorado School of Mines Press, 1977), a college-level textbook. Also helpful here is *Our Industry Petroleum*.

Chapter 14 — Offshore

Major sources for this chapter were the *Offshore Technology Yearbook* (Dallas: Energy Communications, 1974); *Our Industry Petroleum*; various articles from *Petroleum Engineer International* and *Sea Technology*; manufacturers' literature; technical conferences; and visits.

Pages 146-47

The Offshore Search for Oil and Gas, Exxon Background Series, September 1980, was most helpful in the discussion of areas of major offshore activity.

Page 149

The May 1983 issue of the *Basic Petroleum Data Book* provided the statistical material used in this chapter.

Chapter 15 — Completing the Well

This chapter owes much to information prepared by Thomas C. Frick, a retired Atlantic Richfield executive and now a petroleum industry consultant in Dallas, Texas. Other sources include the often-cited general books on the industry, each of which has a chapter on well completion; and the current technical literature, particularly *Petroleum Engineer International*. Especially useful was the August 1979 issue, which contained articles on the first BOP, early cementing practices, origins of jet perforating, and early pumping units. L. Douglas Patton and William A. Abbott's multi-part series, "Well Completion and Workover," begun in the April 1979 issue contributed significantly to this chapter.

Pages 157-58

Rex A. Hudson of Halliburton Services Company provided background information on the methods of well completion used in the 1930s and 1950s.

Chapter 16 — Reservoir Engineering

This chapter owes much to Mr. Gene Herbeck of Atlantic Richfield's Reservoir Engineering staff in Dallas, Texas, as well as to Thomas C. Frick, mentioned in the Notes for Chapter 15.

Other sources include those cited in Chapter 10, and L. C. Case, *Water*

Problems in Oil Production (Tulsa: Petroleum Publishing Company, 1970); L. P. Dake, *Fundamentals of Reservoir Engineering* (Amsterdam, Oxford, and New York: Elsevier Scientific Publications, 1978); and Norman J. Clark, *Elements of Petroleum Reservoirs* (Dallas: Society of Petroleum Engineers, 1960).

Others include a compilation of abstracts by the U.S. Department of Energy; *Enhanced Oil Recovery: CO_2 Injection* (Oak Ridge, Tenn.: Technical Information Center, March 1980); numerous abstracts of relevant papers published by the Department of Energy, such as "Penn State to Study Chemicals That Stretch Lifetime of Oil Wells," in *Energy Update*, Dec. 10, 1979, and "Progress Reported on Louisiana Fireflood Oil Recovery Test," *Energy Update*, June 13, 1979.

Other references include an undated paper by R. J. Blackwell, "Status of EOR Recovery" (Houston, Tex.: Exxon Production Research Co.); Peter Nulty, "We Can Wring More Out of the Oil Patch," *Fortune*, Dec. 31, 1979, p. 58; manufacturers' literature such as that for the American Cyanamid Company's Cyanatrol polymers for enhanced oil recovery; and, of course, current technical literature such as *Petroleum Engineer International*.

Chapter 17 — Supplies and Services

The major source for up-to-date information in this chapter was the 1980 Offshore Technology Conference held in Houston, Texas. This annual meeting is the petroleum industry's biggest technical and product get-together, in which every company with the remotest interest in serving the industry displays its wares and services, ranging from microscopes to complete drilling rigs, housing complexes, and jet aircraft, including helicopters. Indeed, the roar of helicopters arriving and departing the grounds, mingled with the sounds of many different types of operating machinery, sometimes made conversation difficult.

A visitor can depart this exhibit with hundreds of pounds of product literature, and samples of smaller items and be assured of a continuing stream of new information in the mail from the many lists on which he is sure to be entered.

"Offshore" is a misnomer. It is, simply, the petroleum technology conference.

Page 178

Gulf Research developed the airborne magnetometer in 1939, just in time to help the Allies find submarines, says James A. Clark's *Chronological History*, citing company sources.

Pages 178-80

As a rule of thumb, the industry uses 11 to 12 tons of tubular goods per thousand feet of drilling each year, says National Supply Company of Houston, Texas. Much of this must be left in the ground whether or not the well is successful. In 1981 the average well drilled 4,530 feet deep used 54 tons of tubular goods. The weight used is not linear, of course, because the deeper well requires heavier pipe and casing to give adequate diameter at the bottom and to support the great weight of the drilling string.

Page 180

For an intriguing account of this operation, with photographs, see Peter J. Brennan, "Some Engineering Aspects of a World's Fair," *Engineer Magazine* (March-April, 1968): 20.

Page 181

The history of the Johnston Drill-Stem Tester comes from a flyer given away at the Offshore Technology Conference and an article by Gene Fuller, "Evolution of Formation Test Tools," *Petroleum Engineer International* (August, 1979): 100.

Page 182

On the giant hovercraft barge, see F. Jay Schempf, "Hover Barge Deemed Viable for Arctic," *Drilling Contractor* (May, 1980): 86; also, product literature distributed by manufacturers, such as British Hovercraft Corporation and Columbia Helicopters.

Pages 183-84

Southwest Research Institute, San Antonio, Texas, has been a leading researcher and developer of underwater systems, including midget working submarines. Oceaneering's one-atmosphere diving suit, "Jim," was featured in *Chevron World*, Summer 1978.

Page 184

An occasional publication called *Pipefreezing News*, published by B. C. B. Pipe Freezing Services, Ltd., England, gives more information than most nonprofessionals will wish to have. The technique is not limited to petroleum operations. A feature article discusses beer.

Chapter 18 — Marine Transportation

Much of the technical information in this chapter comes from the Marine Department at Atlantic Richfield. Every major petroleum company is also a major shipping company. All are at least charterers of large ships, if not also owners who have ships designed and built to their own specifications. Oil companies number among their staffs marine dispatchers, caterers, sea captains and seamen, and naval architects. The author thus had adequate resources in his own company on which to base this chapter.

Pages 186-89

Most standard works on the industry have a lengthy chapter on transportation by sea, including a history that starts with wooden barrels and the *Gluckauf* ("Good Luck" in German). There are many discrepancies. The British Petroleum book *Our Industry Petroleum* says the ship was launched in 1886, while Shell's *The Petroleum Handbook* gives 1885 as the

date. BP says she had a carrying capacity of 3,000 tons, while Shell says a gross tonnage of 2,240. Neither is the deadweight tonnage, which refers to the total lifting capacity of the vessel, not only its cargo carrying capacity. Modern tankers, but not dry-cargo vessels or passenger ships, are rated in deadweight tons. *Gluckauf's* deadweight tonnage was about 3,300. The vessel was built for a German importer of Standard Oil products. And Standard Oil says 1,886.

Controversy surrounds even the status of the *Gluckauf* as the first ship built expressly as an oil tanker. Philip W. Bishop's pamphlet, *Petroleum* (Washington, D.C.: Smithsonian Institution, 1969), says it was the first, but Robert W. Tolf, *The Russian Rockefellers* (Stanford: Hoover Institution Press, 1976), argues conclusively that the Nobels built the first oil tankers. Because the Nobels' tankers never went farther than the Caspian or Mediterranean seas, a technical compromise is possible: the *Gluckauf* was the first ocean-going oil tanker.

Pages 188-89

For a fascinating account of modern tanker operations — a shipboard view — see Noel Mostert, *Supership* (New York: Warner Books, 1976). A less-than-friendly but highly informative fictional account is Justin Scott, *The Shipkiller* (New York, Dial Press, 1978). For the reader who wants to delve into the arcane, cold-bloodedly economic world of shipping, Roy Nersesian's *Ships and Shipping: A Comprehensive Guide*, (Tulsa, Okla.: PennWell Publishing Co., 1981), is an excellent, readable overview. Nersesian doesn't mention the *Gluckauf*, but there's not much else he overlooks in his two chapters on tankers.

Page 191

The wreck of a ship carrying millions of barrels of crude oil is a disaster. But it is not a permanent disaster. Huge as was the spill from the *Amoco Cadiz* on the French coast, a recent study indicates the harmful effects in the worst afflicted areas, tidal estuaries with shellfish beds, lasted only three years. Now, some six years after the disaster, little is left but the memory. This information comes from *The New York Times*, July 19, 1983, C:3, report on an article in *Science*, (July 8, 1983). See also Chapter 24, "Petroleum and the Environment," regarding all the oil spilled from torpedoed tankers in World War II.

However, climate and geography are factors. The ecology in the chilly Straits of Magellan has still not fully recovered from the effects of the wreck of the Shell supertanker *Metulla* in 1973.

Page 192

For more detail on tanker economics, see Nersesian, *Ships and Shipping: A Comprehensive Guide*. For a shorter treatise on ship financing, see Peter J. Brennan, "The Financing of Shipping," *Marine Engineering/Log* (May 1982): 101.

Chapter 19 — Land Transport

Land transportation is also a standard chapter in most works on the industry previously cited here. Clark's *Chronological History* mentions Chinese bamboo gas pipelines in A.D. 900. He cites as his source Dean Hale, ed., "Diary of an Industry," *American Gas Journal* (October 1959): 21-52. The aqueducts and pipelines for water in ancient times remain marvelous monuments to hydraulic engineering. More impressive to an engineer than the many-arched aqueducts that still traverse the European plains are the stone siphons that brought water over mountains near Ephesus in Asia Minor, now the Aegean coast of Turkey. All sources seem to agree that the first modern-style metal pipeline was built by Samuel Van Syckel. It came on line in Pennsylvania in 1865.

The author is indebted to members of the Atlantic Richfield Transportation Department for much of the information contained in this chapter. Those people, after all, must be current with the laws, rules, and regulations as well as the technology governing petroleum transportation.

Page 197

The many general histories of the industry cited in earlier chapters are prolific sources for the battles between the pipelines, railroads, teamsters, and Standard Oil.

Pages 198-99

In 1981, according to the *Pipeline & Gas Journal*, there were more than two million miles of pipeline in the United States. These included all types — crude and gas trunk lines, field gathering lines, gas distribution mains, and gas service lines.

Pages 199-200

For a succinct but thorough account of the Trans Alaska Pipeline, see Brennan, "The Trans-Alaska Pipeline: Startup and Operation."

Page 201

According to *Pipeline & Gas Journal*, there were 477 pipeline companies in the United States alone and nearly 800 worldwide in 1981.

Pages 201-202

Information on the flowmeters came from manufacturers' product literature and from Brennan, "Startup and Operation."

Pages 204-205

The longest offshore pipeline in the North Sea runs 180 miles from producing field to shore. The deepest underwater line lies beneath 2,000 feet of sea between Algeria and Sicily, transporting Algerian gas to Europe. The pressure at that depth is about 900 pounds per square inch.

Chapter 20 — Communications

The standard books cited before have little or nothing to say about communications in the petroleum industry. Before the advent of microwave and satellite communications systems that could be privately owned and operated, oilmen used the same means used by the public at large, from semaphores to carrier pigeons, telephone, and telegraph. The technology of communications is one all its own that remains the same regardless of the user and end use. If any non-communications sector (other than the military) was a major factor in pushing the development of communications technology, it was the railroads and, later, shipping. In this field, the petroleum industry played a relatively small role until very recent times.

Most information in this chapter comes from the equipment manufacturers, the well-known names in electronics and communications, who advertise widely in petroleum publications and are a large presence at such exhibitions as the Offshore Technology Conference.

Pages 208-210

Most of the background about Comsat and Marisat comes from Comsat, which published a monthly journal called *Marifacts*. The publication related new and innovative uses of satellite communication systems and case histories such as "How Marisat Helped the *LNG Aries* Coordinate the Rescue of 21 Men from a Stricken Ship," "When the Question Is How to Communicate with Your Fleet," and "Canadian Offshore Operations Depend on Communications Via Satellite."

Page 210

A good, short article on meteorological satellites is Henry W. Brandli, "Meteorological Satellites and the Oil Industry," *Petroleum Engineer International* (May 1980): 46.

Pages 210-11

On the Trans-Alaska Pipeline, see Brennan, "Startup and Operation."

Page 213

Teleconferencing — the word had not been invented when this section was written — is becoming an increasingly important business tool. Atlantic Richfield was a pioneer (see *Business Week*, July 7, 1980, p. 81), but now there are many imitators. As with any other communications technique or technology, the petroleum industry has no monopoly on this one. But as so often in other areas, the industry just moved a little bit faster to exploit and adapt the system to its own needs.

Chapter 21 — Refining

Many sources provided material for this chapter. All the standard works referenced earlier have chapters on refining, or "manufacturing" as it is called in the industry. *Our Industry Petroleum* has the most complete, smoothly written and lavishly illustrated account. *Petroleum Handbook* is good for descriptions of individual processes. Shell published a vastly expanded edition of *Petroleum Handbook* in 1982, which treats all petroleum processes in exhaustive detail.

The author has of course relied also on internal papers from the company with which he has spent so much time, Atlantic Richfield. Other references include the *Petroleum Technology Handbook, The Chemical Dictionary, The Chemical Engineer's Handbook*, and others. Special issues of *Hydrocarbon Processing*, were most useful, including September 1972 and Market Data 1984. The magazine contains one-page summaries with flowsheets of virtually every process known to the petroleum refiner. Some of those flowsheets were redrawn to illustrate this chapter.

Page 214

More than 200 U.S. refineries have capacities of fewer than 50,000 bpd; only 16 exceed 250,000 bpd. The nation's largest is the 640,000-bpd Exxon refinery at Baytown. Texas has more refining capacity—5.3 million bpd—than the next two largest states combined, Louisiana and California, 2.4 and 2.6 mbpd respectively. This data comes from *Oil and Gas Journal,* March 30, 1981, via National Supply Company's *Oil and Gas Pocket Reference 1982.*

Page 216

The effort to produce more gasoline per barrel of crude continues. In 1980, Ashland Oil surprised the industry by announcing a new process that boosts gasoline output by 40 percent, making it possible to get a 75-percent gasoline yield from a crude. The "Reduced Crude Conversion" process converts heavy bottoms normally sold as residual fuel to higher-value gasoline at an acceptable price in investment and operating costs.

Page 220

Although catalysts have been around since 1913, the first cat cracker went onstream in 1937 at Marcus Hook, Pa. The modern petroleum age really began in 1910 with the advent of thermal cracking as a commercial process. The American Petroleum Institute covers the development in its 1959 publication *One Hundred Years of Oil*, cited by Clark. So does the *Encyclopedia Americana*.

Page 223

As the book was written, during 1983, total demand was under 15 million barrels per day. Gasoline demand continued below 7 million barrels per day.

Refinery product mix figures come from Mobil Oil Corporation. Of course, they vary from one refinery to another, depending on the crude, the refinery design, and market demand. Figures calculated from the Department of Energy's *Weekly Petroleum Status Report* show the following averages per 100,000 barrels for the first half of 1983:

Motor gasoline	43,400
Naphtha-type jet fuel	1,392
Kerosene-type jet fuel	5,388
Distillate fuel oil	17,716
Residual fuel oil	9,691
Other oils	22,241

The amount for residual fuel oil was down 23 percent from the previous year, reflecting a mild winter.

Chapter 22 — Petrochemicals

The petrochemical industry's recent rise to prominence as the leading branch of chemistry and source of many industrial and consumer goods is well documented. There's not much one author can say in general terms that many others have not already said. Processes differ in detail, and volumes have been written about those details, but the technology and industry as a whole are well known and extensively written about. That the industry should be so well covered is not surprising, because the way it displaced many other types of chemistry and dominated the chemical markets is itself astonishing.

As recently as 1950, for example, petrochemicals formed 0 percent of all organic chemicals in Japan and France, only 2 percent in West Germany, 4 percent in Western Europe as a whole, and 10 percent in the United Kingdom of Great Britain and Northern Ireland. Only in the United States did petrochemicals represent as much as half of all organic chemicals. Other nations derived their organic chemicals primarily from coal.

By 1971, these figures had changed enormously. Petrochemicals exceeded 90 percent of all organic chemicals in every nation and region. Today, the proportion is close to 100 percent, though scientists are once again looking to coal as a source of organic chemicals in the future. These figures come from *Our Industry Petroleum*.

The author's general sources for this chapter include the general works already cited many times as well as the monumental *Kirk-Othmer Encyclopedia of Chemical Technology*, (New York: John Wiley, 1978). This work comprises thirty volumes, and is still in progress. Frank W. Long of Atlantic Richfield's Commercial Development Department also provided important information for this chapter.

Those interested in examining the field in depth will find *Information Sources on the Petrochemical Industry*, (New York: United Nations, 1978), a useful source of worldwide information.

Page 274

Reference to Mendeleyev came from *The Petroleum Handbook*. Clark missed this one, but Fanning's *The Rise of American Oil* cites the Russian chemist as an early theorist on the origins and values of petroleum.

Chapter 23 — Natural Gas

The history of natural gas parallels that of crude oil, since the two are found together. Although people put gas to work in a natural way before they did oil, the commercialization of gas lagged behind that of oil. References pertaining to the history of gas are much the same as those for oil. Most will be found in the notes for Chapters 1 through 6.

However, because gas developed separately from oil, as noted in this chapter, different organizations still compile statistics and promote the use of the material. Chief among these is the American Gas Association, whose publications provided much information for this chapter. Internal Atlantic Richfield papers also were used extensively.

Page 236

Today, if you see flame and smoke near a U.S. oil and gas field, it is because a process upset required the operator to vent the gas rather than feed it to a pipeline, or there has been an accident. But in the Middle East, which produces more gas with its oil than it can process or ship, flaring vents are still common. The author recalls that during a recent flight into Dubai at midnight, the existence of a world outside in the stygian blackness could be divined only by means of numerous blazing flares reaching to the invisible horizon. Other passengers on that flight might well have wondered just which world they were about to enter.

Page 238

Figures for consumption came from the American Gas Association, 1978. For 1981, the consumption figure is 58 billion cubic feet per day. However, a recent Department of Energy study, "Natural Gas: Use and Expenditures," shows that in 1983 gas consumption had declined 7 percent since the 1970s (although expenditures for gas have increased by 145 percent). Natural gas consumption in 1983 was at its lowest level since 1966.

Pages 239-40

Geopressured gas and deep gas attracted great interest a few years ago, particularly when price controls were eased on deep gas. *The New York Times* ran an editorial on "Gas Embedded in Rocks and Controls," November 16, 1979. Numerous technical meetings and conferences took place. Showing the sensitivity of markets to both supply and demand today, gas is in surplus along with oil and interest in expensive production has backed off. Nonetheless, the gas remains in the ground and the technology to get it out is waiting for the day when the gas is needed.

The wide variation in reserve estimates from industry and government sources simply reflects the difficulty in determining what lies under the ground and sea. The uncertainty is even greater for gas than for oil.

Chapter 24 — Petroleum and the Environment

Pages 244-246

The National Research Council's report, "Petroleum in the Marine Environment," (National Academy of Sciences: Washington, D.C., 1975), cited the following figures for marine pollution:

Total petroleum in the seas: 6.11 million metric tons per year.
From natural seepage: 0.6 million MT/yr or 10 percent.
From marine transportation: 2.13 million MT/yr or 35 percent.
From atmospheric sources: 0.6 million MT/yr or 10 percent.
From land runoff: 1.9 million MT/yr or 31 percent.
From industrial and municipal waste: 0.8 million MT/yr or 13 percent.
From offshore production: 0.08 million MT/yr, or 1.3 percent, which says a lot for environmental control in offshore operations.

Although these figures date to the 1975 publication, members of the National Academy of Sciences Committee responsible for updating that report suspect that they have not changed that much as of 1984. While the figures they have been collecting had not yet been published when this book went to press, they indicate that pollution from land runoff is higher than in 1975, but that the amount lost by tanker operations is lower, so on average the amount is the same. The area of greatest uncertainty which might affect the overall total is the amount precipitated from the atmosphere.

Page 246

The North Atlantic oil seep is described in *Petroleum Engineer International* (December 1979): 14.

Pages 248-50

On TAPS, see Brennan, "Startup and Operation;" Roscow, *800 Miles to Valdez*, chap. 15; and the booklet, "Oil Spill Prevention Measures for the Trans Alaska Pipeline System," prepared for a 1973 conference in Washington, D.C., and available from the Alyeska Pipeline Service Company.

Page 252

The deep-sea recovery of gasoline is described in *Petroleum Engineer International* (December 1979): 14.

On the TAPS ballast-handling system, see Brennan, "Startup and Operation."

Pages 253-54

On industry expenditures, the source was API Publication No. 4314, "Environmental Expenditures of the United States Petroleum Industry 1969-1978." A revision, Publication No. 84143250, "Environmental Expenditures of the U.S. Petroleum Industry 1970-79, 1980," does not significantly change these figures.

Page 257

The capital investment statement came from a report, "Environmental Factors Affecting Refining," prepared by Atlantic Richfield staff and dated August 31, 1978. The exact quote is: "The capital investment required to build a reliable tail gas scrubber is frequently greater than for the basic Claus unit. Since the basic unit recovers 25 times as much sulfur as the tail gas scrubber, the cost effectiveness of sulfur recovery by such scrubbers is relatively poor." It is not the technology *per se* that makes the scrubber costly, but the fact that it must process a much greater volume of gas to recover an equivalent amount of sulfur. As the chapter says later, it is more expensive to extract the final 5 percent than the first 95 percent. That goes for anything. It is an engineering truism.

Chapter 25 — Who Owns the Business

A major source for this chapter was the *CDE Stock Ownership Directory: Energy*, (New York: Corporate Data Exchange, 1980). Some percentages of specific ownership have changed, but not significantly, (except as noted in the next paragraph), since this chapter was written. In any case, the author intends to illustrate orders of magnitude, to put things in approximate relationship to one another rather than to pursue numerical data to an irrelevant degree.

Some events that occurred in late 1983 and early 1984 reinforce the premise of this chapter, namely, that ownership of the petroleum industry is diverse, not in the powerful hands of a very few extremely wealthy individuals or families. The Getty Foundation and Getty family interests, which had been the largest stockholders in Getty Oil, sold the company to Texaco. Shortly afterwards, the Keck family agreed to sell Superior Oil to Mobil. In the largest takeover of them all, Gulf Oil, one of the Seven Sisters, agreed to be bought by Standard Oil of California.

The effect of these mergers, undertaken to boost at relatively low cost the reserves of the acquiring companies, is to remove from the industry or to further dilute some of the few remaining large family holdings. These buyouts, and others likely to follow, also increase somewhat company concentration in the industry. But they hardly lessen competition. Thousands of producing companies and dozens of integrated ones remain to assure continued strong competition for the consumer and corporate customer.

Other sources for the chapter include the daily and current business press as well as the many historical works previously cited. Especially useful have been the frequently updated *Industry Surveys* published by the financial research and publishing house, Standard & Poor's.

Pages 260-61

Data on state-owned energy industries came from a chart on the scope of state ownership in *The Economist* (London), December 30, 1978.

Page 261

For a story on small wildcatters, see *Venture* (March 1980): 35. See also James P. Roscow, "Wall Street's Oil Pipeline," *Financial World*, December 15, 1979, for more on small oil company financing.

Page 262

The Independent Petroleum Association of America is the source for these figures. But see also John M. Balir, *The Control of Oil* (New York: Vintage, 1978). *Oil and Gas Pocket Reference 1982* has many statistics on the costs of drilling and equipping wells.

Pages 262-66

Percentage ownerships were taken from *CDE Stock Ownership Directory*.

Page 266

Cost figures came from the American Petroleum Institute, Independent Petroleum Refiners Association, and industry sources.

Refinery costs are necessarily estimates. The final price cannot be determined until construction is completed. However, the industry uses a guide called the *Nelson Cost Index of the Oil and Gas Journal for Refinery Construction*, which covers the items that go into such a plant. In late 1982 the *Index* stood at 994.3; in 1962 it was 237.6. The billion dollars, more rather than less, remains a round figure.

As a result of the current glut and shutdown of refineries, the older, less efficient ones are being taken out of service. Those companies that continue successfully in tomorrow's more competitive climate will not only have shut down old and inefficient units but will also have upgraded newer ones and, as demand increases, will build new ones utilizing the latest technology. Indeed, the *Oil and Gas Journal*, March 21, 1983, p. 85, states: "The much reduced U.S. refining industry with its hefty downstream processing capability and growing strength in residual conversion is becoming the most modern and flexible in the world... It is probable [that] downstream closures have run their course."

The National Petroleum Refiners Association estimates that refineries will need to invest between $9.7 billion and $13 billion by the year 2000.

Page 267

High interest rates are bad for all businesses. But interest is a before-taxes deductible whereas taxes are an after-profits excise. The industry gets some immediate, industry-related benefit from interest, for example, use of the money on which the interest is paid. It gets no such direct benefit from a tax.

Page 268

Sources for profit increases were the current business literature, industry publications, and Standard & Poor's *Industry Surveys*.

Page 269

Rig counts came from the *Oil and Gas Journal*.

Chapter 26 — Coordination and Supply

This chapter relies heavily on in-house Atlantic Richfield sources, although the coordination and supply activity is much the same from one integrated oil company to another.

Pages 272-73

In shortage or surplus, the industry remains fiercely competitive. When this chapter was first written, shortage prevailed throughout the industry. As these notes are written, glut is the major problem. The author does not know what the situation will be when the reader peruses this chapter. It doesn't much matter: either way, the industry is competitive and the supply-and-coordination department's job complicated.

The specific reference was to products, with everyone running refineries at 110 percent of capacity and cutting prices to maintain market share. Refineries today are running 65 percent capacity on a reduced base, down from 18.6 mbpd in January 1981 to 17 mbpd in 1983. But too much oil and too few customers means surplus and price cutting. Today, because the surplus is at the crude level, rather than at the product level, price cutting occurs among the primary suppliers. Refining levels are far below those where the incremental barrel becomes a factor in pushing more product onto a satiated market. Those days will return, when supply and demand throughout the chain are in near balance and one or another company sees an advantage in pushing the incremental barrel and undercutting the price. The operative word is "periodic."

Page 277

Not since 1979 has the government tried to allocate petroleum products. It did so then because of shortages, not rising prices. As of March 1983, stocks of crude oil and petroleum products in the United States were far below the normal range calculated for the period 1975-1981. Refinery production is also down. The combined declines reflect great reduction in demand, which has been accompanied by falling prices. In such a market, marginal facilities are closed, marginal operators get squeezed out. The stronger ones compete harder to retain their shares of a shrinking pie that all expect will soon begin to expand again. Cost containment is increasingly important, and thus the job of supply and coordination is even more critical. No one wants to get stuck with a tanker load of $34 crude when the price has dropped to $28 and interest rates are 10 percent.

Chapter 27 — Marketing Oil and Gas

As with coordination and supply, marketing is common to all integrated oil companies. It is, however, less an industry-integrated function, because it provides the opportunity for the competitive edge of the business, the place through which a company gains market shares or loses them. The marketing department of Atlantic Richfield is largely responsible for the nuts and bolts of this section, though the Shell, British Petroleum, and other general works all discuss the marketing function in some detail.

Page 278

World production of petroleum since 1977 has been essentially flat, give or take a few hundred million barrels per year. In both 1977 and 1980, total world production was 59,706,000 barrels per day. Recession and

conservation continue to restrain production to about the same levels. During this period, the OPEC countries have seen their production fall by nearly half, from 31 mbpd in 1979 to 17 mbpd in 1983, and thus their share of world markets.

World figures for consumption and production, numbers of vehicles, etc., given for illustrative purposes only, are from *United Nations Statistical Yearbook, 1977*. That may seem dated, but the figures are many years in preparation and long out of date by the time they are published. The later edition of the *Statistical Yearbook* was not ready as these notes were prepared.

According to the U.S. Department of Commerce Bureau of Economic Analysis, the U.S. GNP for 1982 was $3,059,300,000,000, or more than three trillion dollars in current dollars. The world GNP exceeds ten trillion dollars, an estimate because even the U.N. doesn't have complete figures for later than 1980. Thus the petroleum industry accounts for 10 percent of the world's gross product.

Even if the present glut continues, demand for oil will still increase because: (1) oil is a depleting resource, and (2) demand slipped from January 1981 (when totals of products supplied domestically equalled 18.4 mbpd) to March 1983 (when totals were 14.7 mbpd). However, imports fell even further, from 6.3 mbpd to 2.4 mbpd for the same period. This difference in consumption is coming out of the hides of OPEC. World demand will increase as prices seek a stable level and world economies recover, though U.S. demand may well continue downward as U.S. industry and consumers become more energy efficient or back out of oil. From 1970 to 1981, U.S. energy efficiency increased nearly 25 percent. Over time, with occasional burbles, demand for this diminishing resource constantly increases worldwide.

Pages 279-80

Clark, *Chronological History*, cites Kendall Beaton, *Enterprise in Oil*, as the authority for the first filling station. The station established in 1911 in Detroit boasted many firsts, including being the first gas station ever robbed.

Page 280

The author personally remembers: "Five gals one dollar" — even gals for fifteen cents and less — as late as the 1960s.

National companies often price below market in their home markets. They may do so, surreptitiously in the case of OPEC countries, in world markets, but not by much and only to gain market share. Gasoline is cheaper in Tijuana and Riyadh than in Los Angeles. China trades gasoline for boiler fuel, according to Atlantic Richfield's Marketing Department.

Page 281

The "incremental barrel" was a subject of raging editorials in the industry press during the 1960s in particular. Editorial fulminations had no effect. Only external forces eventually drove the price of crude to a level

364

where the incremental barrel was itself too costly to make, and high prices made it difficult to sell.

Page 282

Aproximately 120,000 gas stations still exist in the United States today. For a broader discussion of retail gasoline marketing, see *Recent Changes in Retail Gasoline Marketing* (Washington, D.C.: American Petroleum Institute, 1981).

Page 284

Spot prices can turn on a single cargo and one operator's desperate need either to buy or to sell. The generally reported official price is the price of only one crude, the benchmark Arabian Light of Saudi Arabia, pegged at $29/bbl in April 1983. However, the Department of Energy calculates and publishes a weighted average international price based on many oils from many countries. In April 1983 that price was $28.62/bbl. In February 1981, when the benchmark was $32, the weighted average was $35.53. These numbers reflect more closely what refiners actually pay for crude. This price and the spot price constantly reflect one another, despite the occasional trade well off the market.

Pages 284-85

Predictions about oil prices are in conflict. A *Business Week* article in 1980 predicted $26/bbl in 1985. Another article in the August 1, 1983, issue mentions a world price of $23. Earlier predictions had prices heading toward $75 and more. Who knows?

The fact remains that the price of oil does not reflect an inability on the part of producers to supply all the market can absorb at a profit to some producers all the way down to $5/bbl or even less. The floor is that point at which only the cheapest producers can supply oil. When the cost to the producer of the incremental sixty millionth barrel per day can be calculated (assuming world consumption of 60 mbpd), it will indicate where the floor lies. A freely competitive market will find that floor. The futures market is a new factor tending toward that freely competitive market.

Chapter 28 — Who Works for Oil

The Standard & Poor's *Industry Surveys* cited earlier were particularly informative for this chapter. There is no point in citing any particular report. They are updated regularly, and the latest available was referenced for this section.

Also invaluable for this and other chapters was a publication of the Independent Petroleum Association of America, *The Oil Producing Industry In Your State: 1982*, (Washington, D.C.). This annual publication shows how petroleum impacts the many sectors of each state's economy.

Other sources include industry suppliers, financial houses, and government and market research organizations.

Page 290

The self-reliant smith appears in *Petrolia*.

Page 292

In the week of March 18, 1983, according to the *Oil and Gas Journal*, 2,030 rigs were active in the United States, down some 1,830 from the previous year and the lowest count since June, 1979.

Demand is way down now from what it was in 1981. But it is still a lot higher than it was in 1978. In mid-1983, domestic production appeared to have increased slightly and rig count was stabilizing. The industry seemed to be bottoming out. Indeed, the Department of Energy publication *Short-Term Energy Outlook*, May, 1983, said so. The huge increase in 1981 was a typical boom overshoot. Though things have fallen back since then, the domestic industry has not gone back anywhere near to where it was in 1971 when there were only 976 rigs active. That number did not top 2,000 until 1977. More wells were completed in 1982 than in any previous year, representing more feet drilled, even though the number of rigs was down from 1981.

Pages 292-93

Information on training was based on conversations at the Shell, National Lead, Dresser, Brown & Root, and other training centers in Houston.

"Studies by one company . . . " The company was Sun, and the study was reported in a paper presented at the API Production Department Meeting in new Orleans, April 1-4, 1979. The pattern reflects an industry that stands still for a long time, employing lots of unskilled or semiskilled people, and then suddenly expands as the U.S. industry has. The same pattern is seen in a manufacturing plant where the average age of the workforce advances year to year until everybody retires at once. "Personnel needs are based on the total number of completed wells and current numbers of producing oil wells." Both these numbers are up, thanks both to more drilling and longer-lived wells, which "outlast the workspan of most individuals."

Pages 293-96

Figures for the petroleum industry impact on the national economy were taken from the Bureau of Labor Statistics, which computes detailed input-output analyses for the economy, the Independent Petroleum Association of America, and the Petroleum Equipment Suppliers Associations.

Page 295

National Supply's *Oil and Gas Pocket Reference 1982* prices offshore rigs as follows:

Jack-up	$22-60 million
Ship shapes	$50-100 million
Semis	$50-180 million

The source for the major companies in the field was Standard & Poor's *Industry Survey* dated October 1982.

Sources for industry expenditures were *Pocket Reference 1982* and the Independent Petroleum Association of America's *United States Petroleum Statistics 1983 Preliminary*.

Pages 296-97

Source for export figures is Research Study No. 1154, *Oilfield Machinery and Equipment in the United States* (New York: Frost & Sullivan, Inc., 1983).

Page 297

The cited survey, Report # A831, was *The Tar Sands and Oil Shale and Associated Plant and Equipment Market in the United States* (New York: Frost and Sullivan, Inc., 1980).

Chapter 29 — Government and the Industry

The reader should refer to the Notes for Chapters 1-6 for more detail on the historical background of regulation. The author is also grateful to the Legal Department of Atlantic Richfield for its contributions to this chapter, notably for the observation that there are nearly 800 bills in Congress that concern themselves with one or another aspect of the energy crisis alone and for the details and ramifications of many laws enacted over the years. Only lawyers deeply immersed in their topics would have such information at their disposal.

Page 298

The cost of regulation is discussed in Steven Rattner, "Regulating the Regulators," *The New York Times Magazine*, June 10, 1979, p. 104. The $100 billion figure, attributed to Washington University economist, Murray Weidenbaum, does not go undisputed. Even economists have axes to grind. Rattner's article contains much history and points out many absurdities of regulation but does not deal specifically with petroleum.

Page 300

A story, "The Great Search for Paper," in *Shell News* recounts the trouble and expense the company encountered in answering a subpoena for documents in a court case. Shell estimated that compliance with the original subpoena would have cost 751-man-years and $20.6 million.

A press release from the Federal Energy Regulatory Agency dated December 4, 1980, discusses conversion of the Florida gas line and the opposition of marine interests.

Page 304

Two useful studies contributed to these paragraphs. They are Paul F. Dickens III, *Effects of Oil Regulation on Prices and Quantities: A Qualitative Analysis* (Washington, D.C.: U.S. Department of Energy, October, 1978), and Martin Taschdjian, *The Effect of Legislative and Regulatory Actions on Competition in Petroleum Markets* (Washington, D.C.: U.S. Department of Energy, October, 1978).

In November 1978 the DOE Department of Public Affairs published a 60-page booklet gathering all information about every aspect of the National Energy Act of 1978.

Page 305

A DOE press release dated April 29, 1980 (R-80-075), discusses the supplier purchaser rule. The deputy energy secretary is quoted: "It is one of the priorities of this Administration to simplify regulations and progress is being made."

Pages 305-307

The strange case of Sohio was reported on in the *Wall Street Journal*, April 2, 1980, p. 27. Shortly thereafter, the DOE issued a press release, dated May 2, 1980 (R-80-080), announcing that the rule was rescinded.

Page 307

A Federal Energy Regulating Commission release of December 5 is the base of the agricultural uses item.

Despite waning enthusiasm for synthetic fuels, the Synthetic Fuels Corporation is far from dead, nor is the concept. Of nineteen synthetic fuel projects, only two have been cancelled outright. The rest continue, though most are vastly scaled back, some to a virtual standstill, according to the Standard & Poor's *Industry Survey* of October 1982. When Exxon gave up on oil shale, everyone got the impression that the industry had given up totally on synfuels. Not so. There is still private R&D as well as SFC money, though on a greatly reduced scale. Eventually the technology will be needed.

Chapter 30 — Taxes

While this section was in preparation, the battle over the windfall profits tax was in full swing. The business and popular press reported daily on the many ramifications of the proposed tax, and editorials foretold the end of the industry. "The Close-the-Wells Tax" (*Wall Street Journal*, January 22, 1980, p. 20) was mild compared to the editorials in the industry press. Concurrent with the furor, many groups compiled statistics intended to prove this point or that one. Thus, a wealth of compiled information exists on the petroleum industry and taxation in 1979-1980. Sources such as the API, Chase Manhattan Bank, financial and accounting houses, and others, published studies are cited in the text.

The windfall profits tax, of course, eventually became law. The industry lives with it. It has been modified from time to time. The furor died. So did all the studies that pulled together tax information from many sources which could then be used to show how heavily the industry is taxed (or not) and why new taxes should not be (or should be) imposed.

The compiled information still abounds. The figures given illustrate the point — the industry is heavily taxed.

A useful source besides those already mentioned was Joseph Bowring, *Selected Federal Tax and Non-Tax Subsidies for Energy Use and Production* (Washington, D.C.: U.S. Department of Energy, January, 1980). Papers prepared by tax specialists within Atlantic Richfield were also extremely helpful.

Page 315

A few details on the windfall profits tax of 1980 may help to illustrate the complexity of this issue. The tax rates shown are levied on the difference between cost and price after deregulation or resulting from OPEC increases. There are exemptions from the law for types of Alaskan oil (which brought a court challenge against the tax as a whole — the challenge lost), tertiary oil, oil from Indian lands, oil owned by tax-exempt organizations. Exemptions, of course, create problems, such as the court case, because inventive people immediately try to get their situation moved into an exempt category, by whatever means.

The tax rates:

Old oil (pre-1979 production)	70%
Stripper oil	60%
Incremental tertiary and heavy oil	30%
Newly discovered oil by 1986 reduced to	15%

As noted in the text, the tax has been modified several times. The Economic Recovery Tax Act of 1981 included a royalty credit for 1981 and exemptions for 1982 and thereafter. The Miscellaneous Revenue Act of 1982 accounted for last-in, first-out inventories. The Tax Equity and Fiscal Responsibility Act of 1982 eliminated the TAPS adjustment. The Technical Corrections Act of 1982 tried to clarify the original tax altogether.

The tax is supposed to phase out between September 1990 and September 1993 or after raising $227.3 billion, whichever comes later.

Despite all the discussions over its passage, the industry accepted the tax with some equanimity. There were bigger issues to consider, notably, deregulation. One senior industry executive was quoted as saying that if it is politically essential that there be a windfall profits tax to accomplish decontrol, "then so be it." He was not assailed in the editorials.

Chapter 31 — The Future of Oil

Page 316

The World Energy Conference in 1978 was discussed in an Exxon paper, "Exploration in Developing Countries," dated May 1978.

World production peaked at 62 mbpd in 1978 and is currently around 60 mbpd, and climbing. Exxon, quoted in a Standard & Poor's *Industry Survey*, estimates that total world energy demand will grow at 2.5 percent per year through the year 2000 but that demand for oil will increase at only 0.4 percent per year for the next twenty years.

Pages 318-19

Statistics on world energy consumption came from *BP Statistical Review of World Energy, 1982,* and those for the United States came from the Department of Energy Information Service.

Pages 319-20

Regarding OPEC, the operative phrase is "so long as they act in concert." OPEC can produce between 29 mbpd and 32 mbpd. Clearly, OPEC can break the market any time it can pull itself together. More significantly, Saudi Arabia alone can control the market. It has the reserves and the production capacity on the upside, has least financial needs on the downside, and has the lowest cost of production. It can undercut any producer at any time in the foreseeable future but also can cause prices to skyrocket by shutting in production. OPEC can keep moving that carrot just beyond reach.

Pages 320-21

Production figures, future estimates, and projected costs come from many sources, including the United Nations, Department of Energy, American Petroleum Institute, and companies in the industry.

Page 322

As has been noted already, it is hard to estimate a flat figure for new refinery construction. A good guide is the *Nelson Cost Index for Refinery Construction* published by the *Oil and Gas Journal*.

Pages 324-25

With private enterprise backing out of synfuels development, only the SFC will provide the funding needed for ongoing technological development against the day when it will be needed.

Chapter 32 — Energy for Tomorrow

Few aspects of energy have so gained popular attention in recent years as non-fossil, renewable, unconventional energy. The United Nations sponsored the Conference on New and Renewable Sources of Energy (UNCNRSE) in Nairobi, Kenya, in 1981. Virtually every nation produced a paper for that conference which discussed that nation's state of technology and its use of various energy forms. Of greater interest, numerous international working groups produced outstanding technical papers on every type of new and renewable energy, from animal power to solar cells. Some of the economic studies were less than supportive to adamant advocates of non-fossil energy, but the studies as a whole should be required materials for anyone deeply interested in this topic.

A good short treatise on the state of non-fossil energy is Peter J. Brennan, "The Unpleasant Realities of New and Renewable Energy" *Industrial World* (March, 1982): 24.

The American Petroleum Institute has a number of publications that touch on the present and future of energy. Among them are *Myths and Misperceptions: Exploring the New World of Energy* (1983) and *Looking for Energy? A Guide to Information Sources,* 3rd ed., (1982).

Pages 333-34

Recent surveys show that the public is more concerned about nuclear waste disposal than any other facet of the nuclear industry. People are asking this very question, though not in these words, and assurances that disposal sites are safe for the foreseeable future do not assure them. They want to know about the unforeseeable future.

Page 336

On ethanol, see *Net Energy Analysis of Alcohol Fuels* (API Publication No. 4312, November, 1979). Ethanol as a replacement for oil so intrigued the United Nations and its non-OPEC countries that the organization mounted a major conference on the subject, "Workshop on Fermentation Alcohol for Use as Fuel and Chemical Feedstock in Developing Countries," held at Vienna, Austria, in March, 1979. Dozens of papers explored the subject in great depth. But alcohol remains a marginal resource, highly sensitive to the world petroleum price and agricultural factors.

ILLUSTRATION
CREDITS

Clark's *Chronological History*, and other sources. **Page 28:** Exxon, *Lamp*, April 1946 and Photographic Archives, University of Louisville, Standard Oil of New Jersey Collection. **Page 32:** War Emergency Pipelines, Inc. **Page 33:** United Press International. **Page 35:** courtesy of Volkswagon of America, Inc. **Page 35:** by permission of Bill Mauldin and Wil-Jo Associates, Inc.

Chapter 4. Page 37: (International News Photos) United Press International. **Page 38:** Scallop Corp., Shell Photo Service. **Page 39:** left and right: Scallop Corp., Shell Photo Service. **Page 40:** British Petroleum. **Page 41:** British Petroleum. **Page 42:** Gulbenkian Foundation. **Page 43:** redrawn from *Staff Report of the Federal Trade Commission*, August 22, 1952. **Page 44:** upper: Amin Rihani; lower: Chevron. **Page 45:** upper: United Press International; lower: Fikes Hall of Special Collections and the DeGolyer Library, Southern Methodist University. **Page 46:** Mexican Eagle Oil Co.

Chapter 5. Page 49: upper: U.S. Army Signal Corps; lower: Wide World Photos. **Page 51:** redrawn with permission from Aramco. **Page 52:** from *The Seven Sisters* by Anthony Sampson. © 1975 by Anthony Sampson. Reprinted by permission of Viking Penguin Inc. **Page 54:** Getty Oil Co. **Page 55:** United Press International. **Page 57:** United Press International. **Page 58:** United Press International. **Page 59:** © 1983 *Los Angeles Times*. Reprinted by permission; by Bozena Syska, published February 27, 1983. **Page 60:** © 1983 *Los Angeles Times*. Reprinted by permission; by Bozena Syska, published February 27, 1983.

Chapter 6. Page 63: reprinted courtesy of *Perspectives* Magazine, Pennzoil Co. **Page 66:** upper: ARCO; lower: Chevron. **Page 67:** ARCO. **Page 69:** Exxon, *Lamp*, April 1944 and Photographic Archives, University of Louisville, Standard Oil of New Jersey Collection. **Page 71:** redrawn with permission from *National Petroleum News 1982 Factbook*. **Page 72:** ARCO. **Page 73:** ARCO.

Chapter 7. Page 75: Continental Oil Co. and the American Petroleum Institute. **Page 79:** from Scruton, 1960, reprinted with permission from the American Association of Petroleum Geologists. **Page 80:** upper: reprinted with permission from PennWell Publishing; lower: from *Geology Today*. © 1973 by Ziff-Davis Publishing Company. Reprinted by permission of CRM Books, a Division of Random House, Inc. **Page 81:** left and right: reproduced from A. Holmes, *Principles of Physical Geology*, 3rd ed., © D. L. Holmes, by permission of Van Nostrand Reinhold (UK). **Page 82:** ARCO.

Chapter 8. Page 85: reproduced with permission from Brown & Root, Inc. **Page 86:** upper and lower: reprinted with permission from *Scientific American*, from "The Continental Drift" by J. Tuzo Wilson, April, 1963. **Page 88:** reprinted with permission from William

Chapter 1. Page 3: Drake Well Museum. **Page 5:** Upper left: British Museum; upper right: Museum Teheran; bottom: Sem Presser. **Page 7:** upper left: from *The History of Oil Well Drilling* by J. E. Brantly. © 1971 by Gulf Publishing Company, Houston, Tex. Used with permission. All rights reserved. Upper right: *The Geneva Pamphlet*, 1490. **Page 9:** upper and lower: Drake Well Museum. **Page 10:** American Petroleum Institute Historical Photo Library **Page 11:** Spindletop Museum and the American Petroleum Institute.

Chapter 2. Page 13: Drake Well Museum. **Page 15:** Drake Well Museum. **Page 16:** Drake Well Museum. **Page 17:** upper and lower: Drake Well Museum. **Page 18:** upper: Ida Tarbell, *A History of Standard Oil* and Drake Well Museum; lower: Drake Well Museum. **Page 19:** Ida Tarbell, *A History of Standard Oil*. **Page 21:** upper left: Bettmann Archives; upper right: Exxon and Photographic Archives, University of Louisville, Standard Oil of New Jersey Collection; lower left: Chicago *Journal*, 26 June 1908; lower right: Bettmann Archives. **Page 22:** upper: Hughes Tool Division, Hughes Tool Company; lower: U.S. Department of Commerce and Hughes Tool Division. **Page 23:** Atlantic Richfield Archives. **Page 24:** Texas Mid-Continent Oil and Gas Association and the American Petroleum Institute.

Chapter 3. Page 27: chart based on information from Chevron,

Morrow & Co., from Bryan Cooper, *Alaska: the Last Frontier*, 1973. **Page 90:** redrawn with permission from Exxon. **Page 92:** after D. W. Rockwell and A. G. Rojas, 1953, *Coordination of Seismic and Geological Data in Poza Rica-Golden Lane Area, Mexico* in AAPG *Bulletin*, 37: 2551-2565 and Petroleos Mexicanos.

Chapter 9. Page 95: British Petroleum. **Page 97:** ARCO. **Page 98:** LaCoste and Romberg Gravity Meters, Inc. **Page 99:** Shell Oil Co. **Page 100:** ARCO. **Page 101:** Drake Well Museum. **Page 102:** ARCO. **Page 103:** reprinted with permission from Bryan Cooper, *North Sea Oil: The Great Gamble*.

Chapter 10. Page 105: from *Putnam's Geology*, 4th ed., by Edwin E. Larson and Peter W. Birkeland. © 1964, 1971, 1978, 1982 by Oxford University Press, Inc. Reprinted by permission. **Page 107:** Exxon, *Lamp*, Summer 1981. **Page 108:** Humanities Research Center, the University of Texas at Austin. **Page 109:** left: reprinted by permission of John Wiley & Sons, Ltd. from *Modern Petroleum Geology*, edited by G. D. Hobson, 1973; right: British Petroleum. **Page 110:** reprinted with permission from PennWell Publishing Co. **Page 111:** Exxon, *Lamp*, June 1947 and Photographic Archives, University of Louisville, Standard Oil of New Jersey Collection.

Chapter 11. Page 112: Shell Oil Co. **Page 114:** Drake Well Museum. **Page 115:** reproduced by permission of The Huntington Library, San Marino, Calif. **Page 117:** left: Bartlesville Public Library; right: © 1931, The Oklahoma Publishing Co., from *The Daily Oklahoman*, April 12, 1931. **Pages 118-119:** redrawn with permission from the U.S. Department of Interior.

Chapter 12. Page 121: Drake Well Museum. **Page 122:** Exxon, *Lamp*, November 1948 and Photographic Archives, University of Louisville, Standard Oil of New Jersey Collection. **Page 124:** Meyers Photo Shop, Barney Hillerman, photographer. **Page 125:** reprinted with permission from McGraw Hill. **Page 126:** Andrew J. Maslowski. **Page 128:** reprinted with permission from Elsevier Scientific Publishing Co. from R. Chapman, *Petroleum Geology: A Concise Study*, 1973. **Page 129:** Long Beach Public Library and Information Center, the Petroleum Industry Collection.

Chapter 13. Page 131: reprinted with permission from the Petroleum Extension Service, the University of Texas at Austin. **Page 132:** ARCO. **Page 133:** Long Beach Public Library and Information Center, the Petroleum Industry Collection. **Page 134:** Long Beach Public Library and Information Center, the Petroleum Industry Collection. **Page 135:** from *The History of Oil Well Drilling*, by J. E. Brantly. © 1971 by Gulf Publishing Company, Houston, Tex. Used with permission. All rights reserved. **Page 137:** left: from *The History of Oil Well Drill-*

ing, by J. E. Brantly. © 1971 by Gulf Publishing Company, Houston, Tex. Used with permission. All rights reserved; right: Hughes Tool Co. **Page 138:** left: Chevron; right: Hydril Co. **Page 139:** ARCO. **Page 141:** upper: reprinted by permission of Colorado School of Mines, from *Subsurface Geology*, by L. W. LeRoy, 4th ed., 1977; lower: ARCO. **Page 142:** left: NL Sperry Sun: right: British Petroleum.

Chapter 14. Page 145: left: Chevron; right: California Historical Society/Ticor Title Insurance (Los Angeles) and Long Beach Public Library and Information Center, the Petroleum Collection. **Pages 146-47:** redrawn with permission from *Offshore* Magazine. **Page 149:** upper: drawn based on information from *Basic Petroleum Data Book*, May 1983; lower: ARCO. **Pages 150-51:** courtesy of Kerr-McGee's *Resources* Magazine and the Beatrice Field Consortium. **Pages 152-53:** Global Marine Deepwater Drilling Co. and Chevron. **Page 154:** ARCO. **Page 155:** ARCO.

Chapter 15. Page 157: left: Long Beach Public Library and Information Center, the Petroleum Collection; right: Halliburton Co. **Page 158:** Halliburton Co. **Page 159:** Long Beach Public Library and Information Center, the Petroleum Collection. **Page 160:** reprinted with permission from Colorado School of Mines, from *Subsurface Geology*, by L. W. LeRoy, 4th ed., 1977. **Page 161:** left and right: reprinted with permission from Colorado School of Mines, from *Subsurface Geology*, by L. W. LeRoy, 4th ed., 1977. **Page 163:** upper and lower: Drake Well Museum. **Page 164:** upper: Halliburton Co.; lower: from *Oil From Prospect to Pipeline*, 4th ed., by Robert R. Wheeler and Maurine Whited. © 1981 by Gulf Publishing Company, Houston, Tex. Used with permission. All rights reserved. **Page 165:** ARCO. **Page 166:** ARCO.

Chapter 16. Page 169: Long Beach Public Library and Information Center, the Petroleum Collection. **Page 170:** upper: redrawn with permission from University of Oklahoma Press; lower: British Petroleum's *Our Industry Petroleum*. **Page 171:** upper and lower: British Petroleum's *Our Industry Petroleum*. **Page 173:** upper: redrawn with permission from University of Oklahoma Press; lower: redrawn with permission from Exxon. **Page 174:** upper and lower: redrawn with permission from Amoco. **Page 175:** left: redrawn with permission from Amoco; right: redrawn with permission from Gulf Oil Corp. **Page 176:** upper and lower: redrawn with permission from Amoco.

Chapter 17. Page 179: ARCO. **Page 181:** ARCO. **Page 182:** Columbia Helicopters. **Page 183:** Chevron. **Page 184:** LOOP. **Page 185:** ARCO.

Chapter 18. Page 187: British Petroleum's *Statistical Review of World Energy*, 1982. **Page 188-189:** ARCO. **Page 189:** upper: ARCO: lower: redrawn with permission from Northwestern University Press. **Page**

303: upper: ARCO; lower: © Tony Auth, *Philadelphia Inquirer*. Page 304: reprinted with permission from the *Oil and Gas Journal*, November 28, 1983. **Page 305:** reprinted with permission from the *Oil and Gas Journal*, November 28, 1983. **Page 306:** reprinted with permission from William W. Talley, II, Ph.D., the RAM Group Ltd.

Chapter 30. Page 309: left: based on information provided by Nicholas P. Gal, American Petroleum Institute Analysis Department; right: © 1981 by The New York Times Company. Reprinted by permission. July 12, 1981. **Page 311:** *National Petroleum News 1983 Factbook*. **Page 313:** upper and lower: *National Petroleum News 1983 Factbook*. **Page 314:** based on information provided by Nicholas P. Gal, American Petroleum Institute Analysis Department.

Chapter 31. Page 317: left: American Petroleum Institute; right: reprinted with permission from PennWell Publishing Co. **Page 318:** based on statistics from British Petroleum's *Statistical Review of World Energy*, 1982. **Page 319:** reprinted courtesy of *Perspectives* Magazine, Pennzoil Co. and American Petroleum Institute. **Page 321:** from *Putnam's Geology*, 4th ed., by Edwin E. Larson and Peter W. Birkeland. © 1964, 1971, 1978, 1982 by Oxford University Press, Inc. Reprinted by permission. **Page 323:** upper: ARCO; lower: Union Oil Co. **Page 325:** Energy Research and Development Administration, Office of Public Affairs, Photo Office.

Chapter 32. Page 327: reprinted with permission from PennWell Publishing Co. **Page 329:** Fluor Corp. **Page 331:** upper: Bureau of Reclamation. Photo by J. C. Dahilig (DOE); lower left: Bureau of Reclamation. Photo by J. C. Dahilig (DOE); lower right: Department of Energy Information, Office of Public Affairs. **Page 332:** Department of Energy Information, Office of Public Affairs. **Page 333:** United Press International. **Page 334:** upper and lower: from *Putnam's Geology*, 4th ed., by Edwin E. Larson and Peter W. Birkeland. © 1964, 1971, 1978, 1982 by Oxford University Press, Inc. Reprinted by permission. **Page 335:** ARCO. **Page 337:** Department of Energy, Office of Public Affairs Photo Office.

INDEX

Compiled by Anna Marie and Everett G. Hager

378

384

FUNDAMENTALS OF THE PETROLEUM INDUSTRY

Fundamentals of the Petroleum Industry by Robert O. Anderson is written from the point of view of a man who has been intimately involved at the highest levels of the petroleum industry for more than 40 years. Mr. Anderson has seen the business grow to become the largest single industry in the world. As a participant observer, he is in a unique position to see it whole and to speculate about its future.

This wide-ranging survey of the modern petroleum industry is written in clear, non-technical language for the non-expert, and covers not only the historical development of the petroleum age but explains the industry in the context of current economic, social, and scientific trends.

The book discusses all aspects of the business, including such subjects as petroleum geology; the chemical nature of hydrocarbons; petrochemicals; drilling and exploration; reservoir engineering; marketing; taxation and the relationships of the oil industry to government; and the outlook for alternative energies. Present-day practices are thoroughly grounded in history, and at the same time they are interrelated to form a larger view of the marvelously complex apparatus that supplies energy from hydrocarbons to our modern industrial society.

ABOUT THE AUTHOR

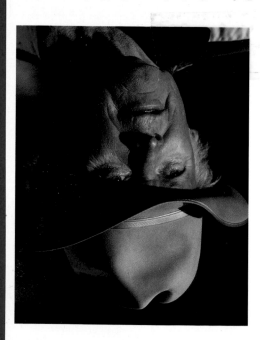

Robert O. Anderson is Chairman of the Board and Chairman of the Executive Committee of Atlantic Richfield Company. He has been active on all levels of the oil industry for over 40 years, beginning with his own small company in New Mexico shortly after his graduation from the University of Chicago. A resident of New Mexico since 1941, Mr. Anderson is an officer of numerous civic, charitable, educational, and cultural organizations. He has served on the National Petroleum Council since 1951.